DATE DUE

Demco, Inc. 38-293

Digital Avionics Handbook
SECOND EDITION

AVIONICS

ELEMENTS, SOFTWARE AND FUNCTIONS

The Electrical Engineering Handbook Series

Series Editor
Richard C. Dorf
University of California, Davis

Titles Included in the Series

The Handbook of Ad Hoc Wireless Networks, Mohammad Ilyas

The Biomedical Engineering Handbook, Third Edition, Joseph D. Bronzino

The Circuits and Filters Handbook, Second Edition, Wai-Kai Chen

The Communications Handbook, Second Edition, Jerry Gibson

The Computer Engineering Handbook, Vojin G. Oklobdzija

The Control Handbook, William S. Levine

The CRC Handbook of Engineering Tables, Richard C. Dorf

The Digital Avionics Handbook, Second Edition Cary R. Spitzer

The Digital Signal Processing Handbook, Vijay K. Madisetti and Douglas Williams

The Electrical Engineering Handbook, Second Edition, Richard C. Dorf

The Electric Power Engineering Handbook, Leo L. Grigsby

The Electronics Handbook, Second Edition, Jerry C. Whitaker

The Engineering Handbook, Third Edition, Richard C. Dorf

The Handbook of Formulas and Tables for Signal Processing, Alexander D. Poularikas

The Handbook of Nanoscience, Engineering, and Technology, William A. Goddard, III, Donald W. Brenner, Sergey E. Lyshevski, and Gerald J. Iafrate

The Handbook of Optical Communication Networks, Mohammad Ilyas and Hussein T. Mouftah

The Industrial Electronics Handbook, J. David Irwin

The Measurement, Instrumentation, and Sensors Handbook, John G. Webster

The Mechanical Systems Design Handbook, Osita D.I. Nwokah and Yidirim Hurmuzlu

The Mechatronics Handbook, Robert H. Bishop

The Mobile Communications Handbook, Second Edition, Jerry D. Gibson

The Ocean Engineering Handbook, Ferial El-Hawary

The RF and Microwave Handbook, Mike Golio

The Technology Management Handbook, Richard C. Dorf

The Transforms and Applications Handbook, Second Edition, Alexander D. Poularikas

The VLSI Handbook, Wai-Kai Chen

Digital Avionics Handbook
SECOND EDITION

AVIONICS
ELEMENTS, SOFTWARE AND FUNCTIONS

Edited by
CARY R. SPITZER

AvioniCon, Inc.
Williamsburg, Virginia, U.S.A.

CRC Press
Taylor & Francis Group
Boca Raton   London   New York

CRC Press is an imprint of the
Taylor & Francis Group, an informa business

CRC Press
Taylor & Francis Group
6000 Broken Sound Parkway NW, Suite 300
Boca Raton, FL 33487-2742

International Standard Book Number-10: 0-8493-8438-9 (Hardcover) 0-8493-8441-9 (Hardcover)
International Standard Book Number-13: 978-0-8493-8438-7 (Hardcover) 978-0-8493-8441-7 (Hardcover)

Library of Congress Cataloging-in-Publication Data

Spitzer, Cary R.
    Avionics : elements, software, and functions / Cary R. Spitzer.
        p. cm.
    Includes bibliographical references and index.
    ISBN 0-8493-8438-9
    1. Avionics. I. Title.

TL695.S748 2006
629.135--dc22
                                                                                2006050555

Visit the Taylor & Francis Web site at
http://www.taylorandfrancis.com

and the CRC Press Web site at
http://www.crcpress.com

Preface

Avionics is the cornerstone of modern aircraft. More and more vital functions on both military and civil aircraft involve electronic devices. After the cost of the airframe and the engines, avionics is the most expensive item on the aircraft, but well worth every cent of the price.

Many technologies emerged in the last decade that are used in this millennium. After proof of soundness in design through ground application, advanced microprocessors are finding their way onto aircraft to provide new capabilities that were unheard of a decade ago. The Global Positioning System has enabled satellite-based precise navigation and landing, and communication satellites are now capable of supporting aviation services. Thus, the aviation world is changing to satellite-based communications, navigation, and surveillance for air traffic management. Both the aircraft operator and the air traffic services provider are realizing significant benefits.

Familiar technologies in this book include data buses, one type of which has been in use for over 30 years, head mounted displays, and fly-by-wire flight controls. New bus and display concepts are emerging that may displace these veteran devices. An example is a retinal scanning display.

Other emerging technologies include speech interaction with the aircraft and synthetic vision. Synthetic vision offers enormous potential for both military and civil aircraft for operations under reduced visibility conditions or in cases where it is difficult to install sufficient windows in an aircraft.

This book offers a comprehensive view of avionics, from the technology and elements of a system to examples of modern systems flying on the latest military and civil aircraft. The chapters have been written with the reader in mind by working practitioners in the field. This book was prepared for the working engineer and his or her boss and others who need the latest information on some aspect of avionics. It will not make one an expert in avionics, but it will provide the knowledge needed to approach a problem.

Editor

Cary R. Spitzer is a graduate of Virginia Tech and George Washington University. After service in the Air Force, he joined NASA Langley Research Center.

During the last half of his tenure at NASA he focused on avionics. He was the NASA manager of a joint NASA/Honeywell program that made the first satellite-guided automatic landing of a passenger transport aircraft in November 1990. In recognition of this accomplishment, he was nominated jointly by ARINC, ALPA, AOPA, ATA, NBAA, and RTCA for the 1991 Collier Trophy "for his pioneering work in proving the concept of GPS aided precision approaches." He led a project to define the experimental and operational requirements for a transport aircraft suitable for conducting flight experiments and to acquire such an aircraft. Today, that aircraft is the NASA Langley B-757 ARIES flight research platform.

Mr. Spitzer was the NASA representative to the Airlines Electronic Engineering Committee. In 1988 he received the Airlines Avionics Institute Chairman's Special Volare Award. He is only the second federal government employee ever so honored.

He has been active in the RTCA, including serving as chairman of Special Committee 200 Integrated Modular Avionics (IMA), chairman of the Airport Surface Operations Subgroup of Task Force 1 on Global Navigation Satellite System Transition and Implementation Strategy, and as Technical Program Chairman of the 1992 Technical Symposium. He was a member of the Technical Management Committee.

In 1993 Mr. Spitzer founded *AvioniCon*, an international avionics consulting firm that specializes in strategic planning, business development, technology analysis, and in-house training.

Mr. Spitzer is a Fellow of the Institute of Electrical and Electronics Engineers (IEEE) and an Associate Fellow of the American Institute of Aeronautics and Astronautics (AIAA). He received the AIAA 1994 Digital Avionics Award and an IEEE Centennial Medal and Millennium Medal. He is a Past President of the IEEE Aerospace and Electronic Systems Society. Since 1979, he has played a major role in the highly successful Digital Avionics Systems Conferences, including serving as General Chairman.

Mr. Spitzer presents one-week shortcourses on digital avionics systems and on satellite-based communication, navigation, and surveillance for air traffic management at the UCLA Extension Division. He has also lectured for the International Air Transport Association.

He is the author of *Digital Avionics Systems,* the first book in the field, published by McGraw-Hill, and Editor-in-Chief of *The Digital Avionics Handbook,* published by CRC Press.

Contributors

Kathy H. Abbott
Federal Aviation
 Administration

Daniel G. Baize
NASA Kennedy Space Center

John G.P. Barnes
John Barnes Informatics

Gregg F. Bartley
Federal Aviation
 Administration

Douglas W. Beeks
Beeks Engineering and
 Design

Barry C. Breen
Honeywell

Dominique Briere
Airbus—Retired

Samuel P. Buckwalter
ARINC

Chris deLong
Honeywell

James L. Farrell
VIGIL, Inc.

Christian Favre
Airbus

Thomas K. Ferrell
Ferrell and Associates
 Consulting

Uma D. Ferrell
Ferrell and Associates
 Consulting

Lee Harrison
Galaxy Scientific Corp.

Steve Henely
Rockwell Collins

Peter J. Howells
Rockwell Collins Flight
 Dynamics

Mirko Jakovljevic
TTTech Computertechnik AG

Myron Kayton
Kayton Engineering Company

Michael S. Lewis
The Boeing Company

Thomas M. Lippert
Microvision Inc.

Daniel A. Martinec
ARINC

James E. Melzer
Rockwell Collins Optronics

Roy T. Oishi
ARINC

Russell V. Parrish
MVP Technologies, Inc.

Paul J. Prisaznuk
ARINC

Philip A. Scandura, Jr.
Honeywell International

Dennis L. Schmickley
The Boeing Company

Pascal Traverse
Airbus

David G. Vutetakis
Concorde Battery Corporation

Randy Walter
Smiths Industries

Robert B. Wood
Rockwell Collins Flight
 Dynamics

Contents

SECTION I Elements and Functions

SECTION II   Functions

Section I

Elements and Functions

1

AS 15531/MIL-STD-1553B Digital Time Division Command/Response Multiplex Data Bus

Chris deLong
Honeywell

1.1 Introduction

MIL-STD-1553 is a standard which defines the electrical and protocol characteristics for a data bus. SAE AS-15531 is the commercial equivalent to the military standard. A data bus is similar to what the personal computer and office automation industry have dubbed a "Local Area Network (LAN)." In avionics, a data bus is used to provide a medium for the exchange of data and information among various systems and subsystems.

1.1.1 Background

In the 1950s and 1960s, avionics were simple standalone systems. Navigation, communications, flight controls, and displays were analog systems. Often, these systems were composed of multiple boxes interconnected to form a single system. The interconnections between the various boxes were accomplished with point-to-point wiring. The signals mainly consisted of analog voltages, synchro-resolver signals, and relay/switch contacts. The location of these boxes within the aircraft was a function of operator need, available space, and the aircraft weight and balance constraints. As more and more systems

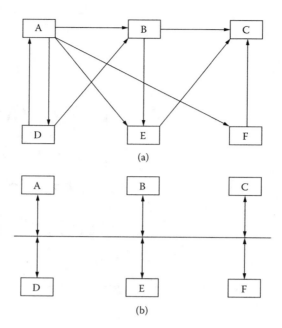

FIGURE 1.1   Systems configurations.

were added, the cockpits became crowded due to the number of controls and displays, and the overall weight of the aircraft increased.

By the late 1960s and early 1970s, it was necessary to share information between various systems to reduce the number of black boxes required by each system. A single sensor providing heading and rate information could provide those data to the navigation system, the weapons system, the flight control system, and pilot's display system (see Figure 1.1a). However, the avionics technology was still basically analog, and while sharing sensors did produce a reduction in the overall number of black boxes, the interconnecting signals became a "rat's nest" of wires and connectors. Moreover, functions or systems that were added later became an integration nightmare as additional connections of a particular signal could have potential system impacts, plus since the system used point-to-point wiring, the system that was the source of the signal typically had to be modified to provide the additional hardware needed to output to the newly added subsystem. As such, intersystem connections were kept to the bare minimums.

By the late 1970s, with the advent of digital technology, digital computers had made their way into avionics systems and subsystems. They offered increased computational capability and easy growth, compared to their analog predecessors. However, the data signals — the inputs and outputs from the sending and receiving systems — were still mainly analog in nature, which led to the configuration of a small number of centralized computers being interfaced to the other systems and subsystems via complex and expensive analog-to-digital and digital-to-analog converters.

As time and technology progressed, the avionics systems became more digital. And with the advent of the microprocessor, things really took off. A benefit of this digital application was the reduction in the number of analog signals, and hence the need for their conversion. Greater sharing of information could be provided by transferring data between users in digital form. An additional side benefit was that digital data could be transferred bidirectionally, whereas analog data were transferred unidirectionally. Serial rather than parallel transmission of the data was used to reduce the number of interconnections within the aircraft and the receiver/driver circuitry required with the black boxes. But this alone was not enough. A data transmission medium which would allow all systems and subsystems to share a single and common set of wires was needed (see Figure 1.1b). By sharing the use of this interconnect, the various subsystems could send data among themselves and to other systems and subsystems, one at a time, and in a defined sequence. Enter the 1553 Data Bus.

1.1.2 History and Applications

MIL-STD-1553(USAF) was released in August of 1973. The first user of the standard was the F-16. Further changes and improvements were made and a tri-service version, MIL-STD-1553A, was released in 1975. The first user of the "A" version of the standard was again the Air Force's F-16 and the Army's new attack helicopter, the AH-64A Apache. With some "real world" experience, it was soon realized that further definitions and additional capabilities were needed. The latest version of the standard, 1553B, was released in 1978.

Today the 1553 standard is still at the "B" level; however, changes have been made. In 1980, the Air Force introduced Notice 1. Intended only for Air Force applications, Notice 1 restricted the use of many of the options within the standard. While the Air Force felt this was needed to obtain a common set of avionics systems, many in industry felt that Notice 1 was too restrictive and limited the capabilities in the application of the standard. Released in 1986, the tri-service Notice 2 (which supersedes Notice 1) places tighter definitions upon the options within the standard. And while not restricting an option's use, it tightly defines how an option will be used if implemented. Notice 2, in an effort to obtain a common set of operational characteristics, also places a minimum set of requirements upon the design of the black box. The military standard was converted to its commercial equivalent as SAE AS 15531, as part of the government's effort to increase the use of commercial products.

Since its inception, MIL-STD-1553 has found numerous applications. Notice 2 even removed all references to "aircraft" or "airborne" so as not to limit its applications. The standard has also been accepted and implemented by NATO and many foreign governments. The U.K. has issued Def Stan 00-18 (Part 2) and NATO has published STANAG 3838 AVS, both of which are versions of MIL-STD-1553B.

1.2 The Standard

MIL-STD-1553B defines the term Time Division Multiplexing (TDM) as "the transmission of information from several signal sources through one communications system with different signal samples staggered in time to form a composite pulse train." For our example in Figure 1.1b, this means that data can be transferred between multiple avionics units over a single transmission media, with the communications among the different avionics boxes taking place at different moments in time, hence time division. Table 1.1 is a summary of the 1553 data bus characteristics. However, before defining how the data are transferred, it is necessary to understand the data bus hardware.

1.2.1 Hardware Elements

The 1553 standard defines certain aspects regarding the design of the data bus system and the black boxes to which the data bus is connected. The standard defines four hardware elements: transmission media, remote terminals, bus controllers, and bus monitors; each of which is detailed as follows.

1.2.1.1 Transmission Media

The transmission media, or data bus, is defined as a twisted shielded pair transmission line consisting of the main bus and a number of stubs. There is one stub for each terminal (system) connected to the bus. The main data bus is terminated at each end with a resistance equal to the cable's characteristic impedance. This termination makes the data bus behave electrically like an infinite transmission line. Stubs, which are added to the main bus to connect the terminals, provide "local" loads, and produce an impedance mismatch where added. This mismatch, if not properly controlled, produces electrical reflections and degrades the performance of the main bus. Therefore, the characteristics of both the main bus and the stubs are specified within the standard. Table 1.2 is a summary of the transmission media characteristics.

The standard specifies two stub methods: direct and transformer coupled. This refers to the method in which a terminal is connected to the main bus. Figure 1.2 shows the two methods, the primary

TABLE 1.1   Summary of the 1553 Data Bus Characteristics

Data Rate	1 MHz
Word Length	20 bits
Data Bits per Word	16 bits
Message Length	Maximum of 32 data words
Transmission Technique	Half-Duplex
Operation	Asynchronous
Encoding	Manchester II Bi-phase
Protocol	Command-Response
Bus Control	Single or Multiple
Message Formats	Controller-to-Terminal (BC-RT)
	Terminal-to-Controller (RT-BC)
	Terminal-to-Terminal (RT-RT)
	Broadcast
	System Control
Number of Remote Terminals	Maximum of 31
Terminal Types	Remote Terminal (RT)
	Bus Controller (BC)
	Bus Monitor (BM)
Transmission Media	Twisted Shielded Pair Cable
Coupling	Transformer or Direct

TABLE 1.2   Summary of Transmission Media Characteristics

Cable Type	Twisted Shielded Pair
Capacitance	30.0 pF/ft max — wire to wire
Characteristic Impedance	70.0 to 85.0 ohms at 1 MHz
Cable Attenuation	1.5 dbm/100 ft at 1 MHz
Cable Twists	4 twists per foot maximum
Shield Coverage	90% minimum
Cable Termination	Cable impedance ($\pm$2%)
Direct Coupled Stub Length	Maximum of 1 ft
Transformer Coupled Stub Length	Maximum of 20 ft

difference between the two being that the transformer coupled method utilizes an isolation transformer for connecting the stub cable to the main bus cable. In both methods, two isolation resistors are placed in series with the bus. In the direct coupled method, the resistors are typically located within the terminal, whereas in the transformer coupled method, the resistors are typically located with the coupling transformer in boxes called data bus couplers. A variety of couplers are available, providing single or multiple stub connections.

Another difference between the two coupling methods is the length of the stub. For the direct coupled method, the stub length is limited to a maximum of 1 ft. For the transformer coupled method, the stub can be up to a maximum length of 20 ft. Therefore for direct coupled systems, the data bus must be routed in close proximity to each of the terminals, whereas for a transformer coupled system, the data bus may be up to 20 ft away from each terminal.

1.2.1.2   Remote Terminal

A remote terminal is defined within the standard as "All terminals not operating as the bus controller or as a bus monitor." Therefore if it is not a controller, monitor, or the main bus or stub, it must be a remote terminal — sort of a "catch all" clause. Basically, the remote terminal is the electronics necessary to transfer data between the data bus and the subsystem. So what is a subsystem? For 1553 applications, the subsystem is the sender or user of the data being transferred.

In the earlier days of 1553, remote terminals were used mainly to convert analog and discrete data to/from a data format compatible with the data bus. The subsystems were still the sensor which provided

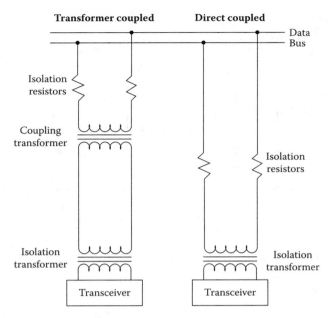

FIGURE 1.2 Terminal connection methods.

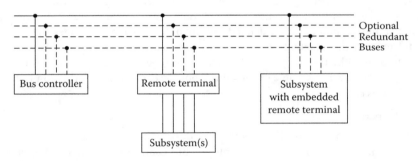

FIGURE 1.3 Simple multiplex architecture.

the data and computer which used the data. As more and more digital avionics became available, the trend has been to embed the remote terminal into the sensor and computer. Today it is common for the subsystem to contain an embedded remote terminal. Figure 1.3 shows the different levels of remote terminals possible.

A remote terminal typically consists of a transceiver, an encoder/decoder, a protocol controller, a buffer or memory, and a subsystem interface. In a modern black box containing a computer or processor, the subsystem interface may consist of the buffers and logic necessary to interface to the computer's address, data, and control buses. For dual redundant systems two transceivers and two encoders/decoders would be required to meet the requirements of the standard.

Figure 1.4 is a block diagram of a remote terminal and its connection to a subsystem. In short, the remote terminal consists of all the electronics necessary to transfer data between the data bus and the user or originator of the data being transferred.

But a remote terminal is more than just a data formatter. It must be capable of receiving and decoding commands from the bus controller, and respond accordingly. It must also be capable of buffering a message-worth of data, be capable of detecting transmission errors and performing validation tests upon the data, and reporting the status of the message transfer. A remote terminal must be capable of performing a few

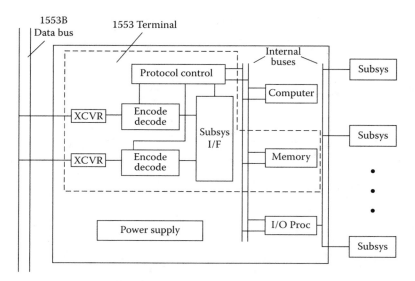

FIGURE 1.4 Terminal definition.

of the bus management commands (referred to as mode commands), and for dual redundant applications it must be capable of listening to and decoding commands on both buses at the same time.

A remote terminal must strictly follow the protocol as defined by the standard. It can only respond to commands received from the bus controller (i.e., it only speaks when spoken to). When it receives a valid command, it must respond within a defined amount of time. If a message does not meet the validity requirements defined, then the remote terminal must invalidate the message and discard the data (not allow the data to be used by the subsystem). In addition to reporting status to the bus controller, most remote terminals today are also capable of providing some level of status information to the subsystem regarding the data received.

1.2.1.3 Bus Controller

The bus controller is responsible for directing the flow of data on the bus. While several terminals may be capable of performing as the bus controller, only one bus controller is allowed to be active at any one time. The bus controller is the only device allowed to issue commands onto the data bus. The commands may be for the transfer of data, or the control and management of the bus (referred to as mode commands).

Typically, the bus controller is a function that is contained within some other computer, such as a mission computer, a display processor, or a fire control computer. The complexity of the electronics associated with the bus controller is a function of the subsystem interface (the interface to the computer), the amount of error management and processing to be performed, and the architecture of the bus controller. There are three types of bus controllers architectures: a word controller, a message controller, and a frame controller.

A *word controller* is the oldest and simplest type. Few word controllers are built today and they are only mentioned herein for completeness. For a word controller, the terminal electronics transfers one word at a time to the subsystem. Message buffering and validation must be performed by the subsystem.

Message controllers output a single message at a time, interfacing with the computer only at the end of the message or perhaps when an error occurrs. Some message controllers are capable of performing minor error processing, such as transmitting once on the alternate data bus, before interrupting the computer. The computer will inform the interface electronics of where the message exists in memory and provide a control word. For each message the control word typically informs the electronics of the message type (e.g., an RT-BC or RT-RT command), which bus to use to transfer the message, where to read or write the data words in memory, and what to do if an error occurs. The control words are a function of the hardware design of the electronics and are not standardized among bus controllers.

A *frame controller* is the latest concept in bus controllers. A frame controller is capable of processing multiple messages in a sequence defined by the computer. The frame controller is typically capable of error processing as defined by the message control word. Frame controllers are used to "off-load" the computer as much as possible, interrupting only at the end of a series of messages or when an error it can not handle is detected.

There is no requirement within the standard as to the internal workings of a bus controller, only that it issue commands onto the bus.

1.2.1.4 Bus Monitor

A bus monitor is just that. A terminal which listens to (monitors) the exchange of information on the data bus. The standard strictly defines what bus monitors may be used for, stating that the information obtained by a bus monitor be used "for off-line applications (e.g., flight test recording, maintenance recording or mission analysis) or to provide a back-up bus controller sufficient information to take over as the bus controller." Monitors may collect all the data from the bus or may collect selected data.

The reason for restricting its use is that while a monitor may collect data, it deviates from the command-response protocol of the standard in that a monitor is a passive device that does not transmit a status word, and therefore can not report on the status of the information transferred. Therefore, bus monitors fall into two categories: a recorder for testing, or as a terminal functioning as a back-up bus controller.

In collecting data, a monitor must perform the same message validation functions as the remote terminal and, if an error is detected, inform the subsystem of the error (the subsystem may still record the data, but the error should be noted). For monitors which function as recorders for testing, the subsystem is typically a recording device or a telemetry transmitter. For monitors which function as back-up bus controllers, the subsystem is the computer.

Today it is common that bus monitors also contain a remote terminal. When the monitor receives a command addressed to its terminal address, it responds as a remote terminal. For all other commands, it functions as a monitor. The remote terminal portion could be used to provide feedback to the bus controller of the monitor's status, such as the amount of memory or time left, or to reprogram a selective monitor as to what messages to capture.

1.2.1.5 Terminal Hardware

The electronic hardware among a remote terminal, bus controller, and bus monitor does not differ much. Both the remote terminal and bus controller (and bus monitor if it is also a remote terminal) must have the transmitters/receivers and encoders/decoders to format and transfer data. The requirements on the transceivers and the encoders/decoders do not vary between the hardware elements. Table 1.3 lists the electrical characteristics of the terminals.

All three elements have some level of subsystem interface and data buffering. The primary difference lies in the protocol control logic and often this is just a different series of micro-coded instructions. For this reason, it is common to find 1553 hardware circuitry that is also capable of functioning as all three devices.

There is an abundance of "off-the-shelf" components available today from which to design a terminal. These vary from discrete transceivers, encoders/decoders, and protocol logic devices to a single dual redundant hybrid containing everything but the transformers.

1.3   Protocol

The rules under which the transfers occur is referred to as "protocol". The control, data flow, status reporting, and management of the bus is provided by three word types.

TABLE 1.3   Terminal Electrical Characteristics

Requirement	Transformer Coupled	Direct Coupled	Condition
	Input Characteristics		
Input Level	0.86–14.0 V	1.2–20.0 V	p–p, l–l
No Response	0.0–0.2 V	0.0–0.28 V	p–p, l–l
Zero Crossing Stability	±150.0 nsec	±150.0 nsec	
Rise/Fall Times	0 nsec	0 nsec	Sine Wave
Noise Rejection	140.0 mV WGN[a]	200.0 mV WGN	BER 1[b] per 10^7
Common Mode Rejection	±10.0 V peak	±10.0 V peak	line-gnd, DC-2.0 MHz
Input Impedance	1000 ohms	2000 ohms	75 kHz–1 MHz
	Output Characteristics		
Output Level	18.0–27.0 V	6.0–9.0 V	p–p, l–l
Zero Crossing Stability	25.0 nsec	25.0 nsec	
Rise/Fall Times	100–300 nsec	100–300 nsec	10%–90%
Maximum Distortion	±900.0 mV	±300.0 mV	peak, l–l
Maximum Output Noise	14.0 mV	5.0 mV	rms, l–l
Maximum Residual Voltage	±250.0 mV	±90.0 mV	peak, l–l

[a] WGN = White Gaussian Noise.
[b] BER = Bit Error Rate.

1.3.1   Word Types

Three distinct word types are defined by the standard. These are command words, data words, and status words. Each word type has a unique format yet all three maintain a common structure. Each word is 20 bits in length. The first three bits are used as a synchronization field, thereby allowing the decode clock to re-sync at the beginning of each new word. The following 16 bits are the information field and differ among the three word types. The last bit is the parity bit. Parity is based on odd parity for the single word. The three word types are shown in Figure 1.5.

Bit encoding for all words is based on bi-phase Manchester II format. The Manchester II format provides a self-clocking waveform in which the bit sequence is independent. The positive and negative voltage levels of the Manchester waveform is DC balanced (same amount of positive signal as there is negative signal) and as such is well suited for transformer coupling. A transition of the signal occurs at the center of the bit time. A logic "0" is a signal that transitions from a negative level to a positive level. A logic "1" is a signal that transitions from a positive level to a negative level.

The terminal's hardware is responsible for the Manchester encoding and decoding of the word types. The interface that the subsystem sees is the 16-bit information field of all words. The sync and parity fields are not provided directly. However, for received messages, the decoder hardware provides a signal to the protocol logic as to the sync type the word was and as to whether parity was valid or not. For transmitted messages, there is an input to the encoder as to what sync type to place at the beginning of the word, and parity is automatically calculated by the encoder.

1.3.1.1   Sync Fields

The first three bit times of all word types is called the sync field. The sync waveform is in itself an invalid Manchester waveform as the transition only occurs at the middle of the second bit time. The use of this distinct pattern allows the decoder to re-sync at the beginning of each word received and maintain the overall stability of the transmissions.

Two distinct sync patterns are used: the command/status sync, and the data sync. The command/status sync has a positive voltage level for the first one and a half bit times, then transitions to a negative voltage level for the second one and a half bit times. The data sync is the opposite — a negative voltage level for the first one and a half bit times, then transitions to a positive voltage level for the second one and a half bit times. The sync patterns are shown in Figure 1.5.

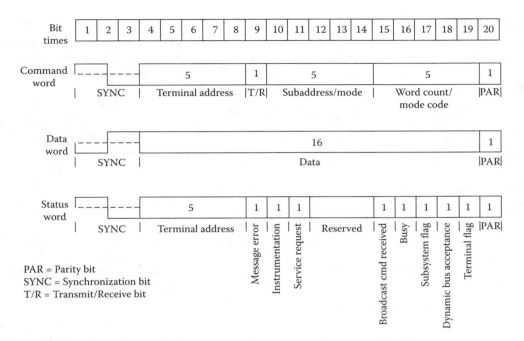

FIGURE 1.5   Word formats.

1.3.1.2   Command Word

The Command Word (CW) specifies the function that a remote terminal(s) is to perform. This word is only transmitted by the active bus controller. The word begins with a command sync in the first three bit times. The following 16-bit information field is as defined in Figure 1.5.

The five-bit Terminal Address (TA) field (bit times 4–8) states to which unique remote terminal the command is intended (no two terminals may have the same address). Note that an address of 00000 is a valid address, and that an address of 11111 is reserved for use as the broadcast address. Also note that there is no requirement that the bus controller be assigned an address, therefore the maximum number of terminals the data bus can support is 31. Notice 2 to the standard requires that the terminal address be wire programmable externally to the black box (i.e., an external connector) and that the remote terminal electronics perform a parity test upon the wired terminal address. The Notice basically states that an open circuit on an address line is detected as a logic "1," that connecting an address line to ground is detected as a logic "0," and that odd parity will be used in testing the parity of the wired address field.

The next bit (bit time 9) is the Transmit/Receive (T/R) bit. This defines the direction of information flow and is always from the point of view of the remote terminal. A transmit command (logic 1) indicates that the remote terminal is to transmit data, while a receive command (logic 0) indicates that the remote terminal is going to receive data. The only exceptions to this rule are associated with mode commands.

The following five bits (bit times 10–14) are the Subaddress (SA)/Mode Command (MC) bits. Logic 00000 or 11111 within this field shall be decoded to indicate that the command is a Mode Code Command. All other logic combinations of this field are used to direct the data to different functions within the subsystem. An example might be that 00001 is position and rate data, 00010 is frequency data, 10010 is display information, and 10011 is self-test data. The use of the subaddresses is left to the designer, however, Notice 2 suggests the use of subaddress 30 for data wraparound.

The next five bit positions (bit times 15–19) define the Word Count (WC) or Mode Code to be performed. If the Subaddress/Mode Code field was 00000 or 11111, then this field defines the mode code to be performed. If not a mode code, then this field defines the number of data words either to be received or transmitted depending on the T/R bit. A word count field of 00000 is decoded as 32 data words.

The last bit (bit time 20) is the word parity bit. Only odd parity shall be used.

1.3.1.3   Data Word

The Data Word (DW) contains the actual information that is being transferred within a message. Data words can be transmitted by either a remote terminal (transmit command) or a bus controller (receive command). The first three bit times contain a data sync. This sync pattern is the opposite of that used for command and status words and therefore is unique to the data word type.

The following 16 bits of information are left to the designer to define. The only standard requirement is that the most significant bit (MSB) of the data be transmitted first. While the standard provides no guidance as to their use, Section 80 of MIL-HDBK-1553A and SAE AS-15532 provides guidance and lists the formats (i.e., bit patterns, resolutions, etc.) of the most commonly used data words.

The last bit (bit time 20), is the word parity bit. Only odd parity shall be used.

1.3.1.4   Status Word

The Status Word (SW) is only transmitted by a remote terminal in response to a valid message. The status word is used to convey to the bus controller whether a message was properly received or the state of the remote terminal (i.e., service request, busy, etc.). The status word is defined in Figure 1.5. Since the status word conveys information to the bus controller, there are two views as to the meaning of each bit — what the setting of the bit means to a remote terminal, and what the setting of the bit means to a bus controller. Each field of the status word, and its potential meanings, is examined below.

1.3.1.4.1   Resetting the Status Word

The Status Word, with the exception of the remote terminal address, is cleared after receipt of a valid command word. The two exceptions to this rule are if the command word received is a Transmit Status Word Mode Code or a Transmit Last Command Word Mode Code. Conditions which set the individual bits of the word may occur at any time. If after clearing the status word, the conditions for setting the bits still exist, then the bits shall be set again.

Upon detection of an error in the data being received, the Message Error bit is set and the transmission of the status word is suppressed. The transmission of the status word is also suppressed upon receipt of a broadcast message. For an illegal message (i.e., an illegal Command Word), the Message Error bit is set and the status word is transmitted.

1.3.1.4.2   Status Word Bits

Terminal Address. The first five bits (bit times 4–8) of the information field are the Terminal Address (TA). These five bits should match the corresponding field within the command word that the terminal received. The remote terminal sets these bits to the address to which it has been programmed. The bus controller should examine these bits to insure that the terminal responding with its status word was indeed the terminal to which the command word was addressed. In the case of a remote terminal to remote terminal message (RT-RT), the receiving terminal should compare the address of the second command word with that of the received status word. While not required by the standard, it is good design practice to insure that the data received are from a valid source.

Message Error. The next bit (bit time 9) is the Message Error (ME) bit. This bit is set to a logic "1" by the remote terminal upon detection of an error in the message or upon detection of an invalid message (i.e., Illegal Command) to the terminal. The error may occur in any of the data words within the message. When the terminal detects an error and sets this bit, none of the data received within the message shall be used. If an error is detected within a message and the ME bit is set, the remote terminal must suppress the transmission of the status word (see Resetting of the Status Word). If the terminal detected an illegal command, the ME bit is set and the status word is transmitted. All remote terminals must implement the ME bit in the status word.

Instrumentation. The Instrumentation bit (bit time 10) is provided to differentiate between a command word and a status word (remember, they both have the same sync pattern). The instrumentation bit in the status word is always set to logic "0." If used, the corresponding bit in the command word is set to a logic "1." This bit in the command word is the most significant bit of the subaddress field, and therefore would limit the subaddresses used to 10000–11110, hence reducing the number of subaddresses available from 30 to 15. The instrumentation bit is also the reason why there are two mode code indentifiers (00000 and 11111), the latter required when the instrumentation bit is used.

Service Request. The Service Request bit (bit time 11) is such that the remote terminal can inform the bus controller that it needs to be serviced. This bit is set to a logic "1" by the subsystem to indicate that servicing is needed. This bit is typically used when the bus controller is "polling" terminals to determine if they require processing. The bus controller upon receiving this bit set to a logic "1" typically does one of the following. It can take a predetermined action such as issuing a series of messages, or it can request further data from the remote terminal as to its needs. The latter can be accomplished by requesting the terminal to transmit data from a defined subaddress or by using the Transit Vector Word Mode Code.

Reserved. Bit times 12–14 are reserved for future growth of the standard and must be set to a logic "0." The bus controller should declare a message in error if the remote terminal responds with any of these bits set in its status word.

Broadcast Command Received. The Broadcast Command Received bit (bit time 15) indicates that the remote terminal received a valid broadcast command. Upon receipt of a valid broadcast command, the remote terminal sets this bit to logic "1" and suppresses the transmission of its status words. The bus controller may issue a Transmit Status Word or Transmit Last Command Word Mode Code to determine if the terminal received the message properly.

Busy. The Busy bit (bit time 16) is provided as a feedback to the bus controller when the remote terminal is unable to move data between the remote terminal electronics and the subsystem in compliance to a command from the bus controller.

In the earlier days of 1553, the Busy bit was required because many of the subsystem interfaces (analogs, synchros, etc.) were much slower compared to the speed of the multiplex data bus. Some terminals were not able to move the data fast enough. So instead of potentially losing data, a terminal was able to set the Busy bit, indicating to the bus controller is could not handle new data at that time, and for the bus controller to try again later. As new systems have been developed, the need for the use of Busy has been reduced. However, there are systems that still need and have a valid use for the Busy bit. Examples of these are radios, where the bus controller issues a command to the radio to tune to a certain frequency. It may take the radio several seconds to accomplish this, and while it is tuning it may set the Busy bit to inform the bus controller that it is doing as it was told.

When a terminal is busy, it does not need to respond to commands in the "normal" way. For receive commands the terminal collects the data, but does not have to pass the data to the subsystem. For transmit commands, the terminal transmits its status word only. Therefore, while a terminal is busy the data it supplies to the rest of the system are not available. This can have an overall effect upon the flow of data within the system and may increase the data latency within time-critical systems (e.g., flight controls).

Some terminals used the Busy bit to overcome design problems, setting the Busy bit whenever needed. Notice 2 to the standard "strongly discourages" the use of the Busy bit. However, as shown in the example above, there are valid needs for its use. Therefore, if used, Notice 2 now requires that the Busy bit may only be set as the result of a particular command received from the bus controller and not due to an internal periodic or processing function. By following this requirement, the bus controller, with prior knowledge of the remote terminal's characteristics, can determine what will cause a terminal to go busy and minimize the effects on data latency throughout the system.

Subsystem Flag. The Subsystem Flag bit (bit time 17) is used to provide "health" data regarding the subsystems to which the remote terminal is connected. Multiple subsystems may logically "OR" their bits together to form a composite health indicator. This single bit is only to serve as an indicator to the bus controller and user of the data that a fault or failure exists. Further information regarding the nature of the failure must be obtained in some other fashion. Typically, a subaddress is reserved for built-in-test (BIT) information, with one or two words devoted to subsystem status data.

Dynamic Bus Control Acceptance. The Dynamic Bus Control Acceptance bit (bit time 18) is used to inform the bus controller that the remote terminal has received the Dynamic Bus Control Mode Code and has accepted control of the bus. For the remote terminal, the setting of this bit is controlled by the subsystem and is based upon passing some level of built-in-test (i.e., a processor passing its power-up and continuous background tests).

The remote terminal upon transmitting its status word becomes the bus controller. The bus controller, upon receipt of the status word from the remote terminal with this bit set, ceases to function as the bus controller and may become a remote terminal or bus monitor.

Terminal Flag. The Terminal Flag bit (bit time 19) is used to inform the bus controller of a fault or failure within the remote terminal circuitry (only the remote terminal). A logic "1" shall indicate a fault condition. This bit is used solely to inform the bus controller of a fault or failure. Further information regarding the nature of the failure must be obtained in some other fashion. Typically, a subaddress is reserved for BIT information, or the bus controller may issue a Transmit BIT Word Mode Code.

Parity. The last bit (bit time 20), is the word parity bit. Only odd parity shall be used.

1.3.2   Message Formats, Validation, and Timing

The primary purpose of the data bus is to provide a common medium for the exchange of data between systems. The exchange of data is based upon message transmissions. The standard defines 10 types of message transmission formats. All of these formats are based upon the three word types just defined. The 10 message formats are shown in Figures 1.6 and 1.7. The message formats have been divided into two groups. These are referred to within the standard as the "information transfer formats" (Figure 1.6) and the "broadcast information transfer formats" (Figure 1.7).

The information transfer formats are based upon the command/response philosophy that all error-free transmissions received by a remote terminal be followed by the transmission of a status word from the terminal to the bus controller. This handshaking principle validates the receipt of the message by the remote terminal.

Broadcast messages are transmitted to multiple remote terminals at the same time. As such, the terminals suppress the transmission of their status words (not doing so would have multiple boxes trying to talk at the same time and thereby "jam" the bus). In order for the bus controller to determine if a terminal received the message, a polling sequence to each terminal must be initiated to collect the status words.

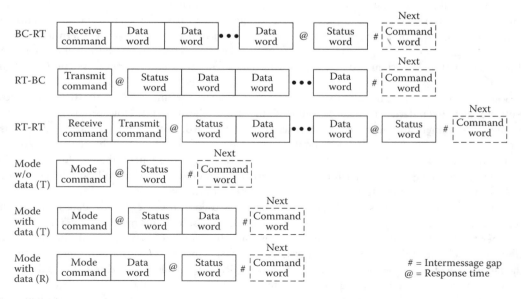

FIGURE 1.6   Information transfer formats.

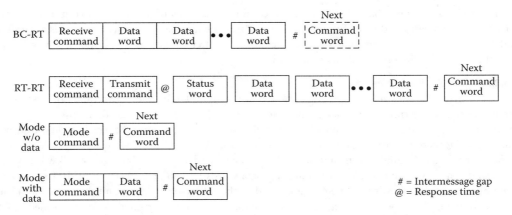

FIGURE 1.7   Broadcast information transfer formats.

Each of the message formats is summarized in the following subsections.

1.3.2.1   Bus Controller to Remote Terminal

The bus controller to remote terminal (BC-RT) message is referred to as the receive command since the remote terminal is going to receive data. The bus controller outputs a command word to the terminal defining the subaddress of the data and the number of data words it is sending. Immediately (without any gap in the transmission), the number of data words specified in the command word are sent.

The remote terminal upon validating the command word and all of the data words will issue its status word within the response time requirements (maximum of 12 μsec).

The remote terminal must be capable of processing the next command that the bus controller issues. Therefore the remote terminal has approximately 56 μsec (status word response time 12 μsec, plus status word transmit time 20 μsec, plus intermessage gap minimum 4 μsec, plus command word transmit time 20 μsec, to either pass the data to the subsystem or buffer the data.

1.3.2.2   Remote Terminal to Bus Controller

The remote terminal to bus controller (RT-BC) message is referred to as a transmit command. The bus controller issues only a transmit command word to the remote terminal. The terminal, upon validation of the command word, will first transmit its status word followed by the number of data words requested by the command word.

Since the remote terminal does not know the sequence of commands to be sent and does not normally operate upon a command until the command word has been validated, it must be capable of fetching from the subsystem the data required within approximately 28 μsec (the status word response time 12 μsec, plus the status word transmission time 20 μsec, minus some amount of time for message validation and transmission delays through the encoder and transceiver).

1.3.2.3   Remote Terminal to Remote Terminal

The remote terminal to remote terminal (RT-RT) command is provided to allow a terminal (the data source) to transfer data directly to another terminal (the data sink) without going through the bus controller. The bus controller may, however, collect and use the data.

The bus controller first issues a command word to the receiving terminal immediately followed by a command word to the transmitting terminal. The receiving terminal is expecting data, but instead of data after the command word it sees a command sync (the second command word). The receiving terminal ignores this word and waits for a word with a data sync.

The transmitting terminal ignored the first command word (it did not contain its terminal address). The second word was addressed to it, so it will process the command as an RT-BC command as described above by transmitting its status word and the required data words.

The receiving terminal, having ignored the second command word, again sees a command (status) sync on the next word and waits further. The next word (the first data word sent) now has a data sync and the receiving remote terminal starts collecting data. After receipt of all of the data words (and validating), the terminal transmits its status word.

1.3.2.3.1 RT-RT Validation

There are several things that the receiving remote terminal of an RT-RT message should do. First, Notice 2 requires that the terminal time is out in 54 to 60 μsec after receipt of the command word. This is required since if the transmitting remote terminal did not validate its command word (and no transmission occurred) then the receiving terminal will not collect data from some new message. This could occur if the next message is either a transmit or receive message, where the terminal ignores all words with a command/status sync and would start collecting data words beginning with the first data sync. If the same number of data words were being transferred in the follow-on message and the terminal did not test the command/status word contents, then the potential exists for the terminal to collect erroneous data.

The other function that the receiving terminal should do, but is not required by the standard, is to capture the second command word and the first transmitted data word. The terminal could compare the terminal address fields of both words to insure that the terminal doing the transmitting was the one commanded to transmit. This would allow the terminal to provide a level of protection for its data and subsystem.

1.3.2.4 Mode Command Formats

Three mode command formats are provided for. This allows for mode commands with no data words and for the mode commands with one data word (either transmitted or received). The status/data sequencing is as described for the BC-RT or RT-BC messages except that the data word count is either one or zero. Mode codes and their use are described later.

1.3.2.5 Broadcast Information Transfer Formats

The broadcast information transfer formats, as shown in Figure 1.8, are identical to the nonbroadcast formats described above with the following two exceptions. First, the bus controller issues commands to terminal address 31 (11111) which is reserved for this function. And secondly, the remote terminals receiving the messages (those which implement the broadcast option) suppress the transmission of their status word.

The broadcast option can be used with the message formats in which the remote terminal receives data. Obviously, multiple terminals cannot transmit data at the same time, so the RT-BC transfer format and the transmit mode code with data format cannot be used. The broadcast RT-RT allows the bus controller to instruct all remote terminals to receive and then instructs one terminal to transmit, thereby allowing a single subsystem to transfer its data directly to multiple users.

Notice 2 allows the bus controller to only use broadcast commands with mode codes (see Broadcast Mode Codes). Remote terminals are allowed to implement this option for all broadcast message formats. The Notice further states that the terminal must differentiate the subaddresses between broadcast and nonbroadcast messages (see Subaddress Utilization).

1.3.2.6 Command and Message Validation

The remote terminal must validate the command word and all data words received as part of the message. The criteria for a valid command word are that the: word begins with a valid command sync, valid terminal address (matches the assigned address of the terminal or the broadcast address if implemented), all bits are in a valid Manchester code, there are 16 information field bits, and there is a valid parity bit (odd). The criteria for a data word are the same except a valid data sync is required and the terminal address field is not tested. If a command word fails to meet the criteria, the command is ignored. After the command has been validated, and a data word fails to meet the criteria, then the terminal shall set the Message Error bit in the status word and suppress the transmission of the status word. Any single error within a message shall invalidate the entire message and the data shall not be used.

1.3.2.7 Illegal Commands

The standard allows remote terminals the option of monitoring for Illegal Commands. An Illegal Command is one that meets the valid criteria for a command word, but is a command (message) that is not implemented by the terminal. An example is if a terminal only outputs 04 data words to subaddress 01 and a command word was received by the terminal that requested it to transmit 06 data words from subaddress 03, then this command, while still a valid command, could be considered by the terminal as illegal. The standard only states that the bus controller shall not issue illegal or invalid commands.

The standard provides the terminal designer with two options. First, the terminal can respond to all commands as usual (this is referred to as "responding in form"). The data received is typically placed in a series of memory locations which are not accessible by the subsystem or applications programs. This is typically referred to as the "bit bucket." All invalid commands are placed into the same bit bucket. For invalid transmit commands, the data transmitted is read from the bit bucket. Remember, the bus controller is not supposed to send these invalid commands.

The second option is for the terminal to monitor for Illegal Commands. For most terminal designs, this is as simple as a look-up table with the T/R bit, subaddress, and word count fields supplying the address and the output being a single bit that indicates if the command is valid or not. If a terminal implements Illegal Command detection and an illegal command is received, the terminal sets the Message Error bit in the status word and responds with the status word.

1.3.2.8 Terminal Response Time

The standard states that a remote terminal, upon validation of a transmit command word or a receive message (command word and all data words) shall transmit its status word to the bus controller. The response time is the amount of time the terminal has to transmit its status word. To allow for accurate measurements, the time frame is measured from the mid-crossing of the parity bit of the command word to the mid-crossing of the sync field of the status word. The minimum time is 4.0 μsec, the maximum time is 12.0 μsec. However, the actual amount of "dead time" on the bus is 2 to 10 μsec since half of the parity bit and sync waveforms are being transmitted during the measured time frame.

The standard also specifies that the bus controller must wait a minimum of 14.0 μsec for a status word response before determining that a terminal has failed to respond. In applications where long data buses are used or where other special conditions exist, it may be necessary to extend this time to 20.0 μsec or greater.

1.3.2.9 Intermessage Gap

The bus controller must provide for a minimum of 4.0 μsec between messages. Again, this time frame is measured from the mid-crossing of the parity bit of the last data word or the status word and the mid-crossing of the sync field of the next command word. The actual amount of "dead time" on the bus is 2 μsec since half of the parity bit and sync waveforms are being transmitted during the measured time frame.

The amount of time required by the bus controller to issue the next command is a function of the controller type (e.g., word, message, or frame). The gap typically associated with word controllers is between 40 and 100 μsec. Message controllers typically can issue commands with a gap of 10 to 30 μsec. But frame controllers are capable of issuing commands at the 4-μsec rate and often must require a time delay to slow them down.

1.3.2.10 Superseding Commands

A remote terminal must always be capable of receiving a new command. This may occur while operating on a command on bus A and after the minimum intermessage gap, a new command appears, or if operating on bus A and a new command appears on bus B. This is referred to as a Superseding Command. A second valid command (the new command) shall cause the terminal to stop operating on the first command and start on the second. For dual redundant applications, this requirement implies that all terminals must, as a minimum, have two receivers, two decoders, and two sets of command word validation logic.

1.3.3 Mode Codes

Mode codes are defined by the standard to provide the bus controller with data bus management and error handling/recovery capability. The mode codes are divided into two groups: with and without data words. The data words that are associated with the mode codes, and only one word per mode code is allowed, contains information pertinent to the control of the bus and do not generally contain information required by the subsystem (the exception may be the Synchronize with Data Word Mode Code). The mode codes are defined by bit times 15–19 of the command word. The most significant bit (bit 15) can be used to differentiate between the two mode code groups. When a data word is associated with the mode code, the T/R bit determines if the data word is transmitted or received by the remote terminal. The mode codes are listed in Table 1.4.

1.3.3.1 Mode Code Identifier

The mode code identifier is contained in bits 10–14 of the command word. When this field is either 00000 or 11111 then the contents of bits 15–19 of the command word are to be decoded as a mode code. Two mode code identifiers are provided such that the system can utilize the Instrumentation bit if desired. The two mode code identifiers shall not convey different information.

1.3.3.2 Mode Code Functions

The following defines the functionality of each of the mode codes.

Dynamic Bus Control. The Dynamic Bus Control Mode Code is used to provide for the passing of the control of the data bus between terminals, thus providing a "round robin" type of control. Using this methodology, each terminal is responsible for collecting the data it needs from all the other terminals. When it is done collecting, it passes control to the next terminal in line (based on some predefined

TABLE 1.4   Mode Code

T/R	Mode Code	Function	Data Word	Broadcast
1	00000	Dynamic Bus Control	No	No
1	00001	Synchronize	No	Yes
1	00010	Transmit Status Word	No	No
1	00011	Initiate Self-Test	No	Yes
1	00100	Transmitter Shutdown	No	Yes
1	00101	Override Transmitter Shutdown	No	Yes
1	00110	Inhibit Terminal Flag Bit	No	Yes
1	00111	Override Inhibit Terminal Flag Bit	No	Yes
1	01000	Reset	No	Yes
1	01001	RESERVED	No	TBD
1	•	•	No	•
1	•	•	No	•
1	01111	RESERVED	No	TBD
1	10000	Transmit Vector Word	Yes	No
0	10001	Synchronize	Yes	Yes
1	10010	Transmit Last Command Word	Yes	No
1	10011	Transmit BIT Word	Yes	No
0	10100	Selected Transmitter Shutdown	Yes	Yes
0	10101	Override Selected Transmitter Shutdown	Yes	Yes
1/0	10110	RESERVED	Yes	TBD
	•	•	Yes	•
	•	•	Yes	•
1/0	11111	RESERVED	Yes	TBD

sequence). This allows the applications program (the end user of the data) to collect the data when it needs it, always insuring that the data collected is from the latest source sample and has not been sitting around in a buffer waiting to be used.

Notices 1 and 2 to the standard forbid the use of Dynamic Bus Control for Air Force applications. This is due to the problems and concerns of what may occur when a terminal, that has passed the control, is unable to perform or does not properly forward the control to the next terminal, thereby forcing the condition of no terminal being in control and having to reestablish control by some terminal. The potential amount of time required to reestablish control could have disastrous effects upon the system (i.e., especially a flight control system).

A remote terminal that is capable of performing as the bus control should be capable of setting the Dynamic Bus Control Acceptance Bit in the terminal's Status Word to logic "1" when it receives the mode code command. Typically, the logic associated with the setting of this bit is based on the subsystem's (computer's) ability to pass some level of confidence test. If the confidence test passes, then the bit is set and the status word is transmitted when the terminal receives the mode command, thereby saying that it will assume the role of bus controller.

The bus controller can only issue the Dynamic Bus Control mode command to one remote terminal at a time. The command obviously is only issued to terminals that are capable of performing as a bus controller. Upon transmitting the command, the bus controller must check the terminal's status word to determine if the Dynamic Bus Control Acceptance Bit is set. If set, the bus controller ceases to function as the controller and becomes either a remote terminal or a bus monitor. If the bit in the status word is not set, the remote terminal which was issued the command is not capable of becoming the bus controller; the current controller must either remain the bus controller or attempt to pass the control to some other terminal.

Synchronize. The synchronize mode code is used to establish some form of timing between two or more terminals. This mode code does not use a data word, therefore the receipt of this command by a terminal must cause some predefined event to occur. Some examples of this event may be the clearing, incrementing, or presetting of a counter; the toggling of an output signal; or the calling of some software routine. Typically, this command is used to time correlate a function such as the sampling of navigation data (i.e., present position, rates, etc.) for flight controls or targeting/fire control systems. Other uses have been for the bus controller to "sync" the back-up controllers (or monitors) to the beginning of a major/minor frame processing.

When a remote terminal receives the Synchronize Mode Command, it should perform its predefined function. For a bus controller, the issuance of the command is all that is needed. The terminal's status word only indicates that the message was received, not that the "sync" function was performed.

Transmit Status Word. This is one of the two commands that does not cause the remote terminal to reset or clear its status word. Upon receipt of this command, the remote terminal transmits the status word that was associated with the previous message, not the status word of the mode code message.

The bus controller uses this command for control and error management of the data bus. If the remote terminal had detected an error in the message and suppressed its status word, then the bus controller can issue this command to the remote terminal to determine if indeed the nonresponse was due to an error. As this command does not clear the status word from the previous message, a detected error by the remote terminal in a previous message would be indicated by having the Message Error bit set in the status word.

The bus controller also uses this command when "polling." If a terminal does not have periodic messages, the RT can indicate when it needs communications by setting the Service Request bit in the status word. The bus controller, by requesting the terminal to transmit only its status word, can determine if the terminal is in need of servicing and can subsequently issue the necessary commands. This "polling" methodology has the potential of reducing the amount of bus traffic by eliminating the transmission of unnecessary words.

Another use of this command is when broadcast message formats are used. As all of the remote terminals will suppress their status words, "polling" each terminal for its status word would reveal whether the terminal received the message by having its Broadcast Command Received bit set.

Initiate Self-Test. This command, when received by the remote terminal, shall cause the remote terminal to enter into its self-test. This command is normally used as a ground-based maintenance function, as part of the system power-on tests, or in flight as part of a fault recovery routine. Note that this test is only for the remote terminal, not the subsystem.

In earlier applications, some remote terminals, upon receipt of this command, would enter self-test and go "offline" for long periods of time. Notice 2, in an effort to control the amount of time that a terminal could be "offline," limited the test time to 100.0 μsec following the transmission of the status word by the remote terminal.

While a terminal is performing its self-test, it may respond to a valid command in the following ways: (a) no response on either bus ("off-line"); (b) transmit only the status word with the Busy bit set; or (c) normal response. The remote terminal may, upon receipt of a valid command received after this mode code, terminate its self-test. As a subsequent command could abort the self-test, the bus controller, after issuing this command, should suspend transmissions to the terminal for the specified amount of time (either a time specified for the remote terminal or the maximum time of 100.0 μsec).

Transmitter Shutdown. This command is used by the bus controller in the management of the bus. In the event that a terminal's transmitter continuously transmits, this command provides for a mechanism to turn the transmitter off. This command is for dual redundant standby applications only.

Upon receipt of this command, the remote terminal shuts down (i.e., turns off) the transmitter associated with the opposite data bus. That is to say if a terminal's transmitter is babbling on the A bus, the bus controller would send this command to the terminal on the B bus (a command on the A bus would not be received by the terminal).

Override Transmitter Shutdown. This command is the complement of the previous one in that it provides a mechanism to turn on a transmitter that had previously been turned off. When the remote terminal receives this command, it shall set its control logic such that the transmitter associated with the opposite bus be allowed to transmit when a valid command is received on the opposite bus. The only other command that can enable the transmitter is the Reset Remote Terminal Mode Command.

Inhibit Terminal Flag. This command provides for the control of the Terminal Flag bit in a terminal's status word. The Terminal Flag bit indicates that there is a error within the remote terminal hardware and that the data being transmitted or the data received may be in error. However, the fault within the terminal may not have any effect upon the quality of the data, and the bus controller may elect to continue with the transmissions knowing a fault exists.

The remote terminal receiving this command shall set its Terminal Flag bit to logic "0" regardless of the true state of this signal. The standard does not state that the built-in-test that controls this bit be halted, but only the results be negated to "0."

Override Inhibit Terminal Flag. This command is the complement of the previous one in that it provides a mechanism to turn on the reporting of the Terminal Flag bit. When the remote terminal receives this command, it shall set its control logic such that the Terminal Flag bit is properly reported based upon the results of the terminal's built-in-test functions. The only other command that can enable the response of the Terminal Flag bit is the Reset Remote Terminal Mode Command.

Reset Remote Terminal. This command, when received by the remote terminal, shall cause the terminal electronics to reset to its power-up state. This means that if a transmitter had been disabled or the Terminal Flag bit inhibited, these functions would be reset as if the terminal had just powered up. Again, remember that the reset applies only to the remote terminal electronics and not to the entire box.

Notice 2 restricts the amount of time that a remote terminal can take to reset its electronics. After transmission of its status word, the remote terminal shall reset within 5.0 μsec. While a terminal is resetting, it may respond to a valid command in the following ways: (a) no response on either bus ("offline"); (b) transmit only the status word with the Busy bit set; or (c) normal response. The remote terminal may, upon receipt of a valid command received after this mode code, terminate its reset function.

As a subsequent command could abort the reset, the bus controller, after issuing this command, should suspend transmissions to the terminal for the specified amount of time (either a time specified for the remote terminal or the maximum time of 5.0 μsec).

Transmit Vector Word. This command shall cause the remote terminal to transmit a data word referred to as the vector word. The vector word shall identify to the bus controller service request information relating to the message needs of the remote terminal. While not required, this mode code is often tied to the Service Request bit in the Status Word. As indicated, the contents of the data word inform the bus controller of messages that need to be sent.

The bus controller also uses this command when "polling." Though typically used in conjunction with the Service Request bit in the status word, wherein the bus controller requests only the status word (Transmit Status Word Mode Code) and upon seeing the Service Request bit set would then issue the Transmit Vector Word Mode Code, the bus controller can always ask for the Vector Word (always getting the status word anyway) and reduce the amount of time required to respond to the terminal's request.

Synchronize with Data Word. The purpose of this synchronize command is the same as the synchronize without data word, except this mode code provides a data word to provide additional information to the remote terminal. The contents of the data word are left to the imagination of the user. Examples from "real world" applications have used this word to provide the remote terminal with a counter or clock value; to provide a backup controller with a frame identification number (minor frame or cycle number); and to provide a terminal with a new base address pointer used in extending the subaddress capability.

Transmit Last Command Word. This is one of the two commands that does not cause the remote terminal to reset or clear its status word. Upon receipt of this command, the remote terminal transmits the status word that was associated with the previous message and the Last Command Word (valid) that it received.

The bus controller uses this command for control and error management of the data bus. When a remote terminal is not responding properly, then the bus controller can determine the last valid command the terminal received and can re-issue subsequent messages as required.

Transmit BIT Word. This mode command is used to provide detail with regards to the Built-in-Test (BIT) status of the remote terminal. Its contents shall provide information regarding the remote terminal only (remember the definition) and not the subsystem.

While most applications associate this command with the Initiate Self Test Mode Code, the standard requires no such association. Typical use is to issue the Initiate Self Test Mode Code, allow the required amount of time for the terminal to complete its tests, and then issue the Transmit BIT Word Mode Code to collect the results of the test. Other applications have updated the BIT word on a periodic rate based on the results of a continuous background test (e.g., as a data wraparound test performed with every data transmission). This word can then be transmitted to the bus controller, upon request, without having to initiate the test and then wait for the test to be completed. The contents of the data word are left to the terminal designer.

Selected Transmitter Shutdown. Like the Transmitter Shutdown Mode Code, this mode code is used to turn off a babbling transmitter. The difference between the two mode codes is that this mode code has a data word associated with it. The contents of the data word specifies which data bus (transmitter) to shutdown. This command is used in systems which provide more than dual redundancy.

Override Selected Transmitter Shutdown. This command is the complement of the previous one in that it provides a mechanism to turn on a transmitter that had previously been turned off. When the remote terminal receives this command, the data word specifies which data bus (transmitter) shall set its control logic such that the transmitter associated with that bus be allowed to transmit when a valid command is received on that bus. The only other command that can enable the selected transmitter is the Reset Remote Terminal Mode Command.

Reserved Mode Codes. As can be seen from Table 1.4, there are several bit combinations that are set aside as reserved. It was the intent of the standard that these be reserved for future growth. It should also be noticed from the table that certain bit combinations are not listed. The standard allows the remote terminal to respond to these reserved and "undefined" mode codes in the following manner: set the message error bit and respond (see Illegal Commands); or respond in form. The designer of terminal hardware or a multiplex system is forbidden to use the reserved mode codes for any purpose.

1.3.3.3   Required Mode Codes

Notice 2 to the standard requires that all remote terminals implement the following four mode codes: Transmit Status Word, Transmitter Shutdown, Override Transmitter Shutdown, and Reset Remote Terminal. This requirement was levied so as to provide the multiplex system designer and the bus controller with a minimum set of commands for managing the multiplex system. Note that the above requirement was placed on the remote terminal. Notice 2 also requires that a bus controller be capable of implementing all of the mode codes, however, for Air Force applications, the Dynamic Bus Control Mode Code shall never be used.

1.3.3.4   Broadcast Mode Codes

Notice 2 to the standard allows the broadcast of mode codes (see Table 1.4). The use of the broadcast option can be of great assistance in the areas of terminal synchronization. Ground maintenance and troubleshooting can take advantage of broadcast Reset Remote Terminal or Initiate Self, but these two commands can have disastrous effects if used while in flight. The designer must provide checks to insure that commands such as these are not issued by the bus controller or operated upon by a remote terminal when certain conditions exists (e.g., in flight).

1.4   Systems-Level Issues

The standard provides very little guidance in how it is applied. Lessons learned from "real world" applications have led to design guides, application notes, and handbooks that provide guidance. This section will attempt to answer some of the systems-level questions and identify implied requirements that, while not specifically called out in the standard, are required nonetheless.

1.4.1   Subaddress Utilization

The standard provides no guidance on how to use the subaddresses. The assignment of subaddresses and their functions (the data content) is left to the user. Most designers automatically start assigning subaddresses with 01 and count upwards. If the Instrumentation bit is going to be used, then the subaddresses must start at 16.

The standard also requires that normal subaddresses be separated from broadcast subaddresses. If the broadcast option is implemented, then an additional memory block is required to receive broadcast commands.

1.4.1.1   Extended Subaddressing

The number of subaddresses that a terminal has is limited to 60 (30 transmit and 30 receive). Therefore, the number of unique data words available to a terminal is 1920 (60 × 32). For earlier applications, where data being transferred were analog sensor data and switch settings, this was more than sufficient. However, in some of today's applications, in which digital computers exchanging data, or for a video sensor passing digitized video data, the number of words is too limited.

Most terminal designs establish a block of memory for use by the 1553 interface circuitry. This block contains a address start pointer and then the memory is offset by the subaddress number and the word count number to arrive at a particular memory address.

A methodology of extending the range of the subaddresses has been successfully utilized. This method uses either a dedicated subaddress and data word, or makes use of the synchronize with data word mode code. The data word associated with either of these contains an address pointer which is used to reestablish the starting address of the memory block. The changing of the blocks is controlled by the bus controller and can be done based on numerous functions. Examples are operational modes, wherein one block is used for startup messages, a different block for take-off and landing, a different block for navigation and cruise, a different block for mission functions (i.e., attack or evade modes), and a different block for maintenance functions.

Another example is that the changing of the start address could also be associated with minor frame cycles. Eight minor frames could have a separate memory block for each frame. The bus controller could synchronize frames and change memory pointers at the beginning of each new minor frame.

For computers exchanging large amounts of data (e.g., GPS almanac tables) or for computers that receive program loads via the data bus at power-up, the bus controller could set the pointers at the beginning of a message block, send 30, 32-word messages, move the memory pointer to the last location in the remote terminals memory that received data, then send the next block of 30, 32-word messages, continuing this cycle until the memory is loaded. The use is left to the designer.

1.4.2 Data Wraparound

Notice 2 to the standard does require that the terminal is able to perform a data wraparound and subaddress 30 is suggested for this function. Data wraparound provides the bus controller with a methodology of testing the data bus from its internal circuitry, through the bus media, to the terminal's internal circuitry. This is done by the bus controller sending the remote terminal a message block and then commanding the terminal to send it back. The bus controller can then compare the sent data with that received to determine the state of the data link. There are no special requirements upon the bit patterns of the data being transferred.

The only design requirements are placed upon the remote terminal. These are that the terminal, for the data wraparound function, be capable of sending the number of data words equal to the largest number of data words sent for any transmit command. This means that if a terminal maximum data transmission is only four data words, it need only provide for four data words in its data wraparound function.

The other requirement is that the remote terminal need only hold the data until the next message. The normal sequence is for the bus controller to send the data, then in the next message it asks for it back. If another message is received by the remote terminal before the bus controller requests the data, the terminal can discard the data from the wraparound message and operate on the new command.

1.4.3 Data Buffering

The standard specifies that the any error within a message shall invalidate the entire message. This implies that the remote terminal must store the data within a message buffer until the last data word has been received and validated before allowing the subsystem access to the data. To insure that the subsystem always has the last message of valid data received to work with would require the remote terminal to, as a minimum, double buffer the received data.

There are several methods to accomplish this in hardware. One method is for the terminal electronics to contain a First-In First-Out (FIFO) memory that stores the data as it is received. Upon validation of the last data word, the terminal's subsystem interface logic will move the contents of the FIFO into memory accessible by the subsystem. If an error occurred during the message, the FIFO is reset.

A second method establishes two memory blocks for each message in common memory. The subsystem is directed to read from one block (block A) while the terminal electronics writes to the other (Block B). Upon receipt of a valid message, the terminal will switch pointers, indicating that the subsystem is to read from the new memory block (block B) while the terminal will now write to block B. If an error occurs within the message, the memory blocks are not switched.

Some of the "off-the-shelf" components available provide for data buffering. Most provide for double buffering, while some provided for multilevels of buffering.

1.4.4   Variable Message Blocks

Remote terminals should be able to transmit any subset of any message. This means that if a terminal has a transmit message at subaddress 04 of 30 data words, it should be capable of transmitting any number of those data words (01–30) if so commanded by the bus controller. The order in which the subset is transmitted should be the same as if the entire message is being transmitted, that being the contents of data word 01 is the same regardless of the word count.

Terminals which implement Illegal Command detection should not consider subsets of a message as illegal. That is to say, if in our example above a command is received for 10 data words, this should not be illegal. But, if a command is received for 32 data words, this would be considered as an illegal command.

1.4.5   Sample Consistency

When transmitting data, the remote terminal needs to ensure that each message transmitted is of the same sample set and contains mutually consistent data. Multiple words used to transfer multiple precision parameters or functionally related data must of the same sampling.

If a terminal is transmitting pitch, roll, and yaw rates, and while transmitting the subsystem updates these data in memory, but this occurs after pitch and roll had been read by the terminal's electronics, then the yaw rate transmitted would be of a different sample set. Having data from different sample rates could have undesirable effects on the user of the data.

This implies that the terminal must provide some level of buffering (the reverse of what was described above) or some level of control logic to block the subsystem from updating data while being read by the remote terminal.

1.4.6   Data Validation

The standard tightly defines the criteria for the validation of a message. All words must meet certain checks (i.e., valid sync, Manchester encoding, number of bits, odd parity, etc.) for each word and each message to be valid. But what about the contents of the data word? MIL-STD-1553 provides the checks to insure the quality of the data transmission from terminal to terminal, sort of a "data in equals data out," but is not responsible for the validation tests of the data itself. This is not the responsibility of the 1553 terminal electronics, but of the subsystem. If bad data are sent, then "garbage in equals garbage out." But the standard does not prevent the user from providing additional levels of protection. The same techniques used in digital computer interfaces (i.e., disk drives, serial interfaces, etc.) can be applied to 1553. These techniques include checksums, CRC words, and error detection/correction codes. Section 80 of MIL-HDBK-1553A which covers data word formats even offers some examples of these techniques.

But what about using the simple indicators embedded within the standard. Each remote terminal provides a status word — indicating not only the health of the remote terminal's electronics, but also that of the subsystem. However, in most designs, the status word is kept within the terminal electronics and not passed to the subsystems. In some "off-the-shelf" components, the status word is not even available to be sent to the subsystem. But two bits from the status word should be made available to the subsystem and the user of the data for further determination as to the validity of the data. These are the Subsystem Flag and the Terminal Flag bits.

1.4.7   Major and Minor Frame Timing

The standard specifies the composition of the words (command, data, and status) and the messages (information formats and broadcast formats). It provides a series of management messages (mode codes), but it does not provide any guidance on how to apply these within a system. This is left to the imagination of the user.

Remote terminals, based upon the contents of their data, will typically state how often data are collected and the fastest rate they should be outputted. For input data, the terminal will often state how often it needs certain data to either perform its job or maintain a certain level of accuracy. The rates are referred to as the transmission and update rates. It is the system designer's job to examine the data needs of all of the systems and determine when data are transferred from whom to whom. These data are subdivided into periodic messages — those which must be transferred at some fixed rate, and aperiodic messages, those which are typically either event driven (i.e., the operator pushes a button) or data driven (i.e., a value is now within range).

A major frame is defined such that all periodic messages are transferred at least once. This is therefore defined by the message with the slowest transmission rate. Typical major frame rates used in today's applications vary from 40 to 640 μsec. There are some systems that have major frame rates in the 1- to 5-sec range, but these are the exceptions, not the norm. Minor frames are then established to meet the requirements of the higher update rate messages.

The sequence of messages within a minor frame is again left undefined. There are two methodologies that are predominately used. In the first method, the bus controller starts the frame with the transmission of all of the periodic messages (transmit and receive) to be transferred in that minor frame. At the end of the periodic messages, the bus controller is either finished (resulting in dead bus time — no transmissions) until the beginning of the next frame, or the bus controller can use this time to transfer aperiodic messages, error handling messages, or transfer data to the back-up bus controller(s).

In the second method (typically used in a centralized processing architecture), the bus controller issues all periodic and aperiodic transmit messages (collects the data), then processes the data (possibly using dead time during this processing), and then issues all the receive messages (outputting the results of the processing). Both methods have been used successfully.

1.4.8   Error Processing

The amount and level of error processing is typically left to the systems designer but may be driven by the performance requirements of the system. Error processing is typically only afforded to critical messages, wherein the noncritical messages just await the next normal transmission cycle. If a data bus is 60% loaded and each message received an error, the error processing would exceed 100% of available time and thereby cause problems within the system.

Error processing is again a function of the level of sophistication of the bus controller. Some controllers (typically message or frame controllers) can automatically perform some degree of error processing. This usually is limited to a retransmission of the message either once on the same bus or once on the opposite bus. Should the retried message also fail, the bus controller software is informed of the problem. The message may then be retried at the end of the normal message list for the minor frame.

If the error still persists, then it may be necessary to stop communicating with the terminal, especially if the bus controller is spending a large amount of time performing error processing. Some systems will try to communicate with a terminal for a predefined number of times on each bus. After this, all messages to the terminal are removed from the minor frame lists, and substituted with a single transmit status word mode code.

An analysis should be performed on the critical messages to determine the effects upon the system if they are not transmitted or the effects of data latency if they are delayed to the end of the frame.

1.5   Testing

The testing of a MIL-STD-1553 terminal or system is not a trivial task. There are a large number of options available to the designer including message formats, mode commands, status word bits, and coupling methodology. In addition, history has shown that different component manufacturers and designers have made different interpretations regarding the standard, thereby introducing products that implement the same function quite differently.

For years, the Air Force provided for the testing of MIL-STD-1553 terminals and components. Today this testing is the responsibility of industry. The Society of Automotive Engineers (SAE), in conjunction

TABLE 1.6   SAE 1553 Test Plans

AS-4111	Remote Terminal Validation Test Plan
AS-4112	Remote Terminal Production Test Plan
AS-4113	Bus Controller Validation Test Plan
AS-4114	Bus Controller Production Test Plan
AS-4115	Data Bus System Test Plan
AS-4116	Bus Monitor Test Plan
AS-4117	Bus Components Test Plan

with the government, has developed a series of test plans for all 1553 elements. These test plans are listed in Table 1.6.

Further Information

In addition to the SAE Test Plans listed in Table 1.6, there are other documents that can provide a great deal of insight and assistance in designing with MIL-STD-1553:

MIL-STD-1553B Digital Time Division Command/Response Multiplex Data Bus
MIL-HDBK-1553A Multiplex Applications Handbook
SAE AS-15531 Digital Time Division Command/Response Multiplex Data Bus
SAE AS-15532 Standard Data Word Formats
SAE AS-12 Multiplex Systems Integration Handbook
SAE AS-19 MIL-STD-1553 Protocol Reorganized
DDC 1553 Designers Guide
UTMC 1553 Handbook

And lastly, there is the SAE 1553 Users Group. This is a collection of industry and military experts in 1553 who provide an open forum for information exchange, and provide guidance and interpretations/clarifications with regard to the standard. This group meets twice a year as part of the SAE Avionics Systems Division conferences.

2

ARINC Specification 429 Mark 33 Digital Information Transfer System

Daniel A. Martinec
ARINC

Samuel P. Buckwalter
ARINC

2.1 Introduction

ARINC specifications 419, 429, and 629 and Project Paper 453 are documents prepared by the Airlines Electronic Engineering Committee (AEEC) and published by Aeronautical Radio Inc. (ARINC). These are among over 300 air transport industry avionics standards published since 1949. These documents, commonly referred to as ARINC 419, ARINC 429, ARINC 453, and ARINC 629, describe data communication systems used primarily on commercial transport airplanes, but which are also used by general aviation and military airplanes. The differences between the systems are described in detail in the subsequent sections.

2.2 ARINC 419

ARINC specification 419, "Digital Data Compendium," provides detailed descriptions of the various interfaces used in the ARINC 500 series of avionics standards prior to 1980. ARINC 419 is often incorrectly assumed to be a standalone bus standard. It provides a summary of electrical interfaces, protocols, and data standards for avionics built prior to the airlines' selection of a single standard (i.e., ARINC 429) for the distribution of digital information aboard aircraft.

2.3 ARINC 429

2.3.1 General

ARINC specification 429, "Digital Information Transfer System (DITS)," was first published in 1977 and has since become the ARINC standard most widely used by the airlines. The title of this airline standard was chosen so as not to describe it as a "data bus." Although ARINC 429 is a vehicle for data transfer, it does not fit the normal definition of a data bus. A typical data bus provides multidirectional transfer of data between multiple points over a single set of wires. ARINC 429's simplistic one-way flow of data significantly limits this capability, but the associated low cost and the integrity of the installations have provided the airlines with a system exhibiting excellent service for more than two decades. Additional information regarding avionics standards may be found at http://www.arinc.com/aeec.

2.3.2 History

In the early 1970s, the airlines recognized the potential advantage of implementation of digital equipment. Some digital equipment had already been implemented to a certain degree on airplanes existing at that time. However, there were three new transport airplanes on the horizon: the Airbus A-310 and the Boeing B-757 and B-767. The airlines, along with the airframe and equipment manufacturers, established a goal to create an all-new suite of avionics using digital technology.

With digital avionics came the need for an effective means of data communications among the avionics units. The airlines recognized that the military was also in the early stages of development of a data bus that could perform the data transfer functions among military avionics. A joint program to produce a data bus common to the air transport industry and the military suggested a potential for significant economical benefits.

The Society of Automotive Engineers (SAE) took on the early work to develop the military's data bus. Participants in the SAE program came from many parts of the military and private sectors of aviation. Considerable effort went into defining all aspects of the data bus with the goal of meeting the needs of both the military and air transport users. That work culminated in the development of the early version of the data bus identified by Mil-Std 1553 (see Chapter 1).

Early in the process of the Mil-Std 1553 development, representatives from the air transport industry realized that the stringent and wide range of military requirements would cause the Mil-Std 1553 to be overly complex for the commercial user and would not exhibit the flexibility to accommodate the varying applications of transport airplanes. Difficulty in certification also was considered a potential problem. The decision was made to abandon a cooperative data bus development program with the military and pursue work on a data bus to more closely reflect commercial airplane requirements.

The numerous single transmitter and multiple receiver data transfer systems used on airplanes built in the early 1970s proved to be reliable and efficient compared to the more complex data buses of the time. These transfer systems, described in ARINC 419, were considered as candidates for the new digital aircraft.

Although none of the systems addressed in the ARINC specification could adequately perform the task, each exhibited desirable characteristics that could be applied to a new design. The result was the release of a new data transfer system exhibiting a high level of efficiency, extremely good reliability, and ease of certification. ARINC 429 became the industry standard. Subsequent to release of the standard, solid-state component manufacturers produced numerous low-cost integrated circuits. ARINC 429 was used widely by the air transport industry and even found applications in nonaviation commercial and military applications. ARINC 429 has been used as the standard for virtually all ARINC 700-series standards for "digital avionics" used by the air transport industry.

ARINC has maintained and provided the necessary routine updates for new data word assignments and formats. There were no significant changes in the basic design until 1980, when operational experience showed that certain shorted wire conditions would allow the bus to operate in a faulty condition with much reduced noise immunity. This condition also proved to be very difficult to locate during routine maintenance. In response, the airlines suggested that the design be changed to ensure that the bus would not continue to operate when this condition occurred, and a change to the receiver voltage thresholds and impedances solved this problem.

No basic changes to the design have been made since that time. ARINC 429 has remained a reliable system and even today is used extensively in the most modern commercial airplanes.

2.3.3 Design Fundamentals

2.3.3.1 Equipment Interconnection

A single transmitter is connected with up to 20 data receivers via a single twisted and shielded pair of wires. The shields of the wires are grounded at both ends and at any breaks along the length of the cable. The shields' connections to the ground should be kept as short as possible.

2.3.3.2 Modulation

Return-To-Zero (RZ) modulation is used. The voltage levels are used for this modulation scheme.

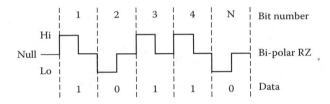

2.3.3.3 Voltage Levels

The differential output voltages across the transmitter output terminal with no load is described in the following table:

	HI(V)	NULL(V)	LO(V)
Line A to line B	+10 ± 1.0	0 ± 0.5	−10 ± 1.0
Line A to ground	5 ± 0.5	0 ± 0.25	−5 ± 0.5
Line B to ground	−5 ± 0.5	0 ± 0.25	5 ± 0.5

The differential voltage seen by the receiver will depend on wire length, loads, stubs, and so on. With no noise present on the signal lines, the nominal voltages at the receiver terminals (A and B) would be the following:

- HI   +7.25V to +11V
- NULL +0.5V to −0.5V
- LO   −7.25V to −11V

In practical installations impacted by noise, and so on, the following voltage ranges will be typical across the receiver input (A and B):

- HI   +6.5V to +13V
- NULL +2.5V to −2.5V
- LO   −6.5V to −13V

Line (A or B) to ground voltages are not defined. Receivers are expected to withstand without damage steady-state voltages of 30 volts alternating current (VAC) root mean square (RMS) applied across terminals A and B, or volts direct current (VDC) applied between terminal A or B and the ground.

2.3.3.4 Impedance Levels

2.3.3.4.1 Transmitter Output Impedance

The transmitter output impedance is 70 to 80 (nominal 75) ohms and is divided equally between lines A and B for all logic states and transitions between those states.

2.3.3.4.2 Receiver Input Impedance

The typical receiver input characteristics are as follows:

- Differential Input Resistance R_I = 12,000 ohms minimum
- Differential Input Capacitance C_I = 50 pF maximum
- Resistance to Ground R_H and $R_G \geq$ 12,000 ohms
- Capacitance to Ground C_H and $C_G \leq$ 50 pF

The total receiver input resistance, including the effects of R_I, R_H, and R_G in parallel, is 8000 ohms minimum (400 ohms minimum for 20 receivers). A maximum of 20 receivers is specified for any one transmitter. See below for the circuit standards.

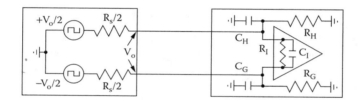

2.3.3.4.3 Cable Impedance

The wire gauges used in the interconnecting cable will typically vary between 20 and 26 depending on desired physical integrity of the cable and weight limitations. Typical characteristic impedances will be in the range of 60 to 80 ohms. The transmitter output impedance was chosen at 75 ohms nominal to match this range.

2.3.3.5 Fault Tolerance

A generator on each engine provides the electrical power on an airplane. The airplane electrical system is designed to take into account any variation in, for example, engine speeds, phase differentials, and power bus switching. However, it is virtually impossible to ensure that the power source will be perfect at all times. Failures within a system can also cause erratic power levels, and the design of the ARINC 429 components takes power variation into account and is not generally susceptible to either damage or erratic operation when those variations occur. The ranges of those variations are provided in the following sections.

2.3.3.5.1 Transmitter External Fault Voltage

Transmitter failures caused by external fault voltages will not typically cause other transmitters or other circuitry in the unit to function outside of their specification limits or to fail.

2.3.3.5.2 Transmitter External Fault Load Tolerance

Transmitters should indefinitely withstand without sustaining damage a short circuit applied (a) across terminals A and B, (b) from terminal A to ground, (c) from terminal B to ground, or (d) b and c simultaneously.

2.3.3.6 Fault Isolation

2.3.3.6.1 Receiver Fault Isolation

Each receiver incorporates isolation provisions to ensure that the occurrence of any reasonably probable internal line replaceable unit (LRU) or bus receiver failure does not cause any input bus to operate outside its specification limits (both undervoltage and overvoltage).

2.3.3.6.2 Transmitter Fault Isolation

Each transmitter incorporates isolation provisions to ensure that it does not undermine any reasonably probable equipment fault condition providing an output voltage in excess of (a) 30 VAC RMS between terminal A and B, (b) +29 VDC between A and ground, or (c) +29 VDC between B and ground.

2.3.3.7 Logic-Related Elements

This section describes the digital transfer system elements considered to be principally related to the logic aspects of the signal circuit.

2.3.3.7.1 Digital Language

Numeric Data — The ARINC 429 accommodates numeric data encoded in two digital languages: BNR expressed in twos complement fractional notation and binary coded decimal (BCD) per the numerical subset of International Standards Organization (ISO) Alphabet No. 5. An information item encoded in both BCD and BNR is assigned unique labels for both BCD and BNR (see Section 2.4.3).

Discrete Data — In addition to handling numeric data as specified above, ARINC 429 is also capable of accommodating discrete items of information either in the unused (pad) bits of data words or, when necessary, in dedicated words.

The rule in the assignment of bits in discrete numeric data words is to start with the least significant bit (LSB) of the word and to continue toward the most significant bit available in the word. There are two types of discrete words: general-purpose discrete words and dedicated discrete words. Seven labels (270 XX to 276 XX) are assigned to the general-purpose discrete words. These words are assigned in ascending label order (starting with 270 XX), where XX is the equipment identifier.

32	31 30	29 28 27 26 25 24 23 22 21 20 19 18 17 16 15 14 13 12 11	10 9	8 7 6 5 4 3 2 1
P	SSM	Data → ← Pad ← Discretes	SDI	Label
		MSB LSB		

Generalized BCD word format

P	SSM	BCD CH #1	BCD CH #2	BCD CH #3	BCD CH #4	BCD CH #5	SDI	8 7 6 5 4 3 2 1
		4 2 1	8 4 2 1	8 4 2 1	8 4 2 1	8 4 2 1		
0	0 0	0 1 0	0 1 0 1	0 1 1 1	1 0 0 0	0 1 1 0	0 0	1 0 0 0 0 0 0 1
Example	2	5	7	8	6	DME distance		

BCD word format example (No discretes)

32	31 30 29	28 27 26 25 24 23 22 21 20 19 18 17 16 15 14 13 12 11	10 9	8 7 6 5 4 3 2 1
P	SSM	Data → ← Pad ← Discretes	SDI	Label
		MSB LSB		

Generalized BCD word format

Maintenance Data (General Purpose) — The general-purpose maintenance words are assigned labels in sequential order, as are the labels for the general-purpose discrete words. The lowest octal value label assigned to the maintenance words is used when only one maintenance word is transmitted. When more than one word is transmitted, the lowest octal value label is used first, and the other labels are used sequentially until the message has been completed. The general-purpose maintenance words may contain discrete, BCD, or BNR numeric data. They do not contain ISO Alphabet No. 5 messages. The general-purpose maintenance words are formatted according to the layouts of the corresponding BCD, BNR, and discrete data words shown in the word formats above.

2.4   Message and Word Formatting

2.4.1   Direction of Information Flow

The information output of a system element is transmitted from a designated port (or ports) to which the receiving ports of other system elements in need of that information are connected. In no case does information flow into a port designated for transmission. A separate data bus (twisted and shielded pair of wires) is used for each direction when data are required to flow both ways between two system elements.

2.4.2   Information Element

The basic information element is a digital word containing 32 bits. There are five application groups for such words: BNR data, BCD data, discrete data, general maintenance data and acknowledgment, ISO Alphabet No. 5, and maintenance (ISO Alphabet No. 5) data (AIM). The relevant data-handling rules are set forth in Section 2.4.6. When less than the full data field is needed to accommodate the information conveyed in a word in the desired manner, the unused bit positions are filled with binary zeros or, in the case of BNR and BCD numeric data, valid data bits. If valid data bits are used, the resolution may exceed the accepted standard for an application.

2.4.3   Information Identifier

A six-character label identifies the type of information contained in a word. The first three characters are octal characters coded in binary in the first eight bits of the word. The eight bits will identify the information contained within BNR and BCD numeric data words (e.g., DME distance, static air temperature, etc.) and identify the word application for discrete, maintenance, and AIM data.

The last three characters of the six-character label are hexadecimal characters used to identify ARINC 429 bus sources. Each triplet of hexadecimal characters identifies a system element with one or more DITS ports. Each three-character code (and black box) may have up to 255 eight-bit labels assigned to it. The code is used administratively to retain distinction between unlike parameters having like label assignments.

Octal label 377 has been assigned for the purpose of electrically identifying the system element. The code appears in the three least significant digits of the 377 word in a BCD word format. Although data encoding is based on the BCD word format, the sign/status matrix (SSM) encoding is per the discrete word criteria to provide enhanced failure warning. The transmission of the equipment identifier word on a bus will permit receivers attached to the bus to recognize the source of the DITS information. Since the transmission of the equipment identifier word is optional, receivers should not depend on that word for correct operation.

2.4.4   Source/Destination Identifier

Bit numbers 9 and 10 of numeric data words are used for a data source and destination identification function. They are not available for this function in alphanumeric (ISO Alphabet No. 5) data words or when the resolution needed for numeric (BNR/BCD) data necessitates their use for valid data. The source and destination identifier function may be applied when specific words need to be directed to a specific system of a multisystem installation or when the source system of a multisystem installation needs to be recognizable from the word content. When it is used, a source equipment encodes its aircraft installation

TABLE 2.1   SDI Encoding

Bit No.		Installation
10	9	See text
0	0	—
0	1	1
1	0	2
1	1	3

Note: In certain specialized applications of the SDI function, the all-call capability may be forfeited so that code "00" is available as an "installation no. 4" identifier.

number in bits 9 and 10 as shown in Table 2.1. A sink equipment will recognize words containing its own installation number code and words containing code "00," the "all-call" code.

Equipment will fall into the categories of source only, sink only, or both source and sink. Use of the SDI bits by equipment functioning only as a source or only as a sink is described above. Both the source and sink texts above are applicable to equipment functioning as both a source and a sink. Such equipment will recognize the SDI bits on the inputs and also encode the SDI bits, as applicable, on the outputs. DME, VOR, ILS (Instrument Landing System), and other sensors are examples of source and sink equipment generally considered to be only source equipment. These are actually sinks for their own control panels. Many other types of equipment are also misconstrued as source only or sink only. If a unit has a 429 input port and a 429 output port, it is a source and sink. With the increase of equipment consolidation (e.g., centralized control panels), the correct use of the SDI bits cannot be overstressed.

When the SDI function is not used, binary zeros or valid data should be transmitted in bits 9 and 10.

2.4.5   Sign/Status Matrix

This section describes the coding of the SSM field. In all cases, the SSM field uses bits 30 and 31; for BNR data words, the SSM field also includes bit 29. The SSM field is used to report hardware equipment condition (fault/normal), operational mode (functional test), or validity of data word content (verified/ no computed data). The following definitions apply:

Invalid data — This is defined as any data generated by a source system whose fundamental characteristic is the inability to convey reliable information for the proper performance of a user system. There are two categories of invalid data: (1) no computed data and (2) failure warning.

No computed data — This is a particular case of data invalidity in which the source system is unable to compute reliable data for reasons other than system failure. This inability to compute reliable data is caused exclusively by a definite set of events or conditions whose boundaries are uniquely defined in the system characteristic.

Failure warning — This is a particular case of data invalidity in which the system monitors have detected one or more failures. These failures are uniquely characterized by boundaries defined in the system characteristic.

Displays are normally "flagged invalid" during a "failure warning" condition. When a "no computed data" condition exists, the source system indicates that its outputs are invalid by setting the SSM of the affected words to the "no computed data" code, as defined in the subsections that follow. The system indicators may or may not be flagged depending on system requirements.

While the unit is in the functional test mode, all output data words generated within the unit (i.e., pass-through words are excluded) are coded with "functional test." Pass-through data words are those words received by the unit and retransmitted without alteration.

When the SSM code is used to transmit status and more than one reportable condition exists, the condition with the highest priority is encoded in bits number 30 and 31. The order of condition priorities is shown in the following table:

Failure warning	Priority 1
No computed data	Priority 2
Functional test	Priority 3
Normal operation	Priority 4

Each data word type has its own unique utilization of the SSM field. These various formats are described in the following sections.

2.4.5.1   BCD Numeric

When a failure is detected within a system that would cause one or more of the words normally output by that system to be unreliable, the system stops transmitting the affected word or words on the data bus. Some avionics systems are capable of detecting a fault condition that results in less than normal accuracy. In these systems, when a fault of this nature (for instance, partial sensor loss which results in degraded accuracy) is detected, each unreliable BCD digit is encoded "1111" when transmitted on the data bus. For equipment having a display, the "1111" code should, when received, be recognized as representing an inaccurate digit and a dash or equivalent symbol is normally displayed in place of the inaccurate digit.

The sign (plus/minus, north/south, etc.) of BCD numeric data is encoded in bits 30 and 31 of the word as shown in Table 2.2. Bits 30 and 31 of BCD numeric data words are zero where no sign is needed.

The "no computed data" code is annunciated in the affected BCD numeric data words when a source system is unable to compute reliable data for reasons other than system failure. When the "functional test" code appears in bits 30 and 31 of an instruction input data word, it is interpreted as a command to perform a functional test.

2.4.5.2   BNR Numeric Data Words

The status of the transmitter hardware is encoded in the status matrix field (bit numbers 30 and 31) of BNR numeric data words as shown in Table 2.3.

A source system annunciates any detected failure that causes one or more of the words normally output by that system to be unreliable by setting bit numbers 30 and 31 in the affected word(s) to the "failure warning" code defined in Table 2.3. Words containing this code continue to be supplied to the data bus during the failure condition.

The "no computed data" code is annunciated in the affected BNR numeric data words when a source system is unable to compute reliable data for reasons other than system failure.

When the "functional test" code appears as a system output, it is interpreted as advice that the data in the word result from the execution of a functional test. A functional test produces indications of one-eighth of positive full-scale values unless indicated otherwise in an ARINC equipment characteristic.

TABLE 2.2   BCD Numeric Sign/Status Matrix

Bit No.		
31	30	Function
0	0	Plus, North, East, Right, To, Above
0	1	No Computed Data
1	0	Functional Test
1	1	Minus, South, West, Left, From, Below

TABLE 2.3   Status Matrix

Bit No. 31	30	Function
0	0	Failure Warning
0	1	No Computed Data
1	0	Functional Test
1	1	Normal Operation

TABLE 2.4   Status Matrix

Bit No. 29	Function
0	Plus, north, east, right, to, above
1	Minus, south, west, left, from, below

If, during the execution of a functional test, a source system detects a failure that causes one or more of the words normally output by that system to be unreliable, it changes the states of bits 30 and 31 in the affected words such that the "functional test" annunciation is replaced with a "failure warning" annunciation.

The sign (plus, minus, north, south, etc.) of BNR numeric data words are encoded in the sign matrix field (bit 29) as shown in Table 2.4. Bit 29 is zero when no sign is needed.

Some avionics systems are capable of detecting a fault condition that results in less than normal accuracy. In these systems, when a fault of this nature (for instance, partial sensor loss, which results in degraded accuracy) is detected, the equipment will continue to report "normal" for the SSM while indicating the degraded performance by coding bit 11 as shown in Table 2.5.

This implies that degraded accuracy can be coded only in BNR words not exceeding 17 bits of data.

2.4.5.3   Discrete Data Words

A source system annunciates any detected failure that could cause one or more of the words normally output by that system to be unreliable. Three methods are defined. The first method is to set bits 30 and 31 in the affected words to the "failure warning" code defined in Table 2.6. Words containing the "failure

TABLE 2.5   Accuracy Status

Bit No. 11	Function
0	Nominal accuracy
1	Degraded accuracy

TABLE 2.6   Discrete Data Words

Bit No. 31	30	Function
0	0	Verified data, normal operation
0	1	No computed data
1	0	Functional test
1	1	Failure warning

warning" code continue to be supplied to the data bus during the failure condition. When using the second method, the equipment may stop transmitting the affected word or words on the data bus. This method is used when the display or utilization of the discrete data by a system is undesirable. The third method applies to data words, which are defined such that they contain failure information within the data field. For these applications, the associated ARINC equipment characteristic specifies the proper SSM reporting. Designers are urged not to mix operational and built-in test equipment (BITE) data in the same word.

The "no computed data" code is annunciated in the affected discrete data words when a source system is unable to compute reliable data for reasons other than system failure. When the "functional test" code appears as a system output, it is interpreted as advice that the data in the discrete data word contents are the result of the execution of a functional test.

2.4.6 Data Standards

The units, ranges, resolutions, refresh rates, number of significant bits, and pad bits for the items of information to be transferred by the Mark 33 DITS are administered by the AEEC and tabulated in ARINC characteristic 429.

ARINC characteristic 429 calls for numeric data to be encoded in BCD and binary, the latter using two's complement fractional notation. In this notation, the most significant bit of the data field represents half of the maximum value chosen for the parameter being defined. Successive bits represent the increments of a binary fraction series. Negative numbers are encoded as the two's complements of positive value and the negative sign is annunciated in the SSM.

In establishing a given parameter's binary data standards, the unit's maximum value and resolution are first determined in that order. The LSB of the word is then given a value equal to the resolution increment, and the number of significant bits is chosen such that the maximum value of the fractional binary series just exceeds the maximum value of the parameter (i.e., it equals the next whole binary number greater than the maximum parameter value less one LSB value). For example, to transfer altitude in units of feet over a range of zero to 100,000 ft with a resolution of 1 ft, the number of significant bits is 17 and the maximum value of the fractional binary series is 131,071 (i.e., 131,072 − 1).

Note that because accuracy is a quality of the measurement process and not the data transfer process, it plays no part in the selection of word characteristics. Obviously, the resolution provided in the data word should equal or exceed the accuracy not to degrade it.

For the binary representation of angular data, the ARINC 429 employs "degrees divided by" as the unit of data transfer and ±1 (semicircle) as the range for two's complement fractional notation encoding (ignoring, for the moment, the subtraction of the LSB value). Thus, the angular range 0 through 359.XXX degrees is encoded as 0 through ±179.XXX degrees, the value of the most significant bit is one half semicircle, and there are no discontinuities in the code.

This may be illustrated as follows. Consider encoding the angular range 0 to 360 in one degree increments. Per the general encoding rules above, the positive semicircle will cover the range 0 to 179 (one LSB less than full range). All the bits of the code will be zeros for 0 and ones for 179, and the SSM will indicate the positive sign. The negative semicircle will cover the range 180 to 359. All the bits will be zeros for 180. The codes for angles between 181 to 359 will be determined by taking the two's complements of the fractional binary series for the result of subtracting each value from 360. Thus, the code for 181 is the two's complement of the code for 179. Throughout the negative semicircle, which includes 180, the SSM contains the negative sign.

2.5 Timing-Related Elements

This section describes the digital data transfer system elements considered to be principally related to the timing aspects of the signal circuit.

2.5.1 Bit Rate

2.5.1.1 High-speed operation — The bit rate for high-speed operation of the system is 100 kilobits per second (kbps) ±1%.

2.5.1.2 Low-speed operation — The bit rate for low-speed operation of the system is within the range 12.0 to 14.5 kbps. The selected rate is maintained within 1%.

2.5.2 Information Rates

The minimum and maximum transmit intervals for each item of information are specific by ARINC 429. Words with like labels but with different SDI codes are treated as unique items of information. Each and every unique item of information is transmitted once during an interval bounded in length by the specified minimum and maximum values. Stated another way, a data word having the same label and four different SDI codes will appear on the bus four times (once for each SDI code) during that time interval.

Discrete bits contained within data words are transferred at the bit rate and repeated at the update rate of the primary data. Words dedicated to discretes should be repeated continuously at specified rates.

2.5.3 Clocking Method

Clocking is inherent in the data transmission. The identification of the bit interval is related to the initiation of either a HI or LO state from a previous NULL state in a bipolar RZ code.

2.5.4 Word Synchronization

The digital word should be synchronized by reference to a gap of four bit times (minimum) between the periods of word transmissions. The beginning of the first transmitted bit following this gap signifies the beginning of the new word.

2.5.5 Timing Tolerances

The waveform timing tolerances are shown below.

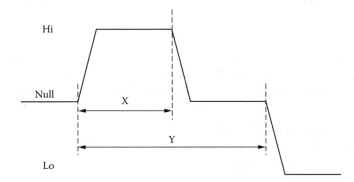

Note: Pulse rise and fall times are measured between the 10% and 90% voltage amplitude points on the leading and trailing edges of the pulse and include time skew between the transmitter output voltages A-to-ground and B-to-ground.

TABLE 2.7

Parameter	High-Speed Operation	Low-Speed Operation
Bit rate	100 kbps ± 1%	12 to 14.5 kbps
Time Y	10 μsec ± 2.5%	Za μsec ± 2.5%
Time X	5 μsec ± 5%	Y/2 ± 5%
Pulse rise time	1.5 ± 0.5 μsec	10 ± 5 μsec
Pulse fall time	1.5 ± 0.5 μsec	10 ± 5 μsec

ªZ 5 1/R where R 5 bit rate selected from 12 to 14.5 kbps range.

2.6   Communications Protocols

2.6.1   Development of File Data Transfer

AEEC adopted ARINC 429 in July 1977. It defined a broadcast data bus with general provisions for file data transfer. In October 1989, AEEC updated a file data transfer procedure with a more comprehensive process that will support the transfer of both bit- and character-oriented data. The new protocol became known as the "Williamsburg Protocol."

2.6.1.1   File Data Transfer Techniques

This "file data transfer techniques" specification describes a system in which an LRU may generate binary extended length messages "on demand." Data is sent in the form of link data units (LDUs) organized in 8-bit octets. System address labels (SALs) are used to identify the recipient. Two data bus speeds are supported.

2.6.1.2   Data Transfer

The same principles of the physical layer implementation apply to file data transfer. Any avionics system element having information to transmit does so from a designated output port over a single twisted and shielded pair of wires to all other system elements having need of that information. Unlike the simple broadcast protocol that can deliver data to multiple recipients in a single transmission, the file transfer technique can be used only for point-to-point message delivery.

2.6.1.3   Broadcast Data

The broadcast transmission technique described above can be supported concurrently with file data transfer.

2.6.1.4   Transmission Order

The most significant octet of the file and LSB of each octet should be transmitted first. The label is transmitted ahead of the data in each case. It may be noted that the label field is encoded in reverse order (i.e., the LSB of the word is the most significant bit of the label). This "reversed label" characteristic is a legacy from past systems in which the octal coding of the label field was, apparently, of no significance.

2.6.1.5   Bit-Oriented Protocol Determination

An LRU will require logic to determine which protocol (character- or bit-oriented) and which version to use when prior knowledge is not available.

2.6.2   Bit-Oriented Communications Protocol

This subsection describes Version 1 of the bit-oriented (Williamsburg) protocol and message exchange procedures for file data transfer between units desiring to exchange bit-oriented data assembled in data files. The bit-oriented protocol is designed to accommodate data transfer between sending and receiving units in a form compatible with the Open Systems Interconnect (OSI) model developed by the ISO. This document directs itself to an implementation of the link layer; however, an overview of the first four layers (physical, link, network, and transport) is provided.

Communications will permit the intermixing of bit-oriented file transfer data words (which contain system address labels (SALs)) with conventional data words (which contain label codes). If the sink should receive a conventional data word during the process of accepting a bit-oriented file transfer message, the sink should accept the conventional data word and resume processing of the incoming file transfer message.

The data file and associated protocol control information are encoded into 32-bit words and transmitted over the physical interface. At the link layer, data are transferred using a transparent bit-oriented data file transfer protocol designed to permit the units involved to send and receive information in multiple word frames. It is structured to allow the transmission of any binary data organized into a data file composed of octets.

Physical medium — The physical interface is described above.

Physical layer — The physical layer provides the functions necessary to activate, maintain, and release the physical link that will carry the bit stream of the communication. The interfacing units use the electrical interface, voltage, and timing described earlier. Data words will contain 32 bits; bits 1 through 8 will contain the SAL and bit 32 will be the parity (odd) bit.

Link layer — The link layer is responsible for transferring information from one logical network entity to another and for enunciating any errors encountered during transmission. The link layer provides a highly reliable virtual channel and some flow control mechanisms.

Network layer — The network layer performs a number of functions to ensure that data packets are properly routed between any two terminals. The network layer expects the link layer to supply data from correctly received frames. The network layer provides for the decoding of information up to the packet level to determine which node (unit) the message should be transferred to. To obtain interoperability, this process, though simple in this application, must be reproduced using the same set of rules throughout all the communications networks (and their subnetworks) on-board the aircraft and on the ground. The bit-oriented data link protocol was designed to operate in a bit-oriented network layer environment. Specifically, ISO 8208 would typically be selected for the subnetwork layer protocol for air and ground subnetworks. There are, however, some applications in which the bit-oriented file transfer protocol will be used under other network layer protocols.

Transport layer — The transport layer controls the transportation of data between a source end-system to a destination end-system. It provides "network independent" data delivery between these processing end-systems. It is the highest order function involved in moving data between systems and it relieves higher layers from any concern specifically with the transportation of information between them.

2.6.2.1 Link Data Units

A LDU contains binary encoded octets. The octets may be set to any possible binary value. The LDU may represent raw data, character data, bit-oriented messages, character-oriented messages, or any string of bits desired. The only restriction is that the bits must be organized into full 8-bit octets. The interpretation of those bits is not a part of the link layer protocol. The LDUs are assembled to make up a data file.

LDUs consist of a set of contiguous ARINC 429 32-bit data words, each containing the SAL (see Section 2.6.2.3) of the sink. The initial data word of each LDU is a start of transmission (SOT). The data described above are contained within the data words that follow. The LDU is concluded with an end of transmission (EOT) data word. No data file should exceed 255 LDUs.

Within the context of this document, LDUs correspond to frames and files correspond to packets.

2.6.2.2 Link Data Unit Size and Word Count

The LDU may vary in size from 3 to 255 ARINC 429 words including the SOT and EOT words. When a LDU is organized for transmission, the total number of ARINC 429 words to be sent (word count) is calculated. The word count is the sum of the SOT word, the data words in the LDU, and the EOT word.

In order to obtain maximum system efficiency, the data is typically encoded into the minimum number of LDUs.

The word count field is 8 bits in length. Thus, the maximum number of ARINC 429 words that can be counted in this field is 255. The word count field appears in the request to send (RTS) and clear to send (CTS) data words. The number of LDUs needed to transfer a specific data file will depend upon the method used to encode the data words.

2.6.2.3 System Address Labels

LDUs are sent point-to-point, even though other systems may be connected and listening to the output of a transmitting system. In order to identify the intended recipient of a transmission, the label field (bits 1–8) is used to carry a SAL. Each on-board system is assigned a SAL. When a system sends an LDU to another system, the sending system (the source) addresses each ARINC 429 word to the receiving system (the sink) by setting the label field to the SAL of the sink. When a system receives any data containing its SAL that is not sent through the established conventions of this protocol the data are ignored.

In the data transparent protocol data files are identified by content rather than by an ARINC 429 label. Thus, the label field loses the function of parameter identification available in broadcast communications.

2.6.2.4 Bit Rate and Word Timing

Data transfer may operate at either high speed or low speed. The source introduces a gap between the end of each ARINC 429 word transmitted and the beginning of the next. The gap should be 4 bit times (minimum), and the sink should be capable of receiving the LDU with the minimum word gap of 4 bit times between words. The source should not exceed a maximum average of 64 bit times between data words of an LDU.

The maximum average word gap is intended to compel the source to transmit successive data words of an LDU without excessive delay. This provision prevents a source that is transmitting a short message from using the full available LDU transfer time. The primary value of this provision is realized when assessing a maximum LDU transfer time for short fixed-length LDUs, such as for automatic dependence surveillance (ADS).

If a Williamsburg source device were to synchronously transmit long-length or full LDUs over a single ARINC 429 data bus to several sink devices, the source may not be able to transmit the data words for a given LDU at a rate fast enough to satisfy this requirement because of other bus activity. In aircraft operation, given the asynchronous burst mode nature of Williamsburg LDU transmissions, it is extremely unlikely that a Williamsburg source would synchronously begin sending a long-length or full LDU to more than two Williamsburg sink devices. A failure to meet this requirement will result in either a successful (but slower) LDU transfer or an LDU retransmission due to an LDU transfer time-out.

2.6.2.5 Word Type

The Word Type field occupies bits 31–29 in all bit-oriented LDU words and is used to identify the function of each ARINC 429 data word used by the bit-oriented communication protocol.

2.6.2.6 Protocol Words

The protocol words are identified with a Word Type field of 100 and are used to control the file transfer process.

2.6.2.6.1 Protocol Identifier

The protocol identifier field occupies bits 28–25 of the protocol word and identifies the type of protocol word being transmitted. Protocol words with an invalid protocol identifier field are ignored.

2.6.2.6.2 Destination Code

Some protocol words contain a Destination Code. The Destination Code field (bits 24–17) indicates the final destination of the LDU. If the LDU is intended for the use of the system receiving the message, the destination code may be set to null (hex 00). However, if the LDU is a message intended to be passed on to another on-board system, the Destination Code will indicate the system to which the message is to be passed. The Destination Codes are assigned according to the applications involved and are used in the Destination Code field to indicate the address of the final destination of the LDU.

In an OSI environment, the link layer protocol is not responsible for validating the destination code. It is the responsibility of the higher-level entities to detect invalid destination codes and to initiate error logging and recovery. Within the pre-OSI environment, the Destination Code provides network layer information. In the OSI environment, this field may contain the same information for routing purposes between OSI and non-OSI systems.

2.6.2.6.3   Word Count

Some protocol words contain a Word Count field. The Word Count field (bits 16–9) reflects the number of ARINC 429 words to be transmitted in the subsequent LDU. The maximum word count value is 255 ARINC 429 words and the minimum is 3 ARINC 429 words. A LDU with the minimum word count value of 3 ARINC 429 words would contain a SOT word, a data word, and an EOT word. A LDU with the maximum word count value of 255 ARINC 429 words would contain a SOT word, 253 data words, and an EOT word.

2.7   Applications

2.7.1   Initial Implementation

ARINC 429 was first used in the early 1980s on the Airbus A-310 and Boeing B-757 and B-767 airplanes. Approximately 150 separate buses interconnecting computers, radios, displays, controls, and sensors accommodated virtually all data transfer on these airplanes. Most of these buses operate at the lower speed; the few that operate at the higher speed of 100 kbps are typically connected to critical navigation computers.

2.7.2   Evolution of Controls

The first applications of ARINC 429 for controlling devices were based on the federated avionics approach used on airplanes that comprised mostly analog interfaces. Controllers for tuning communications equipment used an approach defined as two-out-of-five tuning. Each digit of the desired radio frequency was encoded on each set of five wires. Multiple digits dictated the need for multiple sets of wires for each radio receiver.

The introduction of ARINC 429 proved to be a major step toward reduction of wires. A tuning unit needed only one ARINC 429 bus to tune multiple radios of the same type. An entire set of radios and navigation receivers could be tuned with a few control panels, using approximately the same number of wires previously required to tune a single radio.

As cockpit space became more critical, the need to reduce the number of control panels became critical. The industry recognized that a single control panel, properly configured, could replace most of the existing control panels. The multipurpose control/display unit (MCDU), which came from the industry effort, was derived essentially from the control and display approach used by the rather sophisticated controller for the flight management system. For all intents and purposes, the MCDU became the cockpit controller.

A special protocol had to be developed for ARINC 429 to accommodate the capability of addressing different units connected to a single ARINC 429 bus from the MCDU. The protocol employed two-way communications using two pairs of wires between the controlling unit and the controlled device. An addressing scheme provided for selective communications between the controlling unit and any one of the controlled units. Only one output bus from the controller is required to communicate addresses and commands to the receiving units. With the basic ARINC 429 design, up to 20 controlled units could be connected to the output of the controller. Each of the controlled units is addressed by an assigned SAL.

2.7.3 Longevity of ARINC 429

New airplane designs in the twenty-first century continue to employ the ARINC 429 bus for data transmission. The relative simplicity and integrity of the bus, as well as the ease of certification, are characteristics that contribute to the continued selection of the ARINC 429 bus when the required data bandwidth is not critical. The ARINC 629 data bus developed as the replacement for ARINC 429 is used in applications where a large amount of data must be transferred or where many sources and sinks are required on a single bus.

2.8 ARINC 453

ARINC Project Paper 453 was developed by the AEEC in response to an anticipated requirement for data transfer rates higher than achievable with ARINC 429. The original drafts of Project Paper 453 were based on techniques already employed at that time. The electrical characteristics, including the physical medium, voltage thresholds, and modulation techniques were based on Mil-Std 1553 (see Chapter 1). The data protocols and formats were based on those used in ARINC Specification 429.

During the preparation of the drafts of Project Paper 453, Boeing petitioned AEEC to consider the use of the digital autonomous terminal access communications (DATAC) bus developed by Boeing to accommodate higher data throughput. AEEC accepted Boeing's recommendation for the alternative. ARINC 629 was based on the original version of the Boeing DATAC bus. The work on Project 453 was then curtailed. The latest draft of Project Paper 453 is maintained by ARINC for reference purposes only.

3

Commercial Standard Digital Bus

Lee Harrison
Galaxy Scientific Corp.

3.1   Introduction

The Commercial Standard Digital Bus (CSDB) is one of several digital serial integration data buses that currently predominate in civilian aircraft. The CSBD finds its primary implementations in the smaller business and private general aviation (GA) aircraft but has also been used in retrofits of some commercial transport aircraft.

Rockwell Collins developed CSDB, a unidirectional data bus. The bus used in a particular aircraft is determined by which company the airframe manufacturer chooses to supply the avionics.

CSDB is an asynchronous linear broadcast bus, specifying the use of a twisted, shielded pair cable for device interconnection. Two bus speeds are defined in the CSDB specification: a low-speed bus operates at 12,500 bits per second (bps) and a high-speed bus operates at 50,000 bps. The bus uses twisted, unterminated, shielded pair cable and has been tested to lengths of 50 m.

The CSDB standard also defines other physical characteristics such as modulation technique, voltage levels, load capacitance, and signal rise and fall times. Fault protection for short-circuits of the bus conductors to both 28 volts direct-current (VDC) and 115 volts alternating curent (VAC) is defined by the standard.

3.2   Bus Architecture

Only one transmitter can be attached to the bus, but it can accommodate up to ten receivers. Figure 3.1 illustrates the unidirectional bus architecture.

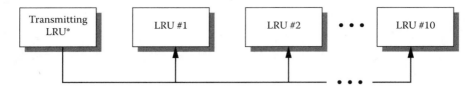

FIGURE 3.1 Unidirectional CSDB communication. *LRU: line replacement unit.

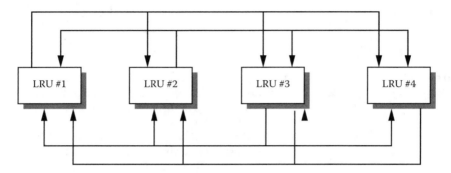

FIGURE 3.2 Bidirectional CSDB communication.

Bidirectional transmission can take place between two bus users. If a receiving bus user is required to send data to any other bus user, a separate bus must be used. Figure 3.2 shows how CSDB may implement bidirectional transmissions between bus users. It can be seen that if each bus user were required to communicate with every other bus user, a significantly greater amount of cabling would be required. In general, total interconnectivity has not been a requirement for CSDB-linked bus users.

It is possible to interface CSDB to other data buses. When this is done, a device known as a gateway interfaces to CSDB and the other bus. If the other bus is ARINC 429 compliant, then messages directed through the gateway from CSDB are converted to the ARINC 429 protocol (see Chapter 2), and vice versa. The gateway would handle bus timing, error checking, testing, and other necessary functions. The system designers would ensure that data latency introduced by the gateway would not cause a "stale data" problem, resulting in a degradation of system performance. Data are stale when they do not arrive at the destination line replaceable unit (LRU) when required, as specified in the design.

3.3 Basic Bus Operation

In Section 2.1.4 of the CSDB standard, three types of transmission are defined: (1) continuous repetition, (2) noncontinuous repetition, and (3) burst transmissions. Continuous repetition transmission refers to the periodic updates of certain bus messages. Some messages on CSDB are transmitted at a greater repetition rate than others. The CSDB standard lists these update rates, along with the message address and message block description. Noncontinuous repetition is used for parameters that are not always valid or available, such as mode or test data. When noncontinuous repetition transmission is in use, it operates the same as continuous repetition. Burst transmission initiates an action (such as radio tuning) or may be used to announce a specific event. Operation in this mode initiates 16 repetitions of the action in each of 16 successive frames, using an update rate of 20 per second.

For CSDB, bytes consist of 11 bits: a start bit, 8 data bits, a parity bit, and a stop bit. The least significant bit (bit 0) follows the start bit. The CSDB standard defines the message block as "a single serial message consisting of a fixed number of bytes transmitted in a fixed sequence" (GAMA CSDB 1986). Essentially, a message block consists of a number of bytes concatenated together, with the first byte always being an

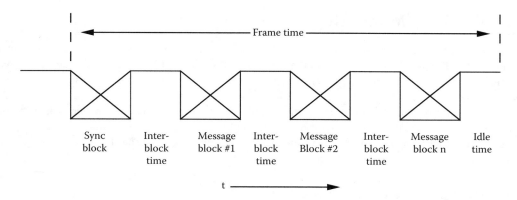

FIGURE 3.3 CSDB data frame structure.

address byte. A status byte may or may not be included in the message block. When it is, it immediately follows the address byte. The number of data bytes in a message block varies.

Data are sent as frames consisting of a synchronization block followed by a number of message blocks. A particular frame is defined from the start of one synchronization block to the start of the next synchronization block. A "sync" block consists of N bytes of the sync character, which is defined as the hexadecimal character "A5." The sync character is never used as an address. While the data may contain a sync character, it may occur in the data a maximum of N − 1 times. Frames consist of message blocks preceded by a sync block. The start of one sync block to the start of the next sync block is one frame time. Figure 3.3 shows what transpires during a typical frame time.

3.4 CSDB Bus Capacity

The CSDB is similar to ARINC 429 in that it is an asynchronous broadcast bus and operates using character-oriented protocol. Data are sent as frames consisting of a synchronization block followed by a number of message blocks. A particular frame is defined from the start of one synchronization block to the start of the next synchronization block. A message block contains an address byte, a status byte, and a variable number of data bytes. The typical byte consists of one start bit, eight data bits, a parity bit, and a stop bit.

The theoretical bus data rate for a data bus operating at 50,000 bps with an 11-bit data byte, is 4,545 bytes per second. For CSDB, the update rate is reduced by the address byte and synchronization block overhead required by the standard.

The CSDB interblock and interbyte times also reduce bus throughput. According to the specification, there are no restrictions on these idle times for the data bus. These values, however, are restrained by the defined update rate chosen by the designer. If the update rate needs to be faster, the interblock time and the Interbyte time can be reduced as required, until bus utilization reaches a maximum.

3.5 CSDB Error Detection and Correction

Two methods of error detection are referenced in the standard. They are the use of parity and checksums. A parity bit is appended after each byte of data in a CSDB transmission. The burst transmission makes use of the checksum for error detection. As the General Aviation Manufacturers Association (GAMA) specification states, "It is expected that the receiving unit will accept as a valid message the first message block that contains a verifiable checksum" (GAMA CSDB 1986).

3.6   Bus User Monitoring

Although the CSDB specification defines many parameters, there is no suggestion that receivers should monitor them. The bus frame, consisting of the synchronization block and message block, may be checked for proper format and content. A typical byte, consisting of start, stop, data, and parity bits, may be checked for proper format.

The bus hardware should include the functional capability to monitor these parameters. Parity, frame errors, and buffer overrun errors are typically monitored in the byte format of character-oriented protocols. The message format can be checked and verified by the processor if the hardware does not perform these checks.

3.7   Integration Considerations

The obvious use of a data bus is for integrating various LRUs that need to share data or other resources. The following sections examine various levels of integration considerations for CSDB, including physical, logical, software, and functional considerations.

3.7.1   Physical Integration

The standardization of the bus medium and connectors addresses the physical integration of LRUs connected to the CSDB. These must conform to the Electronic Industries Association (EIA) Recommended Standard (RS)-422-A (1978), "Electrical Characteristics of Balanced Voltage Digital Interface Circuits." The CSDB standard provides for the integration of up to ten receivers on a single bus, which can be up to 50 m in length. No further constraints or guidelines on the physical layout of the bus are given.

Each LRU connected to a CSDB must satisfy the electrical signals and bit timing that are specified in the EIA RS-422-A. Physical characteristics of the CSDB are given in Table 3.1. Figure 3.4 shows the nonreturn to zero (NRZ) data format used by CSDB LRUs. NRZ codes remain constant throughout a bit interval and either use absolute values of the signal elements or differential encoding in which the polarity of adjacent elements is compared to determine the bit value.

Typical circuit designs for transmitter and receiver interfaces are given in the CSDB standard. Protection against short-circuits is also specified for receivers and transmitters. Receiver designs should include protection against bus conductor shorts to 28 VDC and to 115 VAC, and transmitter designs should afford protection against faults propagating to other circuits of the LRU in which the transmitter is located.

TABLE 3.1   CSDB Physical Characteristics

Modulation Technique	Nonreturn to Zero
Logic sense for logic "0"	Line B positive with respect to line A
Logic sense for logic "1"	Line A positive with respect to line B
Bus receiver	High impedance, differential input
Bus transmitter	Differential line driver
Bus signal rates	Low speed: 12,500 bps
	High speed: 50,000 bps
Signal rise time and fall time	Low speed: 8 μs
	High speed: 0.8 to 1.0 μs
Receiver capacitance loading	Typical: 600 pF
	Maximum: 1,200 pF
Transmitter driver capability	Maximum: 12,000 pF

Source: From *Commercial Standard Digital Bus*, 8th ed., Collins General Aviation Division, Rockwell International Corporation, Cedar Rapids, IA, January 30, 1991.

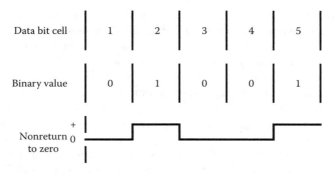

FIGURE 3.4 Nonreturn to zero data example.

To ensure successful integration of CSDB LRUs and avoid potential future integration problems, the electrical load specification must be applied to a fully integrated system, even if the initial design does not include a full complement of receivers. As a result, additional receivers can be integrated at a later time without violating the electrical characteristics of the bus.

3.7.2 Logical Integration

The logical integration of the hardware is controlled by the CSDB standard which establishes the bit patterns that initiate a message block and the start bit, data bits, parity bit, and stop bit pattern that comprises each byte of the message. The system designer, however, must control the number of bytes in each message and ensure that all the messages on a particular bus are of the same length.

3.7.3 Software Integration

Many software integration tasks are left to the system designer for implementation; hence, CSDB does not fully specify software integration. The standard is very thorough in defining the authorized messages and in constraining their signaling and update rates. The synchronization message that begins a new frame of messages is also specified. However, the determination of which messages are sent within a frame for a particular bus is unspecified. Also, there are no guidelines given for choosing the message sequence or frame loading. The frame design is left to the system designer.

In general, the sequencing of the messages does not present an integration problem since receivers recognize messages by the message address, not by the sequence. However, this standard does not disallow an LRU from depending on the message sequence for some other purpose. The system designer must be aware of whether any LRU is depending on the sequence for something other than message recognition, because once the sequence is chosen, it is fixed for every frame.

The bus frame loading is more crucial. There are three types of messages that can occur within a frame: continuous repetition, noncontinuous repetition, and burst transmissions. The system designer must specify which type of transmission to use for each message and ensure that the worst maximum coincidence of the three types within one frame does not exhaust the frame time. The tables of data needed to support this system design are provided, but the system designer must ensure that no parts of the CSDB standard are violated.

3.7.4 Functional Integration

The CSDB standard provides much of the data needed for functional integration. The detailed message block definitions give the interpretation of the address, status byte, and data words for each available message. Given that a particular message is broadcast, the standard completely defines the proper interpretation of the message. The standard even provides a system definition consisting of a suite of predefined buses that satisfy the integration needs of a typical avionics system.

If this predefined system is applicable, most of the system integration questions are already answered. But if there is any variation from the standard, the designer of a subsystem in a CSDB integrated system must inquire to find out which LRUs are generating the messages that the subsystem needs, on which bus each message is transmitted, at what bus speed the messages are transmitted, and the type of transmission. The designer must also ensure that the subsystem provides the messages required by other LRUs. The system designer needs to coordinate this information accurately and comprehensively. The system design must ensure that all the messages on a particular bus are of the same length and must also control the data latencies that may result as data are passed from bus to bus by various LRUs. All testing is left to the system designer.

There are no additional guidelines published for the CSDB. Whatever problems are not addressed by the standard are addressed by Collins during system integration. Furthermore, Collins has not found the need to formalize its integration and testing in internal documents since CSDB-experienced engineers do this work.

3.8 Bus Integration Guidelines

The CSDB, like the ARINC 429 bus, has only one LRU that is capable of transmitting with (usually) multiple LRUs receiving the transmission. Thus, the CSDB has few inherent subsystem integration problems. However, the standard does not address them. The preface to the CSDB standard clearly states its position concerning systems integration: "This specification pertains only to the implementation of CSDB as used in an integrated system. Overall systems design, integration, and certification remain the responsibility of the systems integrator" (GAMA CSDB 1986).

Although this appears to be a problem for the reliability of CSDB-integrated systems, the GA scenario is quite different from the air transport market. The ARINC standards are written to allow any manufacturer to independently produce a compatible LRU. In contrast, the GAMA standard states in its preface that "This specification ... is intended to provide the reader with a basic understanding of the data bus and its usage" (CSDB 1986).

The systems integrator for all CSDB installations is the Collins General Aviation Division. That which is not published in the standard is still standardized and controlled because the CSDB is a sole source item.

Deviations from the standard are allowed, however, for cases in which there will be no further interfaces to other subsystem elements. When variations are made, the change first must be approved in a formal design review and the product specification is then updated accordingly. Integration standards and guidelines for CSDB include the CSDB standard and EIA RS-422-A by the Electronic Industries Association.

3.9 Bus Testing

The CSDB connects avionics LRUs point-to-point to provide an asynchronous broadcast method of transmission. Before the bus was used in the avionics environment, it was put through validation tests similar to those used on other commercial data buses. These included the environmental tests presented in RTCA DO-160 and failure analyses. Most environmental tests were done transparently on the bus after it was installed in an aircraft.

As with other avionics data buses, Collins had to develop external tests to show that the bus satisfied specifications in the standard. Test procedures of this nature are not included in the CSDB standard.

Internal bus tests that the CSDB standard describes include a checksum test and a parity check. Both of these are used to ensure the integrity of the bus's data. Care should be taken when using these tests because their characteristics do not allow them to be used in systems of all criticality levels.

Simulation is used for development and testing of LRUs with a CSDB interface. To simulate an LRU connection to the bus, manufacturers make black box testers to generate and evaluate messages according to the electrical and logical standards for the bus. These consist of a general-purpose computer connected

TABLE 3.2   Aircraft and Their Use of the CSDB

Boeing 727	Retrofit
Boeing 737	Retrofit
McDonnell-Douglas DC-8	Retrofit
Saab 340, Saab 2000	Primary integration bus
Embraer	Primary integration bus
Short Brothers SD330 and SD360	Primary integration bus
ATR42 and ATR72	Primary integration bus
De Haviland Dash 8	Primary integration bus
Canadair Regional	Primary integration bus

Source: From Rockwell Collins, Cedar Rapids, IA.

to bus interface cards. The simplest ones may simulate a single LRU transmitting or receiving. The more complex ones may be able to simulate multiple LRUs simultaneously.

These are not the only external and internal tests that the CSDB manufacturer can perform. Many more characteristics that may require testing are presented in the CSDB specification. Again, it remains the manufacturer's responsibility to prove that exhaustive validation testing of the bus and its related equipment has met all the requirements of the Federal aviation regulations.

3.10   Aircraft Implementations

This section gives a sampling of the aircraft in which the CSDB is installed. Table 3.2 lists some of the commercial transport aircraft and regional airliners using CSDB. CSDB is used both in retrofit installations and as the main integration bus. Additionally, a number of rotorcraft use the CSDB to communicate between the Collins-supplied LRUs.

Defining Terms

Asynchronous: Operating at a speed determined by the circuit functions rather than by timing signals

Checksum: An error detection code produced by performing a binary addition, without carry, of all the words in a message

Frame: A formatted block of data words or bits used to construct messages

Gateway: A bus user that is connected to more than one bus for the purpose of transferring bus messages from one bus to another, where the buses do not follow the same protocol

Linear Bus: A bus where users are connected to the bus medium, one on each end, with the rest connected in between

Parity: An error detection method that adds a bit to a data word based on whether the number of "one" bits is even or odd

Synchronization Block: A special bus pattern, consisting of a certain number of concatenated "sync byte" data words, used to signal the start of a new frame

References

1. GAMA, Commercial Standard Digital Bus (CSDB), General Aviation Manufacturers Association, Washington, D.C., June 10, 1986.

2. Eldredge, D. and E. F. Hitt, "Digital System Bus Integrity," DOT/FAA/CT-86/44, Federal Aviation Administration Technical Center, Atlantic City International Airport, NJ, March 1987.

3. Elwell, D., L. Harrison, J. Hensyl, and N. VanSuetendael, "Avionic Data Bus Integration Technology," DOT/FAA/CT-91-19, Federal Aviation Administration Technical Center, Atlantic City International Airport, NJ, December 1991.

4. Collins, "Serial Digital Bus Specification," Part No 523-0772774, Collins General Aviation Division/ Publications Dept, 1100 West Hibiscus Blvd., Melbourne, FL 32901.

Further Information

The most detailed information available for CSDB is the GAMA CSDB Standard, Part Number 523-0772774. It is available from Rockwell Collins, Cedar Rapids, IA.

References

ARINC Specification 600-7, "Air Transport Avionics Equipment Interfaces," Aeronautical Radio, Inc., Annapolis, MD, January 1987.

ARINC Specification 607, "Design Guidance for Avionics Equipment," Aeronautical Radio, Inc., Annapolis, MD, February 17, 1986.

ARINC Specification 607, "Design Guidance for Avionics Equipment," Supplement 1, Aeronautical Radio, Inc., Annapolis, MD, July 22, 1987.

ARINC Specification 617, "Guidance for Avionics Certification and Configuration Control," Draft 4, Aeronautical Radio, Inc., Annapolis, MD, December 12, 1990.

Card, M. Ace, "Evolution of the Digital Avionics Bus," Proceedings of the IEEE/AIAA 5th Digital Avionics Systems Conference, Institute of Electrical and Electronics Engineers, New York, NY, 1983.

Eldredge, Donald and Ellis F. Hitt, "Digital System Bus Integrity," DOT/FAA/CT-86/44, U.S. Department of Transportation, Federal Aviation Administration, March 1987.

Eldredge, Donald and Susan Mangold, "Digital Data Buses for Aviation Applications," Digital Systems Validation Handbook, Volume II, Chapter 6, DOT/FAA/CT-88/10, U.S. Department of Transportation, Federal Aviation Administration, February 1989.

GAMA, "Commercial Standard Digital Bus (CSDB)," General Aviation Manufacturers Association, Washington, DC, June 10, 1986.

Hubacek, Phil, "The Advanced Avionics Standard Communications Bus," Business and Commuter Aviation Systems Division, Honeywell, Inc., Phoenix, Arizona, July 10, 1990.

Jennings, Randle G., "Avionics Standard Communications Bus — Its Implementation And Usage," Proceedings of the IEEE/AIAA 7th Digital Avionics Systems Conference, Institute of Electrical and Electronics Engineers, New York, NY, 1986.

RS-422-A, "Electrical Characteristics of Balanced Voltage Digital Interface Circuits," Electronic Industries Association, Washington, DC, December 1978.

RTCA DO-160C, "Environmental Conditions and Test Procedures for Airborne Equipment," Radio RTCA DO Technical Commission for Aeronautics, Washington, DC, December 1989.

RTCA DO-178A, "Software Considerations in Airborne Systems and Equipment Certification," Radio RTCA DO Technical Commission for Aeronautics, Washington, DC, March 1985.

Runo, Steven C., "Gulfstream IV Flight Management System," Proceedings of the 1990 AIAA/FAA Joint Symposium on General Aviation Systems, DOT/FAA/CT-90/11, U.S. Department of Transportation, Federal Aviation Administration, May 1990.

Spitzer, Cary R., "Digital Avionics Architectures — Design and Assessment," Tutorial of the IEEE/AIAA 7th Digital Avionics Systems Conference, Institute of Electrical and Electronics Engineers, New York, NY, 1986.

Spitzer, Cary R., Digital Avionics Systems, Prentice Hall, Englewood Cliffs, NJ, 1987.

Thomas, Ronald E., "A Standardized Digital Bus For Business Aviation," Proceedings of the IEEE/AIAA 5th Digital Avionics Systems Conference, Institute of Electrical and Electronics Engineers, New York, NY, 1983.

"WD193X Synchronous Data Link Controller," Western Digital Corporation, Irvine, CA, 1983.

"WD1931/WD1933 Compatibility Application Notes," Western Digital Corporation, Irvine, CA, 1983.

"WD1993 ARINC 429 Receiver/Transmitter and Multi-Character Receiver/Transmitter," Western Digital Corporation, Irvine, CA, 1983.

4

Time-Triggered Protocol

Mirko Jakovljevic
TTTech Computertechnik AG

4.1 Introduction

Time-triggered protocol (TTP) is a high-speed, masterless, multicast, dual-channel 25 Mbit/s field bus communication protocol for safety-critical embedded applications in the transportation industry. TTP communication controllers provide built-in health monitoring, system synchronization, and redundancy services for straightforward development of fault-tolerant embedded systems. A network designed with TTP guarantees manageable modular system design, simplified application development with minimized integration effort, strictly deterministic communication and new levels of safety at lower total life-cycle costs. TTP enables the development of physically distributed, but fully integrated, time-triggered architectures (TTAs) for modern avionics or aerospace control systems. Hence, the development of distributed safety-critical, hard real-time computing and networking for smart control systems (see Figure 4.1) in "more electric" or "all electric" aircraft is fully supported.

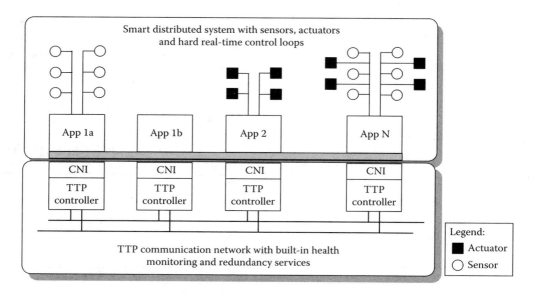

FIGURE 4.1 Distributed TTP-based system with line replaceable units or modules connected by the dual-channel communication network.

4.2 History and Applications

TTP has been developed over the last 25 years as a fully distributed and strictly deterministic safety-critical computing and networking platform. TTP is an open industry solution, and the TTP specification is publicly available on the TTA Group Web site [1]. TTP is designed to provide levels of safety required for the aerospace applications as well as to support its wider cross-industry use.

TTP is deployed in a variety of aerospace applications such as full authority digital engine control systems for Lockheed Martin F-16 and Aermacchi M-346, cabin pressure control systems in Airbus A380 and electric and environmental control systems in Boeing 787 Dreamliner. This makes TTP a key contender for safety-critical subsystems in aircraft under development. TTP is also targeted for on-board space control applications and is applied in development and series projects in other industries such as automotive, off-highway, and railway.

TTP is suited for a range of applications in which the following objectives must be accomplished:

- Configurable, table-driven integrated modular platforms with built-in health monitoring and fault localization suitable for "all electric" aircraft functions
- Optimization of aircraft systems through replacement of centralized functions with smaller physically distributed but tightly integrated line replaceable units (LRUs) and corresponding sensors and actuators found in "more electric" and "all electric" aircraft
- Simplified design of complex distributed systems with straightforward integration of fault-tolerant hard real-time subsystems delivered by different suppliers or teams

4.3 TTP and TTA

In event-triggered systems, the system environment drives temporal behavior. The system reacts to external demand for computing or communication resources when requested. Delays, hazards, and jitter are the most likely consequences if the system is not able to satisfy demand for resouces. This may lead to rare system failures that cannot be easily analyzed. The systems based on the TTA paradigm anticipate resource demand in advance and provide just-in-time access to resources. Instead of being driven by events, TTA is guided by the progression of time with an exact resource use plan tailored to avoid adverse operating conditions. Time-triggered systems make optimal use of resources beyond the saturation point, where an event-driven system spends its useful time working on the resource-sharing conflict resolution or recovery operations. As a core component of the TTA, TTP uses the time-triggered approach to system communication. TTA needs a common time base to provide precise temporal specification of communication interfaces with static distribution of available communication bandwidth to the nodes. Hence, the communication conflict is ruled out by the protocol design.

The deterministic and predictable behavior of distributed fault-tolerant systems is not only dependent on the availability of resources and precise temporal interfaces. Every subsystem should have the same notion of operational status of all other subsystems to carry out a correct action compatible with actions of other subsystems. Furthermore, asymmetric failure conditions among subsystems can create casual ambiguity about failure root-causes or inconsistent notion of system status. As a result, health diagnosis, failure localization, and fault tolerance mechanisms can be difficult to maintain, and the system can become unstable, which can impede the development, integration, and maintenance of complex systems. Manageable and scalable safety-critical systems should completely separate the communication system from the fault-tolerant behavior and application software functionality. Robust partitioning must prevent any latent failure propagation path.

During the creation of TTP all these issues were addressed to reduce system complexity, simplify integration, and enable the development of a fully deterministic, fault-tolerant, hard real-time platform for distributed computing and networking. This chapter begins with fundamentals of TTP communication and basic LRU structure and continues with a more detailed presentation of available safety, redundancy, and communication services. It also provides a concise description of application development for leveraging key strengths of TTA. Finally, this chapter discusses the reasons for straightforward integration, composability, interoperability, and scalability of TTP-based systems and gives a future outlook for this communication technology.

4.4 TTP Fundamentals

4.4.1 Time-Triggered Communication Principles

TTP implements a time division multiple access (TDMA) scheme that avoids collisions on the bus; therefore, communication is organized into TDMA rounds of the same length. A TDMA round is divided into slots with flexible length. Each LRU in the communication system has one slot — its sending slot — and sends frames in every round (Figure 4.2).

The cluster cycle is a recurring sequence of TDMA rounds. In different rounds, different messages can be transmitted in the frames, but in each cluster cycle, the complete set of state messages is repeated. The messages are broadcast to every node on the bus at predefined times with known latency, thus guaranteeing hard real-time arrival of the message at the target LRU. Safety-critical features of the TTP are implemented at higher abstraction levels and do not impose limitations on the physical layer. Therefore, the physical layer is not a part of the protocol specification and different well-established media, transformer coupling, and coding schemes are supported by available TTP controllers [2].

A TTP cluster consists of a maximum of 64 LRUs or LRMs (line replaceable modules) with a TTP controller. Typically, all relevant data in distributed control loops is sent periodically, once in a TDMA round. This means that the update frequency can be as short as one millisecond or shorter, depending on the message length and number of LRUs in the TTP cluster. Future generations of TTP controllers can offer much higher communication speeds range and faster control loops.

A TTP communication system autonomously establishes a fault-tolerant global time reference and coordinates all communication activities based on the globally known message schedule specified at the design time. It requires all communication participants to comply with an exactly specified and rigidly enforced temporal communication schedule that serves as a strict communication interface definition.

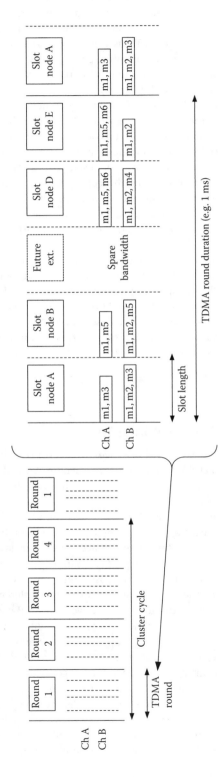

FIGURE 4.2 TTP communication cluster cycle with scheduled slots and rounds. Different sets of messages can be sent in different rounds in the same slot.

TTP-based systems can be designed to tolerate any single LRU fault. As a result of redundant communication channels, TTP tolerates a single channel failure. A correct receiver will consider the sender correct if the sender is correct on at least one of its replicated network interfaces (provided the respective channel is correct). The fault assumptions in TTP cover communication faults such as incorrect timing, inconsistent communication, and state notion differences in different controllers caused by asymmetric (Byzantine) transmission errors.

Incorrect outputs from a single data source (logic faults) caused by the host software or external I/O as an out-of-range sensor data or single-event upset (SEU) are tolerated by voting from replicated data sources supported by the fault-tolerant communication (FT-COM) layer and TTP's precise notion of time and state. In special cases other application-specific diagnostics or specific architectural or design measures such as redundant topologies, error correcting memories, fault-tolerant host and software design, and many others can be used to meet the safety requirement of an application.

4.4.2   Data Frame Format

The frame size allocated to a node can vary from 2 to 240 bytes in length, each frame usually carrying application messages (see Figure 4.3). A 4-bit frame header data is sent before the data frame that is protected by a 24-bit cyclic redundancy checksum (CRC). The first bit in the frame header shows whether or not the frame explicitly carries protocol state information. Status information can be transmitted implicitly as a part of the CRC or explicitly within the message data field. The other three bits represent the code of the requested mode change. Due to the static definition of the TDMA schedule frame and known frame arrival time, message and target identifiers are not required.

4.4.3   Line Replaceable Unit with TTP

A TTP cluster consists of a set of LRUs or LRMs connected by a dual replicated communication channel. Each LRU contains hardware components such as a TTP communication controller and a host controller with software modules (application, FT-COM, and Real-Time Operating System [RTOS]) running on the host (see Figure 4.4).

The TTP field bus connects different nodes and transmits data frames. Frames contain application data and state messages directed to different distributed applications running on a node or subsystem. An application consists of application tasks executable on one or several hosts. Depending on the application software and I/O hardware, the host can acquire and log data from sensors, control the actuators, provide data processing capabilities for a distributed control-loop, or perform a combination of these tasks.

4.4.4   Configurable Table-Driven Communication System

The configuration of message schedule and system networking architecture, as well as other safety-related properties, rely on the Message Descriptor List (MEDL). The MEDL is defined at design time and used by the TTP controller to autonomously communicate with its host and other TTP cluster nodes (LRUs). Overall temporal behavior of the communication system is governed by the information stored in the MEDL. System design tools are used to configure and verify the configuration of all LRUs in a system.

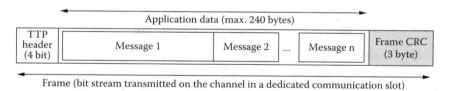

FIGURE 4.3 TTP frame format with low communication overhead (N-frame).

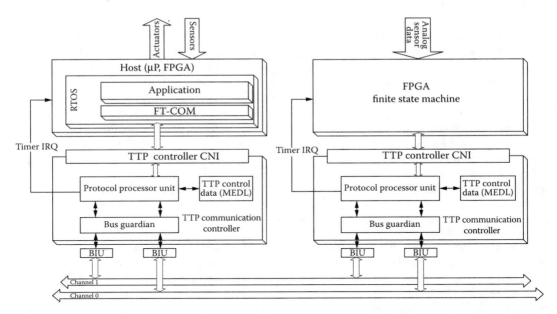

FIGURE 4.4 Electronic modules (LRU or LRM) consisting of the host processor (left) or FPGA state machine (right, e.g., smart sensor configuration), TTP controller and bus guardian.

The configuration of communication system and host application can be simply uploaded over the TTP network. If some communication bandwidth and slots are reserved for the system upgrades (e.g., new subsystems and LRUs), changes in the MEDL are not needed if such upgrades are later implemented.

MEDL stores information such as clock setup data for global timing and communication rate; communication schedules with slots, rounds, and cycles, transmission delays taking into account distances between nodes; bus guardian parameters; startup parameters; and various service and identification parameters. The configuration data also contains CRCs used for continuous scrubbing and self-checking of all configuration data structures.

4.4.5 Global Time Base

The decentralized clock synchronization in TTP provides all nodes with an equivalent time base without use of any dedicated external time source. Accurate global time is essential for the design of deterministic fault-tolerant distributed systems. It supports the coordinated operation of redundant systems and prevents a common mode clocking failure by built-in and formally proven fault tolerant mechanisms [3].

Distributed LRUs in a TTP communication system develop a masterless fault-tolerant global network time reference. The globally known schedule stored in the MEDL prescribes expected message arrival time for every LRU in the system. The difference between the expected and the actual message arrival time is used to correct the local time for every correctly received frame. If an LRU is authorized for the time synchronization, the measured time of the arrival of its frame will be used in the synchronization process.

4.4.6 TTP-Based LRU as Fault Containment Region

4.4.6.1 Communication Network Interface

The communication network interface (CNI) of a TTP controller is a dual-port memory used for data exchange with other hosts. Control signals are not required for the communication with other LRUs in the network; therefore, control signal errors cannot propagate to the network. As the communication system decides when to transfer data, the application or host cannot influence the timing of the message transmission stored in the CNI. The communication subsystem reads message data from the TTP controller's CNI at the predefined fetch instant and delivers it to the CNIs of all receiving LRUs of the cluster at the known delivery instant, overwriting the previous version of the frame.

The CNI separates the behavior of the application software from the operation of the communication network and makes them mutually independent. It is a "temporal firewall" that prevents the error propagation that may influence temporal properties of communication system. Thus, the CNI creates a well-defined fault containment region; faulty host hardware or application software will never be able to influence the operation of the TTP communication network or cause data collisions. In fact, CNI behaves as a shared conflict-free data exchange interface. It encapsulates and hides all functional properties that are not relevant for the system level operation, while making globally relevant data available throughout the network (Figure 4.5). The CNI integrates all applications functionality in the network as defined at design time. It represents a clean interface between application subsystems and provides robust partitioning between the application software and the TTP communication network.

4.4.6.2 Bus Guardian

The fault containment of an LRU is maintained by the controller's CNI, but additional protection mechanisms should ensure that the failure of the TTP controller will not propagate to the network. The bus guardian is an independent unit that protects the TTP network from timing failures of the controller. Any observable effect of a faulty node that impacts temporal behavior (e.g., physical faults, syntactic timing errors) of the prescribed communication interface will be masked by the guardians and will not propagate any further. An internal bus guardian has its own oscillator and is not dependent on controller timing. Depending on the system safety requirements, an external bus guardian can be also added. The coupling of bus guardians determines the topology of the system.

4.4.7 Real-Time Health Monitoring and Redundancy Management

The system health monitoring and continuous error prevention using redundancy management are integral parts of the TTP communication network, and they are robustly separated from the application software functionality. Any LRU connected to the system is continuously monitored. Adherence to the interface specification is strictly enforced in real time without any additional measures in the application software. A noncompliant or faulty component is not allowed to interfere with the continued operation of the system. The failures of intelligent sensors and data-processing LRUs that do not result in temporally changed behavior at the application level may simply cause the delivery of false data (e.g., single event upsets (SEUs), sensor failure). Such problems are tolerated by active redundancy in real time and voting.

In some safety-critical systems such as flight control computer (FCC) software, the application software represents only one third of the total lines of source code. At the same time, the portion of code written to support system redundancy management and fault detection monitoring exceeds 55% of the total code (e.g., Boeing 757 FCC software [4]). This part of the software code can be surprisingly complex to develop and verify, thus adding disproportional costs for system development and integration. The TTA, with TTP in its core, significantly reduces the effort linked to the development of redundancy and health monitoring algorithms in distributed systems.

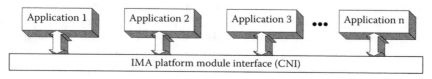

FIGURE 4.5 The CNI enables robust partitioning and independent operation of communication system. The CNI operates as a conflict-free shared memory for data exchange between distributed applications.

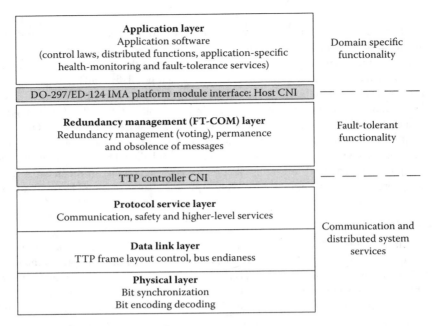

FIGURE 4.6 TTP protocol layers as a conceptual reference architecture.

In contrast to other field buses, many of essential fault-tolerant computing and distributed health monitoring services are implemented in TTP at the protocol level. The application development can be focused on functional issues and use of available services, thus reducing the complexity of software and the development effort for safety-critical distributed systems.

4.5 TTP Communication Protocol Layering

The interaction of different communication services, fault tolerance management, and software applications can be easily represented by using a conceptual reference architecture in the form of a straightforward layered model (Figure 4.6). The most important aspect of layering in TTP is the separation of the communication system, the redundancy and fault tolerance management, and the application development. They are separated by clean data-sharing interfaces configured at design time — Host CNI and TTP Controller CNI. The underlying layers govern their temporally deterministic behavior, without any control signals propagating from the application to the network. Interlayer interfaces such as CNI significantly contribute to the partitioning and fault containment of a single LRU.

4.5.1 Physical Layer

TTP does not prescribe a specific bit encoding or the physical media used. The TTP bus can consist of two independent physical channels that may be based on different physical layers. TTP protocol may work using a shared broadcast medium in bus, a star, or mixed bus-star topologies, which are determined by the location and capacity of the bus guardian devices in the system. The boundaries of the propagation delay must be known and given in the timing schedule stored in MEDL. The choice of a suitable physical layer depends on the application constraints regarding transmission speed, physical environment, and distance among nodes.

4.5.2 Data Link Layer

The data link layer provides the capability to exchange data frames among the LRUs. This layer defines the format of TTP protocol frames with header, status, data, and CRC information. Frames are transmitted on both channels in parallel for safety-critical messages, but if the system design permits relaxed replication of less critical messages, different messages can be sent on both channels. A LRU's status information (i.e., its own perception of the system status) for maintaining system consistency, such as global time, current slot and round position, current mode, pending mode change, and membership vector, is stored in the controller state and broadcast to the network for status and data transmission agreements.

During the synchronized operation, TTP understands three types of data frames:

- **Initialization (I-)frames** are used for the integration of nodes and contains explicit controller state information but do not carry any data. Unused communication slots are filled with these types of frames. These frames are sent when no application data transfer is required in the system.
- **Normal (N-)frames** are used for clock synchronization on correctly transmitted frames and low-overhead data transfer. They carry data and implicit controller state information hidden in the CRC.
- **Extended (X-)frames** are used for integration, data transfer, and high-speed processing of explicit controller state information. An X-frame is, in fact, an I-frame with application data transfer and clock synchronization capability.

During the startup, TTP controller uses another type of frame similar to I-frame:

- **Cold-start frames** contain the global time of the sender and its position in the message schedule (round and slot). At that time, other status data is unknown (e.g., membership) and not a part of the frame.

The frame CRC is calculated by using a configuration table identifier as a seed value common to all MEDLs in the network and uses different parts of its binary value for each channel, thus preventing crossed-out channel connections or communication of LRUs with incompatible configurations.

4.5.2.1 Bus Access Control

Normally, every LRU has its own slot in every round, but different LRUs can use multiplexed slots in different rounds if defined in the MEDL.

4.5.2.2 Physical Distances

The maximal distance between nodes exceeds 130m, which is beneficial for a larger airplane and is not dependent on protocol features such as collision avoidance (e.g., controller area network (CAN) CSMA/collision avoidance (CA)). The maximum distance that can be achieved in a specific TTP architecture depends on the underlying physical layer, coding schemes, shielding, electromagnetic interference (EMI) environment, and driving electronics.

4.5.2.3 Bus Latency, Jitter, and Efficiency

Delays are not introduced by protocol services such as clock synchronization, membership, acknowledgment, or message ordering. The share of control data in the message is reduced to bare minimum. No additional latency is imposed on the system in case of transmission errors, as no retries are attempted and there are no collision checks or resolutions. The predefined communication latency of TTP networks is very low, and the bandwidth can be almost fully utilized.

The message format is kept simple. Theoretically, it provides up to the 98% message data transfer efficiency rate. The remaining 2% includes the bandwidth used for sending CRC and frame header bits. Typically, 50% to 80% of the available bandwidth is used for message data transfer in avionics applications with existing TTP controllers (i.e., AS8202NF). New TTP controllers with faster internal protocol processing may provide much higher data efficiencies close to the theoretical limit of 98% as a result of significantly reduced inter-frame gap (IFG). The overhead caused by the IFG is dependent on the protocol processing speed and the communication speed.

Minimal latency jitter is very important in fast control loops and by-wire applications. Depending on the oscillator and physical layer, the latency jitter can be reduced to the submicrosecond range. Typical values are between 1 and 10 microseconds.

4.5.3 Protocol Layer

The protocol service layer establishes the protocol operation with communication and safety services, and other higher-level services. Communication services cover redundant data transmission, startup and reintegration of nodes, fault-tolerant clock synchronization, and distributed acknowledgments. They simply establish the communication in a distributed time-triggered real-time system. Furthermore, communication services contain all functions required for a temporal firewall and thus completely decouple the communication system from the host.

Safety services include membership for networkwide status notion; clique avoidance for sustaining consistent communication in cases beyond fault assumption; life sign algorithms for the detection of host failures; and network protection against the timing failures of the TTP controller in the form of independent bus guardians. Safety services guarantee fail-silent behavior of a faulty node in the time domain and prevent inconsistent interactions with the existence of different LRU cliques with different controller states. This maintains the consistency of communication and prevents the distribution of faults within the cluster.

Safety and communication services guarantee consistent communication and fail-operational behavior of the TTP network by establishing a LRU as a fault containment region. Therefore, faults cannot propagate from one LRU to the rest of the TTP network. Communication services tolerate single communication faults and detect faulty nodes, while safety services prevent distribution of faults within the cluster. Membership service and distributed acknowledgment from the protocol layer provide distributed real-time health monitoring, diagnosis, and localization of faults in the system without any increase in the bus data bandwidth or delays in communication. Higher-level services contain real-time mode changes of the network schedule, external clock synchronization, and reconfiguration of the node.

4.5.3.1 Controller State as Local Notion of System State

An internal and locally calculated TTP controller status (C-state) represents a notion of the network status from the LRU's perspective based on continuous monitoring of all frame transmissions in the network. Only LRUs with agreed C-state can synchronize with the network and participate in distributed computing and networking. A C-state disagreement between an LRU and the majority of the other LRUs means that the LRU is faulty or it simply has an inconsistent status notion. The C-state contains the global time of the corresponding slot transmission, round slot position, cluster mode data, and membership information for all LRUs in the network. The agreement on C-state is possible among LRUs because of common knowledge of slot transmissions, round positions, and other network architecture design data stored in configuration tables (MEDL).

4.5.3.2 Communication Services

4.5.3.2.1 Redundant Data Transmission

TTP utilizes redundant data transmission on both channels or on the same channel at different times to suppress transient errors. Redundant units can send the same messages at different times in their own slot. The receivers decide which messages will be selected as correct from the set of redundant messages. This is done in the FT-COM layer using different Replica-Determinate Agreement (RDA) algorithms such as majority vote, fault-tolerant average, and high/low wins, among others.

Communication on two channels is protected by the 24-bit CRC unique to the respective TTP channel. In the case of permanent failure on one channel, the second channel still makes data transmission possible. The safety-critical data is always replicated and transmitted on both channels.

Maintenance errors such as false wiring or use of incompatible LRUs are detected by the TTP protocol because the integration into the communication system is not possible for LRUs with incorrect network wiring. If an LRU's channel A and B connectors were mistakenly connected to the network's B and A channels, respectively, this LRU would not be able to integrate into the network communication. Additional user-specific fault detection services can be defined at the application level.

4.5.3.2.2 Cluster Startup and (Re)Integration

Reintegration of individual components or the restart of the whole system are self-stabilizing services that support transition from an unsynchronized into a synchronized state. At power-on or reset, all nodes in a cluster try to become mutually synchronized. After power-up and initialization of the host and TTP controller, the TTP controller listens to all frames with explicit C-state for the duration of the listen timeout (Figure 4.7). The listen timeout is unique for each LRU in the network (Figure 4.8). The reception of such frames starts integration, and LRUs start to send data. Other nodes are added to the membership as the LRU recognizes their activity. This can be done only if the host is alive and ready to send data and the controller has the permission in his MEDL to perform a cold start. The number of allowed cold starts after an activation of the TTP controller is limited by a design parameter to prevent an indefinite number of startups.

A controller expelled from the system communication and membership may want to restart and try to reintegrate on I-frames or X-frames, which contain C-state information to initialize membership and the controller's clock. The following round cycle is closely observed and used to check if the settings acquired from the C-state have resulted in behavior consistent with other participants in the TTP network. After a positive decision, the reintegrating LRU starts communication on the bus.

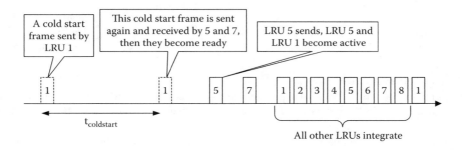

FIGURE 4.7 TTP communication startup.

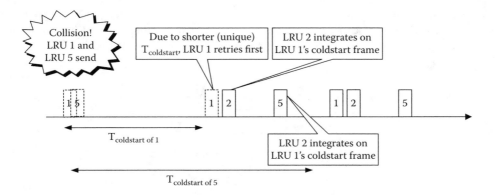

FIGURE 4.8 Startup collision between LRUs 5 and 1 leads to delayed submission.

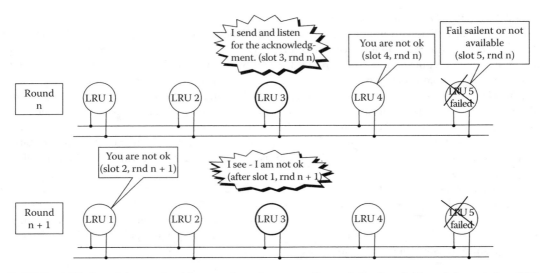

FIGURE 4.9 Update of the membership vector from the stream of mutual distributed acknowledgments for a TTP network consisting of five nodes. First two correctly working successors can provide negative acknowledgment to the LRU3. The same applies for all other sending LRUs in the system. The failure of one or more LRUs does not invalidate the distributed acknowledgment service.

4.5.3.2.3 Distributed Acknowledgment

In order to provide the capability to determine the status of every unit in the network and the agreement on frame transmissions in the network, TTP provides a frame acknowledgment mechanism distributed over all functioning LRUs in the network.

The distributed acknowledgment provides the sending LRU with a confirmation of successful (or failed) data transmission from other correctly working LRUs in the network. A receiving (successor) LRU considers a sending LRU "alive" if at least one of the replicated frames sent on either channel has arrived correctly at the receiver. The correctness of the frame is confirmed by a CRC check. An LRU simply listens to frame transmissions of its successors and compares its own status perception, stored in the membership information submitted in the frame, with their notion of the system status. The sending node expects either a single positive acknowledgment or two mutually reaffirming negative acknowledgments from fault-free LRUs (see Figures 4.9 and 4.10). By listening to the judgment of its successors, a sending LRU determines whether its transmission was successful or unsuccessful.

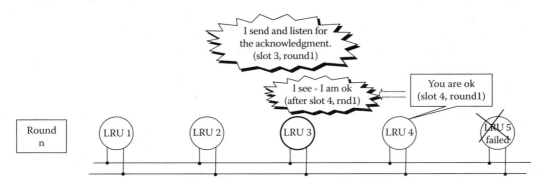

FIGURE 4.10 Positive acknowledgment for LRU3. Only one positive acknowledgment from two operating successors is required to reassure that LRU3 works correctly.

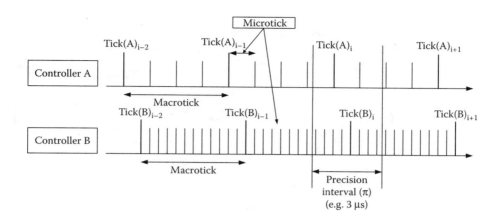

FIGURE 4.11 Sparse time base (macroticks), microticks, and precision interval

4.5.3.2.4 *Fault-Tolerant Clock Synchronization*

The model of clock synchronization in TTP is based on a sparse time base called macroticks, a common time resolution in the TTP network. Every macrotick consists of microticks; the ratio of macroticks to microticks may be different for different TTP controllers depending on their oscillator. Nevertheless, the configuration data in the MEDL considers these local differences so that the sparse time base is consistent throughout the network. The precision interval is determined to enable correction of the clock. This interval is smaller than one macrotick. All nodes within the network should fit the precision interval and global time base accuracy.

The synchronization means that the common understanding of time within the precision interval $\pm\pi/2$ will be continuously adjusted and maintained (see Figure 4.11). This is done by using a fault-tolerant synchronization algorithm at the end of every round. Every TTP controller measures the difference in microticks between the slot arrival time and the expected arrival time stored in the MEDL. The difference is determined by using a fault-tolerant distributed synchronization algorithm, which takes the last four measured values, rejects the maximal and minimal ones, and averages the remaining two. As a result, the correction term for the local time in microticks is calculated. Abrupt (at once) or smooth (over several slots) adjustment can be selected for the correction of time synchronization. The communication controllers periodically check whether their macrotick is within the systemwide precision interval and will raise an error if it is not. At least four LRUs are required for the fault-tolerant masterless time base to tolerate a single asymmetric (Byzantine) timing failure.

The network can be configured to hold a specific number of nodes with spares to contribute to the clock calculation and others excluded from the global time base calculation. This can have an impact on total system cost reduction, because the maximum accuracy in the system may be governed only by a smaller number of LRUs with very accurate internal clocking.

4.5.3.3 Safety Services

4.5.3.3.1 *Membership*

The update of the membership vector from the continuous stream of mutual networkwide distributed acknowledgments shows which units participate in the communication and deliver correct data frames. Only correctly sending LRUs can participate in TTP communication. The membership service informs all nodes in the TTP communication network about the operational state of each LRU in the network within a latency of one TDMA round, thus simplifying the localization of error sources. With each frame submission (broadcast), LRUs provide their own perception of the network status, which can be used by

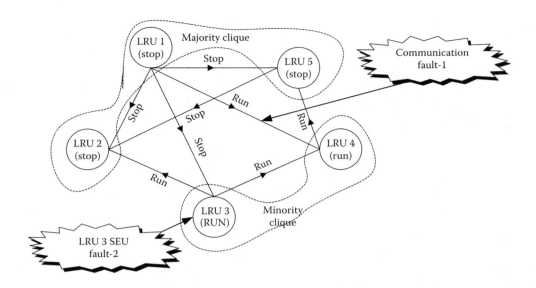

FIGURE 4.12 Development of cliques is resolved by LRU 3 and LRU 4 restart and reintegration. It can be also eliminated by other application-specific recovery operation defined at the development time.

other senders to acknowledge the sender's data transmission. The membership information can be either provided explicitly in the frame or can be hidden in the checksum at the end of each frame.

4.5.3.3.2 Clique Avoidance

If the single-fault assumption is violated (see Figure 4.12), two or more cliques with a different notion of the network status may emerge. Cliques can only occur in multifault scenarios or with asymmetric transmission faults. For example, in a system with the two possible states, RUN and STOP, two faults occur in parallel. LRU 4 misunderstands LRU 1 as a result of a communication fault and believes that the system is in the RUN state. LRU 3, as a result of internal SEU, has a false understanding of the transmission from LRU 1 and firmly believes that everybody else should be in the RUN state. All other LRUs are convinced that the system state is STOP; therefore, there are two groups (cliques) with a different notion of system status. The clique avoidance algorithm solves this kind of problem.

The consistency of the membership vector is continuously agreed upon among all nodes in the system, and conflicts are resolved within two TDMA rounds. Clique avoidance signals to the application whether it is in the majority clique or not, based on the comparison of membership vectors transmitted in the CRC of every node. Before the node sends a frame it checks if it is in a majority clique. If it is part of the minority clique the node signals to the application to decide how to proceed with the operation — by conducting a complete restart and new reintegration or by executing other activities appropriate to the level of safety required by the system.

4.5.3.3.3 Protection against Timing Failures

A bus guardian enforces the temporal fail silence of an LRU in the case of arbitrary faults such as "babbling idiot" or "slightly-off-specification" (SOS). A bus guardian error is immediately reported to the controller and host software. An internal bus guardian in a TTP controller has an independent copy of the communication schedule, and its clock is separate from the TTP controller. The internal bus guardian is synchronized by a start-of-round TTP controller signal.

In TTP mixed topologies may be defined by combining bus and star topologies. Other system architecture decisions, such as parallel use of two or more double-channel TTP buses, can further improve the safety of the systems.

4.5.3.4 Higher-Level Protocol Services

4.5.3.4.1 Mode Changes

In order to provide consistent on-line change in the message schedule timing for all LRUs, mode changes are possible at the end of every cluster cycle. All LRUs will be able to switch at the same time to a different schedule. This permits different operating modes with different communication behavior. If several mode changes are requested within the round, the latest will be used. Any node can prevent the mode change. The current mode request is given in the header of every frame.

4.5.3.4.2 External Clock Synchronization

TTP supports external clock synchronization to provide synchronization among different TTP clusters or provide synchronization with external time sources. By adding drift correction values to the network global time, the differences between global TTP network time base and external time can be equalized.

4.5.3.4.3 Reconfiguration of LRU

This service is part of the communication controller and permits download of the MEDL data and protocol software data independently of the application. It also provides a link for the download client capable of downloading data and new versions of the software to the host.

4.5.4 Fault-Tolerant Layer

In a TTP-based system, the fault tolerance mechanisms are implemented in a dedicated fault tolerant communication layer. The number of replicas, voting, and reintegration remain completely transparent to the application software. FT-COM reduces redundant messages from replicated applications from other subsystems (e.g., LRUs, LRMs) to a single, agreed-upon value before they are presented to the local application but also determines the formatting and endianness of the message data. An application that uses such a fault tolerance communication layer can thus operate with the same functionality and without any modifications in a fault-tolerant system or a standard system.

With a FT-COM layer, TTP implements all services needed to handle complex fault-tolerant distributed software systems effectively and efficiently. For the application developer, it means much simpler and faster application and control algorithm development because there are no mutual interactions between functions and they can be developed independently.

4.5.4.1 Fault-Tolerant Units and Redundant LRUs

Safety-critical, real-time systems impose high demands on the reliability and availability of the system. Active redundancy by replication and voting are the fastest fault tolerance techniques because there is no dead time of the service. The realization of active replication demands a number of mechanisms, such as replica coordination, voting, group membership, internal state alignment, and reintegration of nodes after a transient failure. All of these mechanisms are available in TTP protocol and are separated from the application functionality. The failures of intelligent sensors, actuators, and data-processing LRUs are tolerated by hardware replication in real time by the TTP protocol (including the FT-COM layer). TTP does not set upper limits for the number of redundant LRUs (Figure 4.13) in a fault-tolerant unit (FTU) that behaves functionally and temporally as a single LRU.

4.5.4.2 Clique Detection, Interactive Consistency, and Replica Determinism

The integrity of network communication and fault detection beyond the single-fault hypothesis is maintained by the clique detection mechanism. Clique detection helps prevent subsystems from having an incorrect notion of system status. This mechanism is also of the utmost importance for interactive consistency and replica determinism.

Interactive consistency requires that any transmission results in an identical (atomic) reception at all receivers. With distributed acknowledgment, membership, and clique detection services, interactive consistency guarantees that all nodes receive the same sequence of messages. In basic terms, replica determinism ensures that inputs from one data source received at the same time will be bit-identical to results

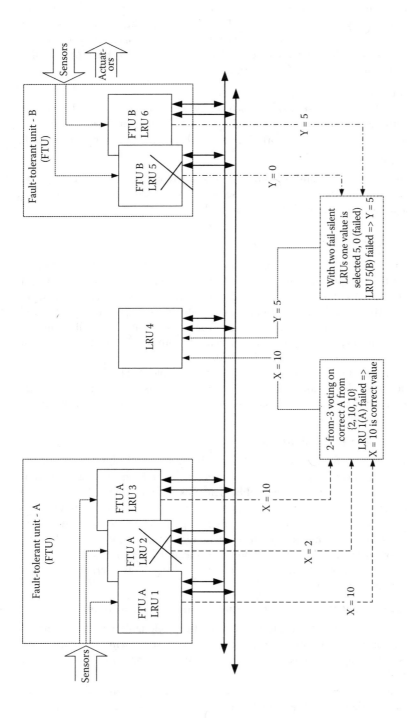

FIGURE 4.13 FTUs improve availability and safety. Active replicated LRUs provide the needed functionality if one LRU in the FTU fails. LRU 4 reads data from FTU-A (voting from triple-redundant LRUs, tolerates an incorrect output from one of the three LRUs 1,2,3) and FTU-B. (LRUs 5 and 6 are designed as fail-silent; in this figure, LRU 5 has failed, and the output is taken from LRU 6.)

received at a later instant from all other working replicas. Based on the global time base and interactive consistency, the same results can be delivered only if all replicas pass through the same system states at the same time. This is possible due to the time-triggered design of LRU application and system tasks synchronized to the global time. Therefore, any event-driven behavior (e.g., interrupts) in replica deterministic FTUs and safety-critical systems in general should be avoided if possible.

Replica determinism guarantees consistent computations and transfer of functionality in the case of transient or permanent failure of one replica in the FTU. Otherwise, the switching from one replica to another can lead to a serious error and upset the controlled system. The lack of those mechanisms can severely increase the design effort of safety-critical distributed systems and add disproportionate costs to the system design.

4.5.5 Application Layer

The application layer contains application tasks running on the LRU's host microcontroller. In distributed applications, tasks are executed on several LRUs and communicate over a TTP communication network to accomplish the desired functionality. A strictly deterministic system behavior can be realized if both the application and communication systems have fully predictable temporal behavior synchronized to global system time. In Figure 4.14, the context of application tasks within the LRU and in relation to the FT-COM layer and RTOS is presented. The application software consists of established manageable software tasks with a task start and deadline during the design.

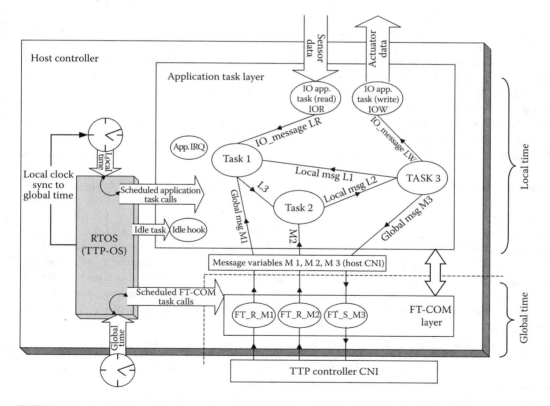

FIGURE 4.14 Application and system software (RTOS) with statically scheduled tasks and polling access to sensors and actuators.

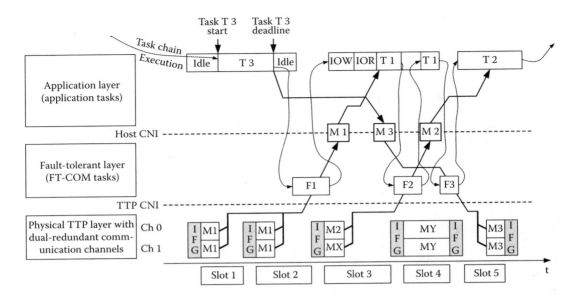

FIGURE 4.15 Statically scheduled application tasks in relation to communication message schedule with conflict-free data transfer. The temporal behavior of an LRU and the system is defined at design time.

Tasks communicate over local messages within a single LRU and over global messages between tasks in different LRUs. Global messages (state variables) are transmitted through the network. The message structure, content, and time of the transmission and access are defined design. Therefore, the message communication as well as all other processing steps are always conflict-free and do not impose any delays in system behavior. Messages in TTP-based systems are transferred just-in-time.

Figure 4.15 shows how this works for every LRU in the system. Global messages (state variables) M1, M2, and M3 are handled as normal variables by application tasks. M1 is taken from two different slots on both channels of the TTP bus and unified into one value in an F1 task in the fault-tolerant layer. Depending on design, averaging or voting on all four correct values can be done in the task F1. The message M2 is taken from one TTP channel and prepared for use by task T2. The message M3 is a result of processing done by task T3 and sent in slot 5 on both TTP channels. All other messages (MX, MY) transmitted through the network are invisible to application tasks as defined at design time. An idle task is executed if no immediate application action is required. This task can be used for self-diagnosis operations or even for running another operating system (e.g., RTAI-Linux). FT-COM tasks and application tasks are executed in an order that permits timely transfer of messages from the communication channels to the processing applications tasks and vice versa. Obviously, the scheduling of the system communication and running of applications on a distributed system is the key issue in the design of hard real-time safety-critical systems with TTP. The FT-COM configuration is generated from network and communication specifications by the design tools. The developer just has to write application code on top of it.

4.5.5.1 Time-Triggered Software Behavior

In order to keep tasks simple and free from event-driven behavior, any temporally indeterministic constructs (e.g., semaphores, blocking read/writes, or variable task execution timing) should be avoided. Furthermore, event-driven functions are not needed for the operation of TTA.

Cyclic sensor and actuator polling scheduled at design time can be accomplished through dedicated lightweight tasks with temporally fixed task execution (e.g., IOW and IOR from Figure 4.14). Therefore, this guarantees the temporal determinism of application execution and very low system latency and jitter in distributed high-speed control loops.

Event-driven software design principles (e.g., interrupts) may be used carefully only if surplus processing capacity is reserved to prevent violation of deadlines for all combinations of interrupts. TTP can also accommodate software applications that are event-driven and do not require fully deterministic system behavior but can profit from deterministic communication, built-in health monitoring, fault containment, system synchronization, and straightforward integration.

4.5.5.2 Error Handling

Even in the case of host failure, the robust separation of the communication system from application software over the CNI guarantees that temporal behavior of the TTP-based system does not change and the failure does not spill over to the network. The most severe failure mode of host computers is thus either to supply wrong data as a result of bit flips or sensor failures or no data at all at the required points in time. The host computer has strict requirements of when to output what data. Its local TTP controller and all other controllers on the network monitor its temporal behavior.

All other LRUs in the network will immediately recognize its incompliance or failure based on controller-driven membership and acknowledgment services. LRUs sending the correct amount of data at the correct time with correct frame CRC, but with faulty message content as a result of host (or sensor) failure, cannot be detected by these mechanisms. It is up to the FT-COM layer or application (for fine-grained application-specific services) to deal with this type of fault. Usually triple modular redundancy (TMR) or lane-switching logic is employed to deal with these faults.

Temporally incorrect task behavior that exceeds the given time budget is captured using error-handling tasks. In this case the TTP controller does not send any data from the failed host. The LRU will have a chance to recover and integrate in one of the following communication rounds after completion of recovery operations. In the case of repetitive failures, the LRU may withdraw permanently from the network communication based on the TTP controller decision as prescribed at design time in the TTP controller's configuration tables.

4.5.5.3 Distributed Application Software Development and Interactive Consistency

Inconsistent communication and different notions of system state are immediately recognized and reported to the application. Failure of one LRU in a fault-tolerant unit will be immediately recognized by safety and communication services and tolerated by voting in the FT-COM layer.

Application developers usually expect interactive consistency in distributed systems and write their software or develop hardware assuming that every node has consistent input data. Without such distributed services based on a global time base, distributed acknowledgment, membership, and clique avoidance, the design of distributed applications becomes a painstaking process. The interactive consistency has to be implemented by the user at the application level which, in turn, requires deep understanding of specific distributed computing issues. The implementation of interactive consistency at the application level contains potential pitfalls and risks that significantly complicate application software structure, certification processes, and system integration. The requirement for consistency may be sometimes neglected or not well-understood during the planning of complex fault-tolerant control systems, but all required support is already built into the TTP communication protocol.

4.5.5.4 Application Development with TTP

The developer can design the system at a higher level of abstraction, simulate physical and control systems using Matlab [5], formally verify its design [6], manually write the application (task) code, or use code generation tools from simulated models [5,6]. As the redundancy and communication layer is completely separated from the application layer, the software application developer can focus solely on functions, sensor data acquisition, control laws, and data formats.

A major part of the development for TTP-based system development is driven by the design of configuration data (e.g., communication, timing and scheduling, fault tolerance, message formatting). This significantly reduces the development effort for the design of distributed safety-critical systems with built-in redundancy and health monitoring. Using built-in TTP services can significantly reduce the design effort for complex distributed safety-critical systems.

4.6   System Integration with TTP

4.6.1   Interoperability

Because the timely delivery of messages and the accurate moment of their submission cannot be easily estimated in all cases, subsystems do not have perfect knowledge of what can be expected from other subsystems. Therefore, the interoperability between subsystems organized to jointly accomplish a function is at risk, which presents a major roadblock to straightforward integration of complex systems. TTP implements global control of network traffic management with clearly defined temporal interfaces and eliminates interoperability issues.

4.6.2   Composability

One of the key factors in guaranteeing temporal composability is the robust separation of temporal control structure from logical control structure, so a subsystem or LRU can be validated in isolation [7]. In TTP the temporal behavior is precisely defined by the design of the communication schedule and the separation of communication system and application software.

An architecture is temporally composable if the system integration does not invalidate the independently established temporal behavior of subsystems or LRUs. This also implies that the separate testing of a standalone LRU, as well as its testing in the system, delivers the same temporal behavior and verification results.

In TTP none of the subsystems designed to work independently in the same network will cause unintended interaction with other subsystems. All subsystems participate in distributed fault-tolerant time synchronization, distributed acknowledgment, membership and clique detection, and jointly support interactive consistency. The fault of one LRU does not influence the networkwide services and network data transport. The applications running on subsystems and LRUs will be able to share data as defined at the design time (see Figure 4.16 and Figure 4.17).

The communication schedule relies on a global time base. A master-free, fault-tolerant time synchronization algorithm is used to prevent any common-mode timing failures. The "babbling idiot" and "slightly-off-specification" errors caused by an erroneous LRU transceiver or oscillators are captured by the bus guardian, which permits submission of frames only in a dedicated slot. The temporal composability would be much harder to achieve without stable interfaces designed to robustly separate a system into fault containment regions (i.e., LRUs or FTUs), which prevent the fault propagation to other subsystems.

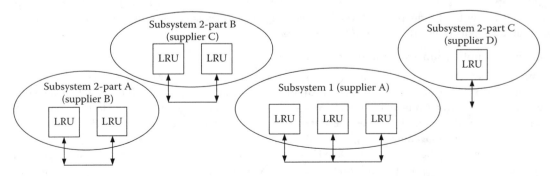

FIGURE 4.16 TTP-based subsystems can be tested separately by different suppliers without composability and interoperability challenges that multiply the integration effort for complex systems.

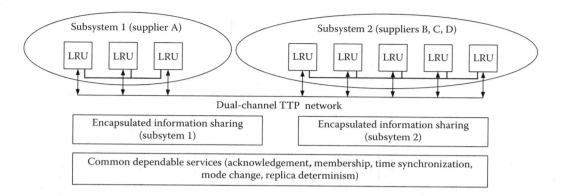

FIGURE 4.17 TTP-based distributed network management services established at the subsystem level work smoothly after system integration.

The logical composability of the fault tolerant system is supported by the robust separation of fault-tolerant functions from application software and their inclusion in the TTP protocol. Interactive consistency and replica determinism are important properties of a composable fault-tolerant system in a temporal and logical domain (see Figure 4.18).

4.6.3 Integration and Cooperative Development

Ultimate responsibility for functional composability and integration lies with suppliers and their cooperative development processes. TTP provides all interfaces (state message formats and timing schedules) needed for accurate and unambiguous common specification and represents the baseline in a joint development effort. Independent communication with predefined temporal behavior supports separate development of the network architecture (integrator), subsystems (supplier), and LRUs (teams). This causes an unparalleled reduction of integration effort.

The physical components of distributed TTP-based systems can be verified separately with the simulated behavior of the remaining system (e.g., other subsystems to accomplish required functionality, aircraft network, flight dynamics, etc.) to provide high predictability of system-level behavior as a result of composability and interoperability of TTP systems (Figure 4.19).

4.7 Modularity and Scalability

Scalable platforms are based on a number of understandable generic elements (distributed services, software/hardware modules, and components) with clean interfaces that support the development of complex systems and scalable architectures. Scalable architectures are open to changes and additional capacity upgrades without exponential increase in development effort.

In general, integration of different subsystems by different suppliers and modular reuse of designed components with sustainable airworthiness throughout several programs represent major challenges, technically as well as regulatory. TTA enables scalable and manageable development of safety-critical fault-tolerant systems and reduces system complexity. Even the increasing system complexity and a rising number of components and system size do not affect the ability to understand the behavior of TTP-based systems. Horizontal layering (abstraction) and vertical layering (partitioning) help to reduce the perceived complexity of large systems [9]. Both principles have been used in TTA platforms.

4.7.1 TTP and DO-297/ED-124 IMA Guidelines

The TTP communication data bus represents an integrated modular avionics (IMA) module and resource in compliance with the terminology and guidelines of DO-297/ED-124. TTP-based platforms incorporate

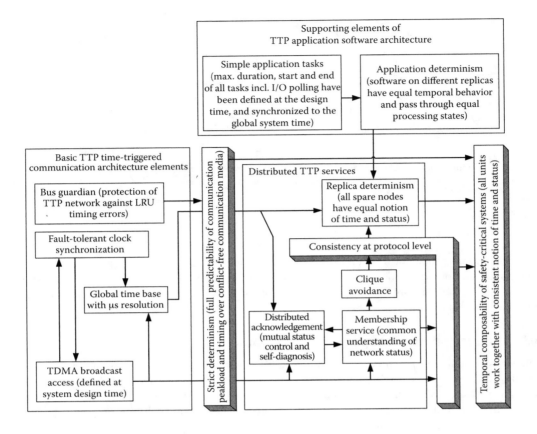

FIGURE 4.18 Origin of temporal composability of a safety-critical TTP-based system. Clique detection, distributed acknowledgment, and membership provide additional support for logical (functional) composability of fault-tolerant systems.

an advanced fault-tolerant distributed IMA concept and also provide full support for forthcoming distributed safety-critical, real-time applications. According to DO-297/ED-124, "IMA is described as a shared set of flexible, reusable, and interoperable hardware and software resources that, when integrated, form a platform that provides services, designed and verified to a defined set of requirements, to host applications performing aircraft functions."

A TTP-based platform supports the IMA concept in compliance with DO-297/ED-124 [10] but offers some unique features for the development of distributed applications with built-in safety and fault-tolerant features.

In order to satisfy DO-297/ED-124 IMA considerations, TTP meets the following requirements:

- Shares resources with multiple applications (i.e., the TTP double channel bus and CNI are shared by an application (a distributed application can consist of one or more tasks executed on different hosts in the network))
- Provides autonomously robust partitioning of shared resources (with spatial and temporal partitioning)
- Allows only hosted applications to interact with the platform and all others through well-defined interfaces (i.e., application software can interact only by using a Host CNI on top of the FT-COM, which interacts directly with CNI and communication hardware)

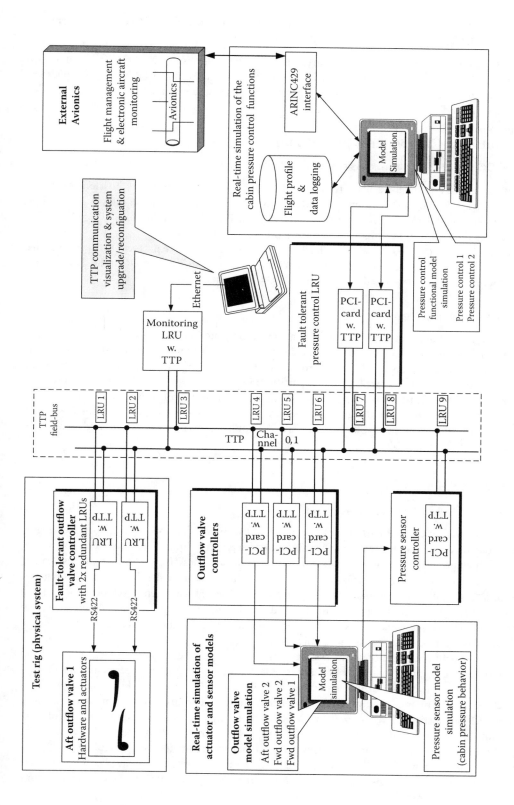

FIGURE 4.19 The development and verification of TTP-based SETTA cabin pressure control system (CPCS) demonstrator architecture using a blend of physical subsystems and simulated system behavior. (Adapted from www.vmars.tuwien.ac.at/projects/setta/index4.htm.)

- Allows configuration of resources to support reuse and modular certification of the platform (i.e., downloadable table-driven FT-COM and MEDL configuration)

The capability to protect shared resources is part of the TTP data bus design based on the use of TDMA bus access and bus guardians, which prevent access at arbitrary times. The TTP-based IMA platform provides distributed fault management and health monitoring using distributed fault-tolerant services, acknowledgment, and membership.

4.7.2 Platform Components and Modules

TTP-based DO-297 IMA platform consists of the following components and modules (Figure 4.20):

- An RTOS (TTP-OS) represents a reusable software module running on the host processor and uses the services of a TTP communication network module. RTOS integrates both the host component and TTP network module. Together with FT-COM running local TTP-OS tasks it provides the Application Program Interface (API) IMA functionality.
- The TTP controller module contains the TTP controller chip and the protocol software executed by the chip. The protocol software and RTOS are developed to support DO-178B Level A; the TTP controllers are developed to support DO-254 certification.

4.7.3 Application Reuse

Robust partitioning at the communication level is guaranteed by the CNI, fault-tolerant time base, static access schedule (defined by the configuration defined in the MEDL), and bus guardian. At the application level, tasks are scheduled during the design to provide required functionality — this guarantees temporal partitioning at the node level, as well as spatial partitioning at the system level. The host resources for every task are predefined at design time, and the application fulfills the constraints set by the resource availability. An application can be designed independently of other applications and unintended interaction with other applications is avoided. Therefore, the application represents a reusable component that is independently modifiable.

4.8 Summary

4.8.1 Aircraft Architecture and TTP

The TTP works well in a serial backplane as well as for distributed field bus applications (see Figure 4.21). TTP-based systems can be physically distributed and separated similarly to federated systems, but the level of integration is comparable to integrated modular avionics (IMA). In many cases, critical avionics systems require real-time reaction times and full synchronization of all subsystems with exchange of all essential data within milliseconds over large distances. The TTP has been developed from scratch for deterministic, safety-critical applications. Reduced weight and improved dispatchability, as well as lower total life-cycle costs, are major reasons for the use of innovative distributed architectures in modern civil aircraft. TTP seamlessly supports those objectives.

TTP enables weight optimization at the system level by placing the fault-tolerant electronics controls in the vicinity of sensors and actuators, which reduces the wiring weight and provides additional options for the replacement of hydraulic or mechanical systems with electrical or electro-hydrostatic systems.

Dispatchability and maintainability are supported by longer and controllable maintenance intervals and simplified diagnostics. TTP has built-in health monitoring and redundancy management with accurate fault localization that prevents casual ambiguity about the error sources. This supports the design of integrated vehicle health management (IVHM) as a system engineering discipline.

A modern airplane may have more complex flight dynamics as a result of more flexible composite materials. Theoretically, the engine and flight controls may be tightly integrated even on the same bus. With TTP, those functions are partitioned and remain separately certifiable. Another example is

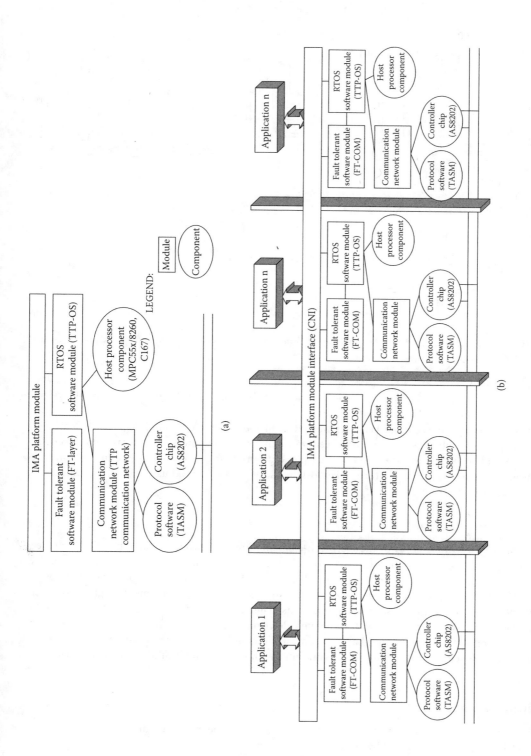

FIGURE 4.20 a. Single TTP-based DO-297 IMA module. b. Distributed IMA platform. All modules and applications are spatially and temporally partitioned.

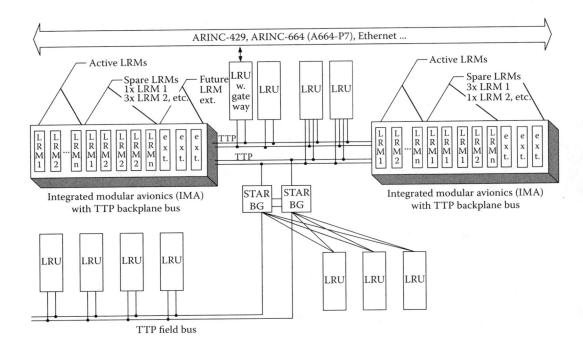

FIGURE 4.21 Integrated and federated architectures in different topologies tailored for specific needs can be created with TTP.

distributed power control in "more electric" aircraft, which ensures synchronized work of all systems with strictly deterministic, fault-tolerant communication.

New relationships between suppliers and integrators include risk sharing and integration of complete subsystems. The suppliers will have more freedom to choose communication systems with lower life-cycle costs and straightforward integration that are appropriate for their application and system-level optimization (Figure 4.22). This, in turn, will boost the use of airworthy, modularly certifiable commercial off-the-shelf (COTS) components.

4.8.2 Future Outlook

The speed of TTP is limited only by the specific implementation of controller technology and the physical layer in use. Therefore, higher speeds on two or more channels in different topologies can be expected in the future. Future generations of TTP controllers can also be extended to provide scalable levels of communication safety and extended support for avionics architecture development. A group of completely independent distributed software applications (Figure 4.23) at different criticality levels can be executed on a fault-tolerant system with completely deterministic behavior, mutually synchronized with microsecond resolution. Hence, TTP-based systems will establish the networked system as a fully partitioned, certifiable, embedded, fault-tolerant computer with reusable applications at different criticality levels.

Defining Terms

Deterministic: A system whose time evolution and behavior can be exactly predicted
Event-Driven: A system triggered by events at arbitrary instants

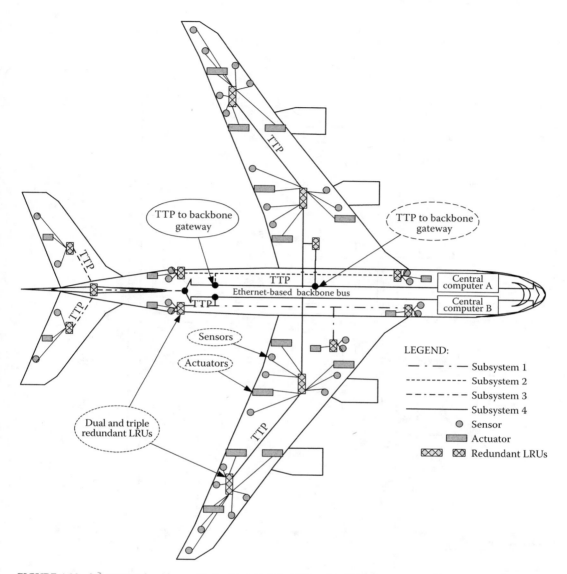

FIGURE 4.22 Subsystems based on strictly deterministic TTAs with TTP support new aircraft architectures for distributed but tightly integrated systems.

Fault-Tolerant Region (FCR): A set of components that is considered to fail as an atomic unit and in a statistically independent way with respect to other FCRs

Fault-Tolerant Unit (FTU): A set of replica-determinate nodes

Jitter: Oscillations in the latency of data transmission from sensors to computing units

Latency: Delay between message transmission and reception

LRM: Line Replaceable Module

LRU: Line Replaceable Unit

Membership: A vector containing a common notion of system status and health information

TDMA: Time Division Multiple Access

Time-Triggered: A system triggered by the progression of time at predefined instants

TMR: Triple modular redundancy, supported naturally by TTP

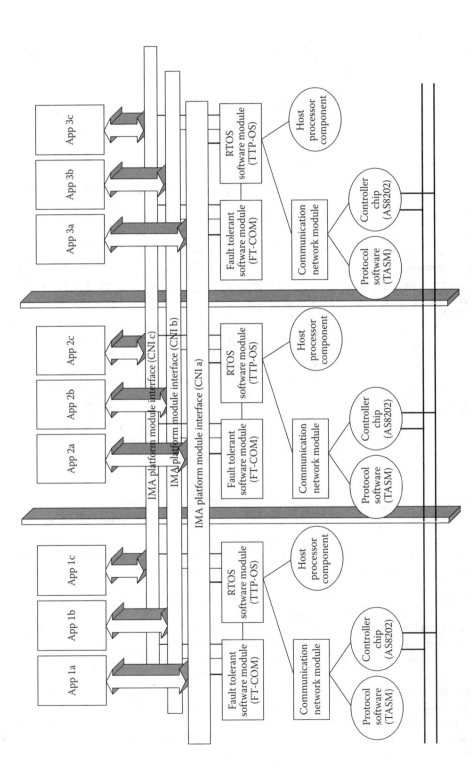

FIGURE 4.23 Distributed modular applications executed on distributed line replaceable modules (LRMs) with full temporal and spatial partitioning at the module and system level.

TTA: Time-triggered architecture; contains TTP and offers full advantage of system-level determinism including the application development approach with TTP tools

TTA Platform: Technology used for development of reusable aerospace control subsystem platforms

TTP: Time-triggered protocol

Acknowledgments

Many thanks to Martin Schwarz, Georg Stoeger, Georg Walkner, and Guenther Bauer, who contributed to this text with their comments.

References

1. TTA-Group, TTP Specification Request, www.ttagroup.org/technology/specification.htm.
2. TTTech, TTP Communication Controllers, www.tttech.com/products/controllers.htm.
3. Pfeifer, H., Schwier, D., and von Henke, F.W., Formal Verification for Time-Triggered Clock Synchronization, Proc. of Dependable Computing for Critical Applications 7 (DCCA 7), IEEE Computer Society, Dependable Computing and Fault-Tolerant Systems Series, January 1999, p. 207–26.
4. Spitzer, C.R, Digital Avionics Systems, Tutorial, 24th DASC Conference, Washington D.C., October 31, 2005.
5. Mathworks, www.mathworks.com.
6. Esterel Technologies, www.esterel-technologies.com.
7. Kopetz, H., *Real time Systems: Design Principles for Distributed Embedded Applications*, Kluwer Academic Publishers, Norwell, MA, 1997.
8. SETTA, SETTA Downloads, www.vmars.tuwien.ac.at/projects/setta/index4.htm.
9. Maier, R., Bauer, G., Stöger, G., and Poledna, S., Time triggered architecture: A consistent computing platform, *IEEE Micro*, July–August 2002.
10. DO-297/ED-124, Integrated Modular Certification Avionics (IMA) Design Guidelines and Certification Considerations, RTCA and European Organization for Civil Aviation Electronics (EURO-CAE), 2005.

5

Head-Mounted Displays

James E. Melzer
Rockwell Collins Optronics

5.1 Introduction

Head-mounted displays (HMD)* are personal information-viewing devices mounted on the head that can provide information in a way that no other display can. The information is always projected into the user's eyes, and it can be made reactive to head and body movements, replicating the way we view, navigate, and explore the world. This unique capability lends itself to the following applications:

- Virtual reality for creating artificial environments[1]
- Medical visualization as an aid in surgical procedures[2,3]
- Military vehicles for viewing sensor imagery[4]
- Airborne workstation applications, reducing size, weight, and power over conventional displays[5]
- Aircraft simulation and training[6,7,8]
- Fixed and rotary wing avionics display applications, as explored in this chapter[9,10]

In some applications, such as the medical and soldier's displays in Figure 5.1, the HMD is used solely as a hands-off information source for viewing endoscopic video, text, maps or graphics. But to truly reap the benefits of the HMD as part of an avionics application it must be part of a visually coupled system (VCS), which includes the HMD, a head position and orientation tracker, and a graphics engine or video source.[11,12] As the pilot turns his or her head, the tracker relays the orientation data to the mission computer, which updates the displayed information accordingly. This gives the pilot access to a myriad of real-time data that is *linked to head orientation*. In a fixed-wing fighter, a missile's sensor can be slaved to the line-of-sight from the pilot's head, allowing the pilot to designate targets away from the forward line-of-sight of the aircraft. In a helicopter, the pilot can point sensors such as forward-looking infrared (FLIR)** and fly at night. Because the aircraft mission computer knows the pilot's head orientation, the

---

*The term head-mounted display is used in this chapter as a more generic term than helmet-mounted display, which most often refers to military-oriented hardware. Helmet-mounted sight (HMS) is another term that often occurs, usually referring to an HMD that provides only a simple targeting reticle.

**Forward-looking infrared (FLIR) is a sensor technology that views differences in black-body emissions — heat differences — from objects.

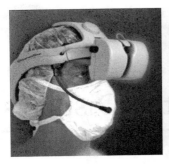

FIGURE 5.1 Three different applications for HMDs. a. The CardioView® for minimally invasive cardiac surgery. (Courtesy of Vista Medical Technologies, Inc.) b. A prototype of the U.S. Army's Land Warrior HMD. (Photo courtesy of Program Manager, Soldier, U.S. Army.) c. The SIM EYE XL100 for aviation simulation and training. (Photo courtesy of Rockwell Collins Optronics.)

HMD can also display real-time data that are either aircraft- or earth-referenced, such as runways, threats, friendly aircraft, and horizon lines.

The U.S. military introduced HMDs into fixed-wing aircraft in the early 1970s for targeting air-to-air missiles. Several hundred of the visual targeting acquisition systems (VTAS) were fielded on F-4 Phantom fighter jets between 1973 and 1979.[10,13] This program was eventually abandoned because the HMD capabilities were not matched by missile technology of the day.* HMDs were given new life when a Soviet MiG-29 was photographed in 1985 showing a simple helmet-mounted sight for off-axis targeting of the Vympel R-73 missile, also called the AA-11 Archer. With this revelation, the Israelis initiated a fast-paced program that deployed the Elbit DASH HMD for off-axis targeting of the Rafael Python 4 missile in 1993 and 1994.[14]

Two domestic simulation studies — Vista Sabre[15] and Vista Sabre II[16] — demonstrated the clear advantages for a pilot equipped with an HMD for missile targeting versus a pilot using only a head-up display (HUD). Encouraged by these studies and by a post-Berlin Wall examination of the close-combat capabilities of the HMD-equipped MiG-29,[17] the U.S. military initiated their own off-boresight missile targeting program. The result is the Joint Helmet Mounted Cueing System (JHMCS, built by Vision Systems International), currently being deployed on the U.S. Navy's F/A-18, the U.S. Air Force's F-15 and F-22, and on both domestic and international versions of the F-16. The JHMCS gives pilots off-axis targeting symbology for the AIM-9X missile and aircraft status;[18] providing them with improved situational awareness of the airspace around the aircraft. Figure 5.2 shows the U.S. Air Force and Navy's JHMCS helmet-mounted display, which is going into service on almost all fixed wing fighter aircraft flown by the U.S. military.

The U.S. Army has taken a more aggressive approach by putting HMD technology on rotary wing aircraft starting with the AH-1S Cobra helicopter gunship in the 1970s. A turreted machine gun is slaved to the pilot's head orientation via a mechanical linkage attached to his helmet. The pilot aims the weapon by superimposing a small helmet-mounted reticle on the target.[19]

In the 1980s, the Army adopted the Integrated Helmet and Display Sighting System (IHADSS) for the AH-64 Apache helicopter. This monocular helmet-mounted display gives the pilot the ability — similar to the Cobra gunship — to target head-slaved weapons (see Figure 5.3). The IHADSS has the added ability to display head-tracked FLIR imagery for nighttime flying. Honeywell** has delivered over 5000 of these cathode ray tube (CRT) based, monochrome systems to the Army for this very successful program.[10]

---

*There was also a Memorandum of Understanding signed in 1980 that relegated the development of short-range missile technology (and therefore HMDs) to the Europeans.

**The IHADSS system is now owned and produced by Elbit, Fort Worth, Inc. (EFW).

FIGURE 5.2 The U.S. Air Force and Navy's Joint Helmet Mounted Cueing System helmet-mounted display, which is going into service on almost all fixed-wing fighter aircraft flown by the U.S. military. (Photo courtesy of Vision Systems International, used with permission.)

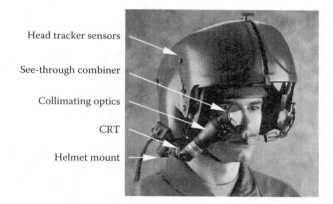

Head tracker sensors

See-through combiner

Collimating optics

CRT

Helmet mount

FIGURE 5.3 The Honeywell IHADSS is a monocular, monochrome, CRT-based, head-tracked, see-through helmet-mounted display used on the U.S. Army's AH-64 Apache helicopter. (Photo courtesy of Honeywell Electronics, used with permission.)

The U.S. Army also has extensive experience using helmet-mounted night vision goggles (NVGs) in aviation environments. These devices have their own unique set of performance, interface, and visual issues[20,21,22,23] and are discussed in more detail in Chapter 8. An interesting hybrid technique being used by the U.S. Army injects aircraft symbology into the front objective lens of the night vision goggles. Though not head-tracked, it gives the pilot "eyes up and out" access to aircraft status data. The ANVIS* HUD is a CRT-based design that has been successfully used on many of the Army's helicopters.[30] There is a movement towards replacing the CRT image source with a small flat-panel image source such as a

---

*ANVIS stands for aviator's night vision system—see Chapter 8.

FIGURE 5.4 The EyeHUD®, a hybrid HMD and night vision goggle technology in which an HMD (shown in a direct-view configuration on the left) can also be used to inject imagery into the objective lens of the night vision goggle (shown on the pilot's right eye in the photo on the right. (Photo courtesy of Rockwell Collins Display Products, San Jose, used with permission.)

liquid crystal display (LCD) to reduce head-supported weight and forward center of gravity (CG), and to eliminate the stiffer cable from the CRT. An example of one of these devices is shown in Figure 5.4.

In addition to these domestic applications, HMD-based pilotage systems are being adopted throughout the international aviation community on platforms such as Eurocopter's *Tiger* helicopter and the South African *Rooivalk*.[31]

5.2   What Is an HMD?

In its simplest incarnation, an HMD consists of one or more image sources, collimating optics, and a means to mount the assembly on the pilot's head. In the IHADSS HMD shown in Figure 5.3, the image source is a single, high-brightness CRT. The monocular optics create and relay a virtual image of the CRT, projecting the imagery onto the see-through combiner to the pilot's eye. This display module is attached to the right side of the aviator's protective helmet with adjustments that let the pilot position the display to see the entire image.

The early VTAS and Cobra helicopter HMDs used a simple targeting reticle to point weapons similar to the one shown on the left in Figure 5.5. The JHMCS HMD has a more sophisticated targeting capability, including "look-to" and shoot cues (shown on the right side of the figure), as well as altitude, airspeed, compass heading, and artificial horizon data. With the IHADSS in the AH-64 Apache helicopter, the pilot sees a similar symbology set superimposed over imagery from the head-tracked FLIR mounted on the nose of the helicopter.

This collection of components, though deceptively simple, has at its core a complex interaction of system and hardware issues with visual, anthropometric, physical, and display issues, because the HMD is viewed by the *human perceptual system*.[24] The design is complicated further in the aircraft environment, because the HMD — now a *helmet*-mounted display — provides both display and life support for the pilot. Issues of luminance, contrast, alignment, and focus must be considered while not impacting pilotage or crash safety. For all these reasons, HMD design requires a careful balancing — a *suboptimization* — of both display and physical requirements. The next sections will examine the important components or features in an HMD.

5.2.1   Image Sources for HMDs

As of 2006, many of the deployed HMDs use CRTs as image sources, primarily because the technology is the most mature and can provide the required high luminance, and because they can be ruggedized

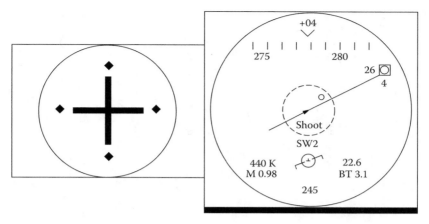

FIGURE 5.5 Comparision of early HMD reticle imagery (left) with a more capable symbology set (right) used with HMDs such as the JHMCS.

to withstand the harsh military environment.[25] Since the mid-1990s, however, small, flat-panel image sources have improved and are being introduced as alternatives to CRTs because of their reduced size, weight, and power requirements.[26,27]

There are two major categories of image sources, *emissive* and *non-emissive* (see Table 5.1). *Nonemissive image sources* modulate a separate illumination on a pixel-by-pixel basis to create the desired imagery. The following are examples:

- *Transmissive Liquid Crystal Displays (LCD)* — The pixel matrix is illuminated from the rear. A modulated electric field controls the transmission of the backlight through the individual liquid crystal cells. High-quality transmissive LCDs are manufactured in large numbers in Japan primarily for video projection applications, though in limited quantity domestically.
- *Reflective Liquid Crystal on Silicon Displays (LCOS)* — This is the same as the transmissive device except that the image source is illuminated from the front. The light transmits through the cell and reflects off of a mirror-like surface when the pixel is transmitting and is scattered when the pixel is turned off. This is a fast-growing area of development in the U.S. because the manufacturing technology is similar to silicon wafer fabrication.
- *Scanning Display* — A point source (such as a laser) or line of point sources (such as light-emitting diodes [LEDs]) is modulated in one or more directions using resonance scanners or opto-acoustic modulators to produce imagery. One excellent example is the Scanned Beam Display[28,29] (see Chapter 6).

Emissive devices include a wide range of image sources in which the image plane of the device emits light without the need for supplemental illumination. Such devices include:

- *Cathode Ray Tube (CRT)* — This is a vacuum tube with an electron gun at one end and a phosphor screen at the other. A beam from the electron gun is modulated by deflection grids and directed onto the screen. The incident electrons excite the phosphor, emitting visible light.[25] CRTs can be very bright and very rugged for the aviation environment, though they are larger than flat-panel displays and require high voltage.
- *Organic Light Emitting Diodes (OLED)* — A low-voltage drive across a thin layer of organic material causes it to emit visible light when the charge carriers recombine within the material. This is an excellent image source technology because of its small size, low power, and environmental ruggedness.

The choice of an image source for an HMD is not easy (see Table 5.1). Depending on the application, it may be preferable to have a backlit (i.e., transmissive) LCD over a reflective one for size, power, or packaging

TABLE 5.1 Categories of Miniature Image Sources Suitable for HMDs

Technology	Transmissive	Reflective	Self-Emissive	Scanning
Description	Light source illuminates the display from the rear; pixels are turned on/off or partially on for gray scale; transistors along the sides of the pixels	Light source illuminates the front of the display with a reflective surface under each pixel; pixels are turned on, off, or partially on for gray scale, blanking out the incident light; transistors underneath the pixels	Individual pixels are turned on/off or partially on for gray scale; transistors underneath the pixels (OLEDs); drive electronics are remote from the image source (CRTs)	Image source (LED or laser) scans across the image plane; drive electronics are remote from image source surface
Examples	Active Matrix Liquid Crystal Display (AMLCD)	Reflective Liquid Crystal on Silicon (LCOS); Digital Micromirror Display (DMD)	Cathode Ray Tube (CRT); Organic Light Emitting Diode (OLED)	Scanning Beam Display; Scanning Light Emitting Diode
Advantages	Very simple illumination design; high quality imagery; available commercially in quantity	High luminous efficiency; high fill factor (transistors under the pixel)	Smallest package; lightest weight; high fill factor (transistors under the pixel); wide temperature range (OLED)	High luminance; saturated colors; potential for image plane distortion
Disadvantages	Less efficient fill factor; transmission loss through LCD; requires spatial or temporal integration for color; limited temperature range (LCD); slower response time (LCD)	Front illumination more difficult to package; scattered light management very important; temporal integration for color	Limited luminance; color by temporal integration	Limited availability; limited resolution (LED); packaging limitations

reasons. Or, it may be preferable to have a self-emissive device such as an OLED with its small package size. Another consideration is that liquid crystal-based image sources have a finite area over which the image is observable; collimating optics with a very short focal length may lose part of the image. When considering which image source to use designers must be concerned with numerous issues such as:

- *Size* — What is the size of the image source itself? If a supplemental illumination source is required, how large is it? How large is the active area of the display? What is the size of the required drive electronics?
- *Weight* — What is the weight of the image source and any required supplemental illumination? If electronic components must be within close proximity to the image source (i.e., head-mounted), how much do they weigh? Can they be taken off the head or moved to a more favorable location on the head? (See Section 5.2.3 of this chapter.)
- *Power* — Some image source technologies such as CRTs require a high-voltage drive. Image sources such as LCDs have low transmission, requiring a brighter backlight. How much power will be required to meet the display luminance requirements?
- *Resolution* — How many pixels can be displayed? Is the image generator or sensor video compatible with this resolution? Is the response time of the image source fast enough to meet pilotage performance requirements?[32] If not, can measures be taken to improve the response time?[33,34]
- *Addressability* — CRTs are considered infinitely addressable because the imagery is drawn in calligraphic fashion. Pixilated devices such as LCDs and OLEDs are considered finite addressable displays because the pixel location is fixed. This limits their ability to compensate for image plane distortion.
- *Aspect ratio* — Most miniature CRTs have a circular format, while most of the solid state (pixilated) devices such as LCDs and OLEDs have a rectangular format. Flat panel devices with VGA, SVGA, or XGA resolution have a 4:3 horizontal-to-vertical aspect ratio. SXGA resolution devices have a 5:4 aspect ratio.* This is an important consideration when choosing an image source because it determines the field of view (FOV) of the display.
- *Luminance and contrast* — It is important that the image source be capable of providing a display luminance that is compatible with viewing against bright ambient backgrounds typically found in the aviation environment (see Section 5.4.3).
- *Color* — Is the image source capable of producing color imagery?[35] Because of the advantage that data color-coding provides to the pilot,[36] color is becoming more prevalent in head-down displays. Though color is not in widespread use in head-up and head-mounted displays, it may become more important due to some preliminary indications that high g-forces can alter color perception in the cockpit.[37]

5.2.2 Optical design

The purpose of the optics in an HMD is threefold:

- *Collimate the image source* — Produce a *virtual image* that appears to be farther away than just a few inches from the face.
- *Magnify the image source* — Make the imagery appear larger than the actual size of the image source.
- *Relay the image source* — Create the virtual image away from the image source and away from the front of the face.

There are two optical design approaches common in HMDs. The first is the *nonpupil forming design* — a simple magnifying lens — hence the term *simple magnifier* (Figure 5.6).[38,39] It is the easiest to design, the least expensive to fabricate, and the lightest and smallest, though it does suffer from a short throw

---

*VGA is 640 horizontal pixels by 480 vertical rows. SVGA is 800 horizontal pixels by 600 vertical rows. XGA is 1024 horizontal pixels by 768 vertical rows. SXGA is 1280 horizontal pixels by 1024 vertical rows.

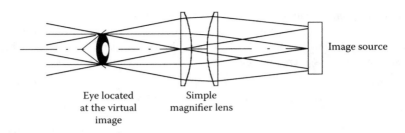

FIGURE 5.6   A simple magnifier, or nonpupil-forming, lens.

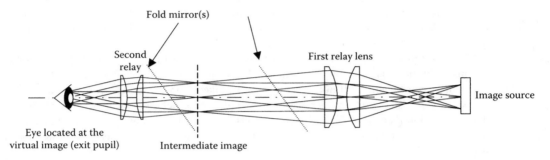

FIGURE 5.7   A pupil-forming optical design is similar to a compound microscope, binoculars, or a periscope.

distance between the image source and the virtual image, which puts the whole assembly on the front of the head, close to the eyes. This is typically used for simple viewing applications such as the medical HMD (Figure 5.1a) and the Land Warrior display (Figure 5.1b). The second optical approach — the *pupil-forming design* — is a bit more complex (Figure 5.7). It is more like the *compound microscope* or a submarine periscope in which a first set of lenses creates an intermediate image of the image source. This intermediate image is *relayed* by another set of lenses to where it creates a pupil, or a hard image of the intermediate image. The advantage is that the pupil-forming design provides more path length from the image plane to the eye. This gives the designer more freedom to insert mirrors as desired to fold the optical train away from the face to a location with a more advantageous weight and CG. The disadvantage is that the additional lenses increase the weight and cost of the HMD and there is no imagery outside the exit pupil — the image of the stop. This approach is typically used when the image source is large (such as a CRT) or when it is desirable to move the weight away from the front of the face, as in Figure 5.1c and Figure 5.3.

In each case, the optical design must be capable of collimating, magnifying, and relaying the image with sufficiently small amounts of residual aberrations,[39] with manual focus (if required) and with proper alignment (if a binocular system).[50] In addition, the optical design must have a sufficiently large exit pupil* (so the user does not lose the image if the HMD shifts on the head) and provide at least 25 mm of eye relief** to allow the user to wear eyeglasses. Table 5.2 summarizes the advantages and disadvantages of pupil-forming and nonpupil-forming approaches.

---

*The exit pupil is found only in pupil-forming designs such as the SIM EYE (Figure 5.1c) and the IHADSS (Figure 5.3). In nonpupil-forming designs of Figures 5.1a and 5.1b, it is more correct to refer to a *viewing eyebox*, because there is a finite unvignetted viewing area.

**There are some differences in terminology usually relating to the writing of specifications. In the classical optical design, the eye relief is the distance along the optical axis from the last optical surface to the exit pupil. In an HMD with angled combiners, eye relief should be measured from the eye to the closest point of the combiner, whether it is on the optical axis or not.

TABLE 5.2   Advantages and Disadvantages of Pupil-Forming and Nonpupil-Forming Optical Designs

	Nonpupil-Forming (Simple Magnifier)	Pupil-Forming (Relayed Lens Design)
Advantages	Simplest optical design; fewer lenses and lighter weight; doesn't "wipe" imagery outside of eye box; less eye box fit problems; mechanically simplest and least expensive	Longer path length means more packaging freedom. Can move away from front of face. More lenses provide better optical correction
Disadvantages	Short path-length puts entire display near eyes/face; short path-length means less packaging design freedom	More complicated optical design; more lenses mean heavier design; loss of imagery outside of pupil; needs precision fitting, more and finer adjustments

5.2.3   Head Mounting

It is difficult to put a precise measurement on the fit or comfort of an HMD, though it is always immediately evident to the wearer. Even if the HMD image quality is excellent, the user will reject it if it does not fit well. Fitting and sizing is especially critical in the case of a helmet-mounted display, which, in addition to being comfortable, must provide a *precision* fit for the display relative to the pilot's eyes. The following are the most important issues for achieving a good fit with an HMD:

- The user must be able to adjust the display to see the imagery.
- The HMD must be comfortable for a long duration of wear without causing "hot spots."
- The HMD must not slip with sweating or under g-loading, vibration, or buffeting.
- The HMD must be retained during crash or ejection.
- The weight of the head-borne equipment must be minimized.
- The mass-moment-of-inertia must be minimized.
- The mass of the head-borne components should be distributed to keep the CG close to that of the head alone.

The human head weighs approximately 9 to 10 lb and sits atop the spinal column. The occipital condyles on the base of the skull mate to the superior articular facets of the first cervical vertebra, the atlas.[40] These two small, oblong mating surfaces on either side of the spinal column are the pivot points for the head (Figure 5.8).

The CG of the head is located at or about the tragion notch, the small cartilaginous flap in front of the ear. Because this is *up* and *forward* of the head and vertebra pivot point, there is a tendency for the head to tip downward were it not for the strong counterforce exerted by the muscles running down the back of the neck — hence when people fall asleep, they "nod off." Adding mass to the head in the form of an HMD can move the CG (now HMD + head) away from this ideal location. High vibration or buffeting, ejection, parachute opening, or a crash will greatly exacerbate the effect of this extra weight and displaced CG, producing effects that can range from fatigue and neck strain to serious or mortal injury.[41] Designers can mitigate the impact of the added head-borne hardware by first minimizing the mass of the HMD and then optimizing the *location* of the mass to restore the head + HMD location to the head alone.

Extensive biomechanics research at the U.S. Army's Aeromedical Research Laboratory (USAARL) supports these conclusions. Figure 5.9 gives a weight versus CG curve in the vertical direction, where the area under the curve is considered crash safe for a helicopter environment. Figure 5.10 defines the weight/ CG combination that will minimize fatigue.[12] It should be noted that these curves are currently under review by USAARL and may change in the near future. Similar work in fixed-wing biomechanics at the Air Force's Wright-Patterson Labs has concluded that the weight of the HMD and oxygen mask cannot exceed 4 pounds and that the resulting CG must also be within a specified region centered about tragion notch.[40]

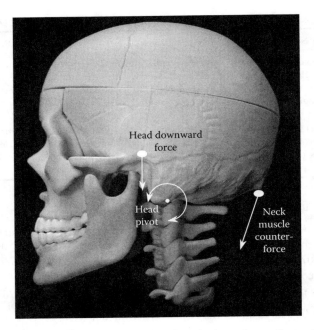

FIGURE 5.8 The human head and neck with the center of gravity located near the tragion notch and the pivot point located at the occipital condyles.

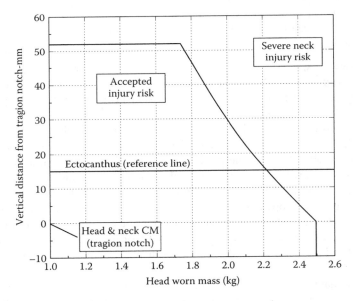

FIGURE 5.9 The USAARL weight and vertical center of gravity curve with the area under the curve considered crash safe in helicopter environments. (Data curve courtesy of U.S. Army Aeromedical Research Labs, used with permission.)

Anthropometry — "the measure of man" — is the compilation of data that define such things as the range of height for males and females, the size of our heads, and how far apart our eyes are. Used judiciously, these data can help the HMD designer achieve a proper fit; however, an over-reliance on these data can be equally problematic. One of the most common mistakes made by designers is to assume

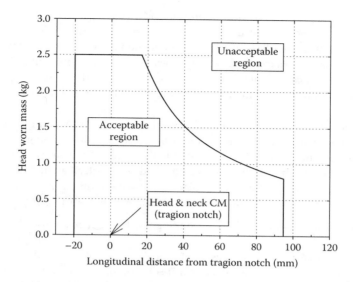

FIGURE 5.10 The USAARL weight and horizontal center of gravity curve with the area under the curve considered acceptable for fatigue in helicopter environments. (Data curve courtesy of U.S. Army Aeromedical Research Labs, used with permission.)

TABLE 5.3 Univariate Anthropometric Data for Key Head Features

Critical Head Dimensions (cm)	5th Percentile Female	95th Percentile Male
Interpupillary distance (IPD)	5.66	7.10
Head length[a]	17.63	20.85
Head width	13.66	16.08
Head circumference	52.25	59.35
Head height (ectocanthus to top of head)[a]	10.21	12.77

[a] These data are head orientation dependent.

a correlation between various anthropometric measurements, because almost all sizing data are *univariate* — that is, they are completely uncorrelated with other data. For example, a person who has a 95th percentile head circumference will not necessarily have a 95th percentile interpupillary distance.[42] One bivariate study did correlate head length and head breadth for male and female aviators, resulting in a rather large spread of data.[43] Table 5.3 shows the univariate (uncorrelated) anthropometric data for key head features. Note the range of sizes for the 5th percentile female up to the 95th percentile male.[44] There are examples in which helmet and HMD developments have been less than successful as a result of an over-emphasis on anthropometric data and an under-emphasis on fitting, resulting in HMDs that don't fit properly or in extraneous helmet sizes (the HGU-53/P).[42]

5.3   The HMD as Part of the Visually Coupled System

In an avionics application, the HMD — a Helmet-Mounted Display or Helmet-Mounted Sight — is part of a VCS consisting of the HMD, head tracker, and mission computer. As the pilot turns his head, the new orientation is communicated to the mission computer, which updates the imagery as required. The information is always with the pilot, always ready for viewing.

Early cockpit-mounted displays — head-down displays — gave the pilot information on aircraft status but required him to return his attention continuously to the interior of the cockpit. This reduced the time he could spend looking outside the aircraft. As jets got faster and the allowable reaction time for pilots got shorter, HUDs provided the next improvement by creating a collimated virtual image that is projected onto a combining glass located on top of the cockpit panel, in the pilot's forward line of sight. This meant that the pilot did not have to redirect his attention away from the critical forward airspace or refocus his eyes to see the image. Because the imagery is collimated, meaning it appears as though from some distant point, it can be superimposed on a distant object. This gives the pilot access to real-time geo- or aircraft-stabilized information such as compass headings, artificial horizons, or sensor imagery.

The HMD expands on this capability by placing the information in front of the pilot's eyes at all times and by linking the information to the orientation of the pilot's line of sight. While the HUD provides information about only the relatively small forward-looking area of the aircraft, the HMD with head tracker can provide information over the pilot's entire field of regard, all around the aircraft with eyes-and head-out viewing. This ability to link the displayed information with the pilot's line of sight increases the area over which the critical aircraft information is available. This new capability can do the following:

- Cue the pilot's attention by providing a pointing reticle to where a sensor has located an object of interest.
- Allow the pilot to slew sensors such as a FLIR for flying at night or in adverse conditions.
- Permit the pilot to aim weapons at targets that are off-boresight from the line of sight of the aircraft.
- Allow the pilot to hand off or receive target information (or location) from a remote platform, wingman, or other crewmember.
- Provide the pilot with aircraft- or geo-stabilized information.

In general, this new ability can provide situational awareness to the pilot by giving him information about the entire space surrounding the aircraft. One excellent example is the U.S. Army's AH-64 Apache helicopter, which is equipped with the IHADSS HMD and head tracker (Figure 5.11). As the pilot moves his head in azimuth or elevation, the tracker communicates the head orientation to the servo system controlling the Pilot Night Vision System (PNVS) FLIR. The sensor follows the head movements, providing the pilot with a viewpoint as though his head were located on the nose of the aircraft. This gives the pilot the ability to "see" at night or in low light in a very intuitive and hands-off manner, similar to the way he would fly during daytime, with the overlay of key flight data such as heading, altitude, and airspeed.

Studies are being conducted to find ways to squeeze even more out of the HMD in high-performance aircraft. A recent simulator study at the Naval Weapons Center explored the use of the HMD to provide "pathway in the sky" imagery to help pilots avoid threats and adverse weather.[45] Another experimental feature compensated for the loss of color and peripheral vision that accompanies g-induced loss of consciousness (g-loc). In this study, as the pilot began to "gray-out," the symbol set was reduced down to just a few critical items, positioned closer to the pilot's central area of vision. Another study provided helicopter pilots with earth-referenced navigation waypoints overlaid on terrain and battlefield engagement areas.[46] The results showed significant improvements in navigation, landing, and the ability to maintain fire sectors, and, most importantly, an overall reduction in pilot workload.

5.4 HMD System Considerations and Tradeoffs

As mentioned in the introduction, good HMD design relies on a suboptimization of requirements, trading off various performance parameters and requirements. The following sections will address some of these issues, including ocularity, FOV and resolution, and luminance and contrast in high ambient luminance environments.

5.4.1 Ocularity

One of the first issues to consider in an HMD is whether it should be biocular, binocular, or monocular. Table 5.4 compares the advantages and disadvantages of all three types.

FIGURE 5.11 The linkage between the IHADSS helmet-mounted display and the Pilot's Night Vision System in the AH-64 Apache helicopter. The PNVS is slaved to the pilot's head line of sight. As he turns his head, the PNVS turns to point in the same direction.

Monocular — *a single video channel viewed by a single eye.* This is the lightest, least expensive, and simplest of all three approaches. Because of these advantages, most of the current HMD systems are monocular, such as the Elbit DASH, the Vision Systems International JHMCS (Figure 5.2), and the EFW IHADSS (Figure 5.3). Some of the drawbacks are the potential for a laterally asymmetric CG and issues associated with focus, eye dominance, binocular rivalry, and ocular-motor instability.[47,48]

Biocular — *a single video channel viewed by both eyes.* The biocular approach is more complex than the monocular design, though it stimulates both eyes, eliminating the ocular-motor instability issues associated with monocular displays. Viewing imagery with two eyes versus one has been shown to yield improvements in detection as well as provide a more comfortable viewing experience.[49,50] However, since it is now a two-eyed viewing system, the designer is subject to a much more stringent set of alignment, focus, and adjustment requirements.[51] The primary disadvantage of the biocular design is that the image source is usually located in the forehead region, making it more difficult to package. In addition, since the luminance from the single image source is split to both eyes, the brightness is cut in half.

Binocular — *each eye views an independent video channel.* This is the most complex, most expensive, and heaviest of all three options, but it has all the advantages of a two-eyed system with the added benefit of providing partial binocular overlap (to enlarge the horizontal FOV), stereoscopic imagery, and more packaging design freedom, such as the SIM EYE shown in Figure 5.1c. A binocular HMD is subject to same alignment, focus, and adjustment requirements as the biocular design, but the designer benefits from the ability to move both the optics and the image sources *symmetrically away* from the face.

5.4.2 Field of View and Resolution

When asked for their initial HMD requirements users will typically ask for more of both FOV *and* resolution. This is not surprising since the human visual system has a total FOV of 200° horizontal by 130° vertical[52] with a grating acuity of 2 minutes of arc[53] in the central foveal region, something that HMD designers have yet to replicate. For daytime air-to-air applications in a fixed wing aircraft, a large FOV is probably not necessary to display the symbology shown in Figure 5.5. The FOV can be approximately 6° for a simple sighting reticle. For an HMD such as the JHMCS system, where the pilot will receive aircraft and weapons status information, a 20° FOV is more effective. If the HMD is intended to display sensor imagery for nighttime pilotage, such as with the IHADSS (a rectangular 30° by 40° FOV), the pilot will "paint" the sky with the HMD, creating a mental map of his surroundings. The larger FOV is advantageous because it provides peripheral cues that contribute to the pilot's sense of self-stabilization, and it lowers pilot workload by reducing the range of head movements needed to fill in the mental map.[54,55,56] Most night vision goggles, such as the ANVIS-6, have a FOV of 40° circular, though most pilots would prefer more.

TABLE 5.4 Advantages and Disadvantages of Monocular, Biocular, and Binocular HMDs

Configuration		Advantages	Disadvantages
Monocular (one image source viewed by one eye)		Lightest weight; simplest to align; least expensive	Potential for asymmetric CG; potential for ocular-motor instability, eye dominance, and focus issues
Biocular (one image source viewed by both eyes)		Simple electrical interface; lightweight; inexpensive	More complex alignment than monocular; difficult to package; difficult for see-through
Binocular (two image sources viewed by both eyes)		Stereo imagery; partial binocular overlap; symmetrical CG	Most difficult to align; heaviest; most expensive

Although display resolution contributes to overall image quality, there is also a direct relationship with performance. The Johnson criteria for image recognition shows that the amount of resolution required is (like most HMD-related issues) task-dependent. For an object such as a tank, increased resolution will allow the pilot to detect ("something is there"), to recognize ("it is a tank"), or to identify ("it is a T-72 tank")[57] at a particular distance.

While more of each is desirable, FOV and resolution in an HMD are linked by the relationship:

$$H = F * \text{Tan } \Theta$$

where *F* is the focal length of the collimating lens.

- If H is the size of the image source, then Θ *is the FOV* or apparent size of the virtual image in space.
- If H is the pixel size, then Θ *is the resolution* or apparent size of the pixel in image space.

Figure 5.12 shows how the focal length of the collimating lens determines the relationship between H, the size of the image source (or pixel size), and Θ, the FOV (or the resolution). Thus, the focal length of the collimating lens *simultaneously* governs the FOV (which you want to be large) *and* the resolution (which you want to be small). For a display with a single image source, the result is either a wide FOV *or* a high resolution but *not both* at the same time.

Given this *F*TanΘ* invariant, there are at least four ways to increase the FOV of a display and still maintain resolution: (1) high-resolution area of interest, (2) partial binocular overlap, (3) optical tiling, and (4) dichoptic area of interest.[58,59] Of these, partial binocular overlap is preferable for binocular flight applications, though optical tiling is under investigation to expand the FOV of night vision goggles.[60]

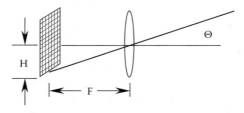

FIGURE 5.12 How the focal length of the collimating lens determines the relationship between H, the size of the image source (or pixel size) and Θ, the field of view (or the resolution).

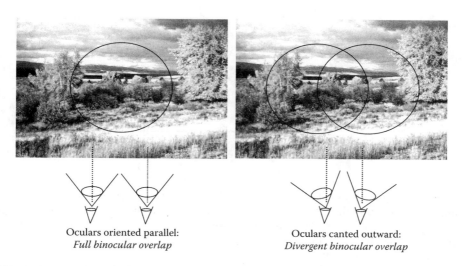

Oculars oriented parallel:
Full binocular overlap

Oculars canted outward:
Divergent binocular overlap

FIGURE 5.13 A comparison of a full binocular overlap and divergent partial binocular overlap. Note the increase in viewable imagery in the horizontal direction with the divergent overlap.

Figure 5.13 shows a comparison of a full binocular overlap and divergent partial binocular overlap. Note the increase in viewable imagery in the horizontal direction with the divergent overlap.

Partial binocular overlap results when the two HMD optical channels are canted either inward (convergent overlap) or outward (divergent overlap), enlarging the horizontal FOV while maintaining the same resolution as the individual monocular channels. Partial overlap requires that two image sources and two video channels are available and the optics and imagery are properly configured to compensate for any residual optical aberrations. Concerns have been voiced about the required minimum binocular overlap as well as the possibility that perceptual artifacts such as binocular rivalry — referred to as "luning" — may have an adverse impact on pilot performance. Although the studies that found image fragmentation did place some workload on the pilot test subjects,[61,62] all were conducted using static imagery. Several techniques are effective in reducing the rivalry effects and their associated perceptual artifacts.[63]

Keep in mind that the resolution of the VCS is a product of the resolution of the HMD and of the imaging sensor. While an HMD with very high resolution may provide a high-quality image, pilotage performance may still be limited by the resolution of the imaging sensor such as the FLIR or camera. It is preferable to match the FOV of the HMD with that of the sensor to achieve a 1:1 correspondence between sensor and display to ensure an optimum flying configuration.

5.4.3 Luminance and Contrast in High Ambient Luminance Environments

In the high ambient luminance environment of an aircraft cockpit daylight readability of displays is a critical issue. The combining element in an HMD is similar to the combiner of a HUD, reflecting the projected imagery into the pilot's eyes. The pilot looks through the combining glass and sees the imagery superimposed on the outside world, so it cannot be 100% reflective — pilots always prefer to have as much see-through as possible. To view the HMD imagery against a bright background, such as sun-lit clouds or snow, this less-than-perfect reflection efficiency means that the image source must be that much brighter. The challenge is to provide a combiner with good see-through transmission and a high luminance image. There are limitations; all image sources have a luminance maximum governed by the physics of the device as well as size, weight, and power of any ancillary illumination. In addition, other factors, such as the transmission of the aircraft canopy and pilot's visor, must be considered when determining the required image source luminance as shown in Figure 5.14.

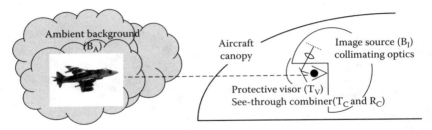

FIGURE 5.14 The contributions for determining image source luminance requirements for an HMD in an aircraft cockpit.

The image source luminance (B_I) is attenuated before entering the eye by the transmission of the collimating optics (T_O) and the reflectance of the combiner (R_C). The pilot views the distant object through the combiner (T_C or $1 - R_C$), the protective visor (T_V), and the aircraft transparency (T_A) against the bright background (B_A). We can calculate the image source luminance for a desired contrast ratio (CR) of 1.3 using the expression:[11]

$$CR = \frac{B_A + B_{Display}}{B_A}$$

where we know that the display luminance to the eye is given by:

$$B_{Display} = B_I * T_O * R_C$$

and as observed by the pilot, the background is given by:

$$B_O = T_C * T_V * T_A * B_A$$

Rewriting, we can see that:

$$CR = \frac{1 + B_I * T_O * R_C}{T_C * T_V * T_A * B_A}$$

We can substitute some nominal values for the various contributions as given in Table 5.5 to achieve a nominal contrast ratio of 1:2.

The first two cases in Table 5.5 compare the difference between a pilot wearing a Class 1 (clear) versus a Class 2 (dark) visor.[64] The dark visor reduces the ambient background luminance, improving HMD

TABLE 5.5 Contributions for the Display Luminance Calculations for Four HMD Configurations

		Case 1 — Clear Visor, 50% Combiner Transmission	Case 2 — Dark Visor, 50% Combiner Transmission	Case 3 — Clear Visor, 80% Combiner Transmission	Case 4 — Dark Visor, 80% Combiner Transmission
Optics transmission	T_O	85%	85%	85%	85%
Combiner reflectance	R_C	50%	50%	20%	20%
Combiner transmission	T_C	50%	50%	80%	80%
Visor transmission	T_v	87%	12%	87%	12%
Aircraft canopy transmission	T_C	80%	80%	80%	80%
Ambient background luminance	B_C	10,000 fL	10,000 fL	10,000 fL	10,000 fL
Required image source luminance	B_I	**2456 fL**	**339 fL**	**9826 fL**	**1355 fL**

image contrast against the bright clouds or snow. These first two cases are relatively simple because they assume a combiner with 50% transmission and 50% reflectance (ignoring other losses). Since pilots need more see-through, this means a reduced reflectance. Cases 3 and 4 assume this more realistic combiner configuration with both clear and dark visors, resulting in a requirement for a much brighter image source. One of the ways to improve both see-through transmission and reflectance is to take advantage of high-reflectance holographic notch filters and V-coats. The problem is that, although these special coatings reflect more of a specific display color, they transmit less of that *same* color, which can alter perceptions of cockpit display color as well as external coloration.

5.5 Summary

HMDs can provide a distinctly unique and personal viewing experience unlike other display technologies. Providing the pilot with display information linked to head orientation frees him from returning his attention to the cockpit interior, allowing the pilot to navigate and fly the aircraft in a more intuitive and natural manner. This is an effective means of providing the pilot with aircraft status as well as information about the surrounding airspace.

However, these capabilities are not without a price. HMDs require careful attention to the complex interactions between hardware and human perceptual issues, made only more complex by the need for the HMD to provide life support in an aviation environment. This will be accomplished only when all factors are considered and the requirements are successfully suboptimized with an understanding of the aviator's tasks and environment.

Recommended Reading

Barfield, W., Furness, T. A., *Virtual Environments and Advanced Interface Design*, New York, Oxford University Press, 1995.

Boff, K. R., Lincoln, J. E., *Engineering data compendium, human perception and performance*, Human Engineering Division, Harry G. Armstrong Aerospace Medical Research Laboratory, Wright-Patterson Air Force Base, OH, 1988.

Kalawsky, R. S., *The Science of Virtual Reality and Virtual Environments*, New York, Addison-Wesley Publishing Company, 1993.

Karim, M. A., Ed., *Electro-Optical Displays*, New York, Marcel Dekker, Inc. 1992.

Lewandowski, R. J., *Helmet- and Head-mounted Displays, Selected SPIE papers on CD-ROM*, SPIE Press, 11, 2000.

Melzer, J. E., and Moffitt, K. W., Eds., *Head-Mounted Displays: Designing for the User*, McGraw-Hill, New York, 1997.

Rash, C. E., Ed., *Helmet-Mounted Displays: Design Issues for Rotary-Wing Aircraft*, U.S. Government Printing Office, 1999.

Velger, M., *Helmet-mounted displays and sights*, Norwood, MA Artech House, Inc. 1998.

References

1. Kalawsky R. S., *The Science of Virtual Reality and Virtual Environments: A Technical, Scientific and Engineering Reference on Virtual Environments*, Addison-Wesley: Wokingham, England, 1996.
2. Schmidt, G. W., Osborn, D. B., Head-mounted display system for surgical visualization, *Proc. SPIE, Biomedical Optoelectronic Instrumentation*, 2396, 345, 1995.
3. Pankratov, M. M., New surgical three-dimensional visualization system, *Proc. SPIE, Lasers in Surgery: Advanced Characterization, Theraputics, and Systems*, 2395, 143, 1995.
4. Casey, C. J., Helmet-mounted displays on the modern battlefield, *Proc. SPIE, Helmet- and Head-Mounted Displays IV*, 3689, 270, 1999.

5. Browne, M. P., Head-mounted workstation displays for airborne reconnaissance applications, *Proc. SPIE, Cockpit Displays V: Displays for Defense Applications*, 3363, 348, 1998.

6. Lacroix, M., Melzer, J., Helmet-mounted displays for flight simulators, *Proceedings of the IMAGE VII Conference, Tucson Arizona, 12–17 June*, 1994.

7. Casey, C. J., Melzer, J. E., Part-task training with a helmet integrated display simulator system, *Proceedings of SPIE, Large-Screen Projection, Avionic and Helmet-Mounted Displays*, 1456, 175, 1991.

8. Thomas, M., Geltmacher, H., Combat simulator display development, *Information Display*, 4&5, 23, 1993.

9. Foote, B., Design guidelines for advanced air-to-air helmet-mounted display systems, *Proc. SPIE, Helmet- and Head-Mounted Displays III*, 3362, 94, 1998.

10. Belt, R. A., Kelley, K., Lewandowski, R., Evolution of helmet-mounted display requirements and Honeywell HMD/HMS systems, *Proc. SPIE, Helmet- and Head-Mounted Displays III*, 3362, 373, 1998.

11. Kocian, D. F., Design considerations for virtual panoramic display (VPD) helmet systems," *AGARD Conference Proceedings No. 425, The man-machine interface in tactical aircraft design and combat automation*, 22–1, 1987.

12. Rash, C. E. (Ed.) *Helmet-Mounted Displays: Design Issues for Rotary-Wing Aircraft*, U.S. Government Printing Office, 1999.

13. Dornheim, M., VTAS sight fielded, shelved in 1970s *Aviation Week & Space Technology*, October 23, 51, 1995.

14. Dornheim, M. A., Hughes, D., U.S. intensifies efforts to meet missile threats, *Aviation Week & Space Technology*, October 16, 36, 1995.

15. Arbak, C., Utility evaluation of a helmet-mounted display and sight, *Proc. SPIE, Helmet-Mounted Displays*, 1116, 138, 1989.

16. Merryman, R. F. K., Vista Sabre II: integration of helmet-mounted tracker/display and high off-boresight missile seeker into F-15 aircraft, *Helmet-and Head-Mounted Displays and Symbology Design Requirements*, 2218, 173, 1994.

17. Lake, J., NATO's best fighter is made in Russia, *The Daily Telegraph*, August, 26, 1991, p. 22.

18. Goodman, Jr., G. W., First look, first kill, *Armed Forces Journal International*, July 2000, p. 32.

19. Braybrook, R., Looks can kill, *Armada International*, 4, 44, 1998.

20. Sheehy, J. B., Wilkinson, M., Depth perception after prolonged usage of night vision goggles, *Aviation, Space and Environmental Medicine*, 60, 573, 1989.

21. Donohue-Perry, M. M., Task, H. L., Dixon, S. A., Visual acuity versus field of view and light level for night vision goggles (NVGs), *Helmet- and Head-Mounted Displays and Symbology Design Requirements*, 2218, 71, 1994.

22. Crowley, J. S., Rash, C. E., Stephens, R. L., Visual illusions and other effects with night vision devices, *Proc. SPIE, Helmet-Mounted Displays III*, 1695, 166, 1992.

23. DeVilbiss, C. A., Ercoline, W. R., Antonio, J. C., Visual performance with night vision goggles (NVGs) measured in U.S. Air Force aircrew members, *Helmet- and Head-Mounted Displays and Symbology Design Requirements*, 2218, 64, 1994.

24. Gibson, J. J., *The ecological approach to visual perception*, Lawrence Erlbaum Associates, Hillsdale, New Jersey, 1986.

25. Sauerborn, J. P., Advances in miniature projection CRTs for helmet displays, *Proc. SPIE, Helmet-Mounted Displays III*, 1695, 102, 1992.

26. Ferrin, F. J., Selecting new miniature display technologies for head mounted applications, *Proc. SPIE, Head-Mounted Displays II*, 3058, 115, 1997.

27. Belt, R. A., Knowles, G. R., Lange, E. H., Pilney, B. J., Girolomo, H.J., Miniature flat panels in rotary wing head mounted displays, *Proc. SPIE, Head-Mounted Displays II*, 3058, 125, 1997.

28. Urey, H., Nestorovic, N., Ng, B., Gross, A. A., Optics designs and systems MTF for laser scanning displays, *Proc. SPIE, Helmet- and Head-Mounted Displays, IV*, 3689, 238, 1999.

29. Urey, H., Optical advantages in retinal scanning displays, *Proc. SPIE, Head- and Helmet-Mounted Displays*, 4021, 20, 2000.

30. Yona, Z., Weiser, B., Hamburger, O., Day/night ANVIS/HUD-24 (day HUD) flight test and pilot evaluations, *Proc. SPIE, Helmet- and Head-Mounted Displays IX: Technologies and Applications*, 5442, 225, 2004.

31. Mace, T.K., Van Zyl, P.H., Cross, T., Integration, development, and qualification of the helmet-mounted sight and display on the Rooivalk Attack Helicopter, *Proc. SPIE., Helmet and Head-Mounted Displays VI*, 4361, 12, 2001.

32. Rabin, J., Wiley, R., Dynamic visual performance: comparison between helmet-mounted CRTs and LCDs, *J. of SID*, 3/3, 97, 1995.

33. Gale, R., Herrmann, F., Lo, J., Metras, M., Tsaur, B., Richard, A., Ellertson, D., Tsai, K., Woodard, O., Zavaracky, M., Presz, M., Miniature 1280 by 1024 active matrix liquid crystal displays, *Proc. SPIE, Helmet- and Head-Mounted Displays IV*, 3689, 231, 1999.

34. Woodard, O. C., Gale, R. P., Ong, H. L., Presz, M. L., Developing the 1280 by 1024 AMLCD for the RAH-66 Comanche, *Proc. SPIE, Head- and Helmet-Mounted Displays*, 4021, 203, 2000.

35. Post, D. L., Miniature color display for airborne HMDs, *Helmet- and Head-Mounted Displays and Symbology Design Requirements*, 2218, 2, 1994.

36. Melzer, J. E., Moffitt, K. W., Color helmet display for the tactical environment: the pilot's chromatic perspective, *Proc. SPIE, Helmet-Mounted Displays III*, 1695, 47, 1992.

37. MacGillis, A., Flying at high G's alters pilot's perception of colors, *National Defense*, 16, September 1999.

38. Task, H. L., HMD image sources, optics and visual interface, in *Head-Mounted Displays: Designing for the User*, Melzer, J.E., and Moffitt, K. W., Eds., McGraw-Hill, New York, 1997, chap. 3.

39. Fischer, R. E., Fundamentals of HMD Optics, in *Head-Mounted Displays: Designing for the User*, Melzer, J.E., and Moffitt, K. W., Eds., McGraw-Hill, New York, 1997, chap. 4.

40. Perry, C. E., Buhrman, J. R., Biomechanics in HMDs, in *Head-Mounted Displays: Designing for the User*, Melzer, J.E., and Moffitt, K. W., Eds., McGraw-Hill, New York, 1997, chap. 6.

41. Guill, F. C., Herd, G. R., An evaluation of proposed causal mechanisms for "ejection associated" neck injuries, *Aviation, Space and Environmental Medicine*, A26, July, 1989.

42. Whitestone, J .J., Robinette, K. M., Fitting to maximize performance of HMD systems, in *Head-Mounted Displays: Designing for the User*, Melzer, J.E., and Moffitt, K. W., Eds., McGraw-Hill, New York, 1997, chap. 7.

43. Barnaba, J.M.,Human factors issues in the development of helmet mounted displays for tactical, fixed-wing aircraft, *Proc. SPIE, Head-Mounted Displays II*, 3058, 2, 1997.

44. Gordon, C.C., Churchill, T., Clauser, C.E., Bradtmiller, B., McConville, J.T., Tebbetts, I., Walker, R.A., *1988 Anthropometric survey of U.S. Army personnel: Summary statistics interim report*. U.S. Army Natick Technical Report TR-89/027, Natick MA, 1989.

45. Procter, P., Helmet displays boost safety and lethality, *Aviation Week & Space Technology*, February 1, 1999, p. 81.

46. Rogers, S. P., Asbury, C. N., Haworth, L. A., Evaluation of earth-fixed HMD symbols using the PRISMS helicopter flight simulator, *Proc. SPIE, Helmet- and Head-Mounted Displays III*, 3389, 54, 1999.

47. Rash, C. E., Verona, R. W., The human factor considerations of image intensification and thermal imaging systems, in Karim, M. A., Ed., *Electro-Optical Displays*, New York, Marcel Dekker, Inc. 1992, chap. 16.

48. Moffitt, K. W., Ocular responses to monocular and binocular helmet-mounted display configurations, *Proc. SPIE, Helmet-Mounted Displays*, 1116, 142, 1989.

49. Boff, K. R., Lincoln, J. E., *Engineering data compendium, human perception and performance*, Human Engineering Division, Harry G. Armstrong Aerospace Medical Research Laboratory, Wright-Patterson Air Force Base, OH, 1988.

50. Moffitt, K. W., Designing HMDs for viewing comfort, in *Head-Mounted Displays: Designing for the User*, Melzer, J.E., and Moffitt, K. W., Eds., McGraw-Hill, New York, 1997, chap. 5.

51. Self, H. C., *Critical tolerances for alignment and image differences for binocular helmet-mounted displays*, Technical Report AAMRL-TR-86-019, Wright-Patterson AFB OH: Armstrong Aerospace Medical Research Laboratory, 1986.

52. U.S. Department of Defense, "MIL-HDBK-141 Optical Design," 1962.

53. Smith, G., Atchison, D. A., *The eye and visual optical instruments*, New York, Cambridge University Press, 1997.

54. Wells, M. J., Venturino, M., Osgood, R. K., Effect of field of view size on performance at a simple simulated air-to-air mission, *Proc. SPIE, Helmet-Mounted Displays*, 1116, 126, 1989.

55. Kasper, E. F., Haworth, L. A., Szoboszlay, Z. P., King, R. D., Halmos, Z. L., Effects of in-flight field-of-view restriction on rotorcraft pilot head movement, *Proc. SPIE, Head-Mounted Displays II*, 3058, 34, 1997.

56. Szoboszlay, Z. P., Haworth, L. A., Reynolds, T. L., Lee, A. G., Halmos, Z. L., Effect of field-of-view restriction on rotocraft pilot workload and performance: preliminary results, *Proc. SPIE, Helmet- and Head-Mounted Displays and Symbology Design Requirements II*, 2465, 142, 1995.

57. Lloyd, J. M., *Thermal Imaging Systems*, Plenum Press, New York, 1975.

58. Melzer, J.E., Overcoming the field of view: resolution invariant in head-mounted displays, *Proc. SPIE, Helmet- and Head-Mounted Displays III*, 3362, 284–293, 1998.

59. Hoppe, M. J., Melzer, J. E., Optical tiling for wide-field-of-view head-mounted displays, *Proc. SPIE, Current Developments in Optical Design and Optical Engineering VIII*, 3779, 146, 1999.

60. Jackson, T. W., Craig, J. L., Design, development, fabrication, and safety-of-flight testing of a panoramic night vision goggle, *Proc. SPIE, Head- and Helmet-Mounted Displays IV*, 3689, 98, 1999.

61. Klymenko, V., Verona, R. W., Beasley, H. H., Martin, J. S., Convergent and divergent viewing affect luning, visual thresholds, and field-of-view fragmentation in partial binocular overlap helmet-mounted displays, *Helmet- and Head-Mounted Displays and Symbology Design Requirements*, 2218, 2, 1994.

62. Klymenko, V., Harding, T. H., Beasley, H. H., Martin, J. S., Rash, C. E., Investigation of helmet-mounted display configuration influences on target acquisition, *Proc. SPIE, Head- and Helmet-Mounted Displays*, 4021, 316, 2000.

63. Melzer, J. E., Moffitt, K., An ecological approach to partial binocular-overlap, *Proc. SPIE, Large Screen, Projection and Helmet-Mounted Displays*, 1456, 124, 1991.

64. U.S. Department of Defense, *MIL-V-85374, Military specification, visors, shatter resistant*, 1979.

6

Display Devices: RSD™ (Retinal Scanning Display)

Thomas M. Lippert

Microvision Inc.

6.1 Introduction

This chapter relates performance, safety, and utility attributes of the retinal scanning display (RSD) as employed in a helmet-mounted pilot-vehicle interface, and by association, in panel-mounted head up display (HUD) and head down display (HDD) applications. Because RSD component technologies are advancing so rapidly, quantitative analyses and design aspects are referenced to permit a more complete description here of the first high-performance RSD system developed for helicopters.

Visual displays differ markedly in how they package light to form an image. The RSD depicted in Figure 6.1, is a relatively new optomechatronic device based initially on red, green, and blue diffraction-limited laser light sources. The laser beams are intensity modulated with video information, optically combined into a single, full-color pixel beam, then scanned into a raster pattern by a roster optical scanning engine (ROSE) comprised of miniature oscillating mirrors, much as the deflection yoke of a cathode-ray tube (CRT) writes an electron beam onto a phosphor screen. RSDs are unlike CRTs in that conversion of electrons to photons occurs prior to beam scanning, thus eliminating the phosphor screen altogether along with its re-radiation, halation, saturation, and other brightness- and contrast-limiting factors. This means that the RSD is fundamentally different from other existing display technologies in that there is no planar emission or reflection surface — the ROSE creates an optical pupil directly. Like the CRT, an RSD may scan out spatially continuous (nonmatrix-addressed) information along

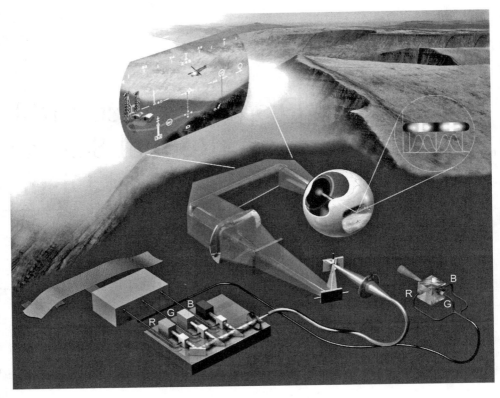

FIGURE 6.1　Functional component diagram of the RSD HMD.

each horizontal scan line, while the scan lines form discrete information samples in the vertical image dimension.*

6.2　An Example Avionic HMD Challenge

Consider the display engineering problem posed by Figure 6.1. An aircraft flying the contour of the earth will transit valleys as well as man-made artifacts: towers, power lines, buildings, and other aircraft. On this flight the pilot is faced with a serious visual obscurant in the form of ground fog, rendered highly opaque by glare from the sun.

The pilot's situational awareness and navigation performance are best when flying "eyes-out" the windshield, in turn requiring "eyes-out" electronic display of his own aircraft attitude and status information. Particularly under degraded visual conditions additional imagery of obstacles (towers, the earth, etc.) synthesized from terrain data bases and mapped into the pilot's ever-changing direction of gaze via Global Positioning System data reduce the hazards of flight. The question has been, which technology can provide a display of adequate brightness, color, and resolution to adequately support pilotage as viewed against the harsh real-world conditions described.

For over 30 years, researchers and designers have improved the safety and effectiveness of HMDs so that mission-critical information would always be available "eyes-out" where the action is, unlike "eyes-in" traditional HDDs.[1] U.S. Army AH-64 Apache helicopter pilots are equipped with such an HMD, enabling nap-of-the-earth navigation and combat at night with video from a visually coupled infrared imager and data computer. This particular pilot-vehicle interface has proven its reliability and effectiveness in over

---

*At the time this volume was published EUROCAE has not approved the publication of ED-124 pending internal review.

1 million hours of flight and was employed with great success in the Desert Storm Campaign. Still, it lacks the luminance required for optimal grayscale display during typical daylight missions, much less the degraded conditions illustrated above.

The low luminance and contrast required for nighttime readability is relatively easy to achieve, but it is far more difficult to develop an HMD bright enough and of sufficient contrast for daylight use. The information must be displayed as a dynamic luminous transparency overlaying the real-world's complex features, colors, and motion. In order to display an image against a typical real-world daytime scene luminance of 3000 fL, the virtual display peak luminance must be about 1500 fL at the pilot's eye. And depending on the efficiency of the specific optics employed, the luminance at the display light source may need to be many times greater. The display technology that provides the best HMD solution might also provide the optimal HUD and HDD approaches.

6.3   CRTs and MFPs

Army aviation is the U.S. military leader in deployed operational HMD systems. The Apache helicopter's monochrome green CRT Helmet Display Unit (HDU) presents pilotage FLIR (forward-looking infrared) imagery overlaid with flight symbology in a 40°(H) × 30°(V) monocular field of view (FOV). The Apache HDU was developed in the late 1970s and early 1980s using the most advanced display technology then available. The RAH-66 Comanche helicopter program expanded the display's performance requirements to include night and day operability of a monochrome green display with a binocular 52° H × 30° V FOV and at least 30° of left/right image overlap.

The Comanche's Early Operational Capability Helmet Integrated Display Sighting System (EOC HIDSS) prototype employed dual miniature CRTs. The addition of a second CRT pushed the total head-supported weight for the system above the Army's recommended safety limit. Weight could not be removed from the helmet itself without compromising safety, so even though the image quality of the dual-CRT system was good, the resulting reduction in safety margins was unacceptable.

The U.S. Army Aircrew Integrated Systems (ACIS) office initiated a program to explore alternate display technologies for use with the proven Aircrew Integrated Helmet System Program (AIHS, also known as the HGU-56/P helmet) that would meet both the Comanche's display requirements and the Army's safety requirements.

Active-matrix liquid-crystal displays (AMLCD), active-matrix electroluminescent (AMEL) displays, field-emission displays (FEDs), and organic light-emitting diodes (OLEDs) are some of the alternative technologies that have shown progress. These postage-stamp size miniature flat-panel (MFP) displays weigh only a fraction as much as the miniature CRTs they seek to replace.

AMLCD is the heir apparent to the CRT, given its improved luminance performance. Future luminance requirements will likely be even higher, and there are growing needs for greater displayable pixel counts to increase effective range resolution or FOV, and for color to improve legibility and enhance information encoding. It is not clear that AMLCD technology can keep pace with these demands.

6.4   Laser Advantages, Eye Safety

The RSD offers distinct advantages over other display technologies because image quality and color gamut are maintained at high luminances limited only by eye-safety considerations.[2,3] The light-concentrating aspect of the diffraction-limited laser beam can routinely produce source luminances that exceed that of the solar disc. Strict engineering controls, reliable safeguards, and careful certification are mandatory to minimize the risk of damage to the operator's vision.[4] Of course, these safety concerns are not limited to laser displays; any system capable of displaying extremely high luminances should be controlled, safeguarded, and certified.

Microvision's products are routinely tested and classified according to the recognized eye safety standard — the maximum permissible exposure (MPE) — for the specific display in the country of delivery. In the U.S. the applicable agency is the Center for Devices and Radiological Health (CDRH)

Division of the Food and Drug Administration (FDA). The American National Standards Institute's Z136.1 reference, "The Safe Use of Lasers," provides MPE standards and the required computational procedures to assess compliance. In most of Europe the IEC 60825-1 provides the standards.

Compliance is assessed across a range of retinal exposures to the display, including single-pixel, single scan line, single video frame, 10-second, and extended-duration continuous retinal exposures. For most scanned laser displays, the worst-case exposure leading to the most conservative operational usage is found to be the extended-duration continuous display MPE. Thus, the MPE helps define laser power and scan-mirror operation-monitoring techniques implemented to ensure safe operation. Examples include shutting down the laser(s) if the active feedback signal from either scanner is interrupted and automatically attenuating the premodulated laser beam for luminance control independent of displayed contrast or grayscale.

6.5   Light Source Availability and Power Requirements

Another challenge to manufacturers of laser HMD products centers on access to efficient, low-cost lasers or diodes of appropriate collectible power (1–100 mW), suitable wavelengths (430–470, 532–580, and 607–660 nm), low video-frequency noise content (<3%), and long operating life (10,000 hr). Diodes present the most cost-effective means because they may be directly modulated up from black, while lasers are externally modulated down from maximum beam power.

Except for red, diodes still face significant development hurdles, as do blue lasers. Operational military-aviation HMDs presently require only a monochrome green, G, display which can be obtained by using a 532-nm diode-pumped solid-state (DPSS) laser with an acoustic-optic modulator (AOM). Given available AOM and optical fiber coupling efficiencies, the 1500-fL G RSD requires about 50 mW of laser beam power. Future requirements will likely include red + green, RG, and full color, RGB, display capability.

6.6   Microvision's Laser Scanning Concept

Microvision has developed a flexible component architecture for display systems (Figure 6.1). RGB video drives AOMs to impress information on Gaussian laser beams, which are combined to form full-color pixels with luminance and chromaticity determined by traditional color-management techniques. The aircraft-mounted photonics module is connected by single-mode optical fiber to the helmet, where the beam is air propagated to a lens, deflected by a pair of oscillating scanning mirrors (one horizontal and one vertical), and brought to focus as a raster format intermediate image. Finally, the image is optically collimated and combined with the viewer's visual field to achieve a spatially stabilized virtual image presentation.

The AIHS Program requires a production display system to be installed and maintained as a helicopter subsystem — designated Aircraft Retained Unit (ARU) — plus each pilot's individually fitted protective helmet, or Pilot Retained Unit (PRU). Microvision's initial concept-demonstration HMD components meet these requirements (Figure 6.2).

Microvision's displays currently employ one horizontal line-rate scanner — the Mechanical Resonant Scanner (MRS) — and a vertical refresh galvanometer. Approaches using a bi-axial microelectro-mechanical system (MEMS) scanner are under development. Also, as miniature green laser diodes become available, Microvision expects to further reduce ARU size, weight, and power consumption by transitioning to a small diode module (Figure 6.1, lower-right) embedded in the head-worn scanning engine, which would also eliminate the cost and inefficiency of the fiber optic link.

For the ACIS project, a four-beam concurrent writing architecture was incorporated to multiply by 4 the effective line rate achievable with the 16-kHz MRS employed in unidirectional horizontal writing mode. The vertical refresh scanner was of the 60-Hz saw-tooth-driven servo type for progressive line scanning. The f/40 writing beams, forming a narrow optical exit pupil (Figure 6.3), are diffraction-multiplied to form a 15-mm circular matrix of exit pupils.

The displayed resolution of a scanned-light-beam display[5] is limited by three parameters: (1) spot size and distribution as determined by cascaded scan-mirror apertures (D), (2) total scan-mirror deflection angles in the horizontal or vertical raster domains (*Theta*), and (3) dynamic scan-mirror flatness under

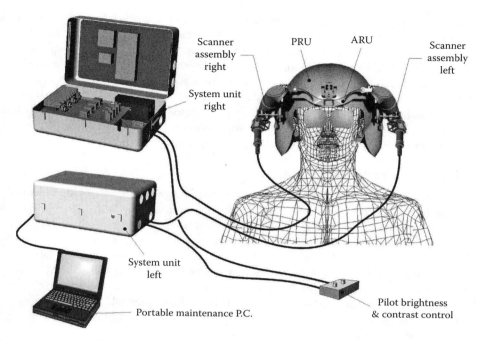

FIGURE 6.2 Microvision's RSD components meet the requirements of the AIHS HIDSS program for an HMD.

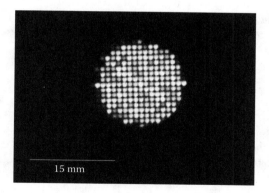

FIGURE 6.3 The far-field beamlet structure of a spot-multiplied (expanded) RSD OEP. The unexpanded 1-mm exit pupil is represented by a single central spot.

normal operating conditions. Microvision typically designs to the full-width/half-maximum Gaussian spot overlap criterion, thus determining the spot count per raster line. Horizontal and vertical displayable spatial resolutions, limited by $(D)^*(Theta)$, must be supported by adequate scan-mirror dynamic flatness for the projection engine to perform at its diffraction limit. Beyond these parameters, image quality is affected by all the components common to any video projection display. Electronics, photonics, optics, and packaging tolerances are the most significant.

6.6.1 Government Testing of the RSD HMD Concept

Under the ACIS program, the concept version of the Microvision RSD HMD was delivered to the U.S. Army Aeromedical Research Laboratory (USAARL) for testing and evaluation in February 1999.[6]

As expected, the performance of the concept-phase system had some deficiencies when compared to the RAH-66 Comanche requirements. However, these deficiencies were few in number and the overall performance was surprisingly good for this initial development phase. Measured performance for exit pupil, eye relief, alignment, aberrations, luminance transmittance, and field-of-view met the requirements completely. The luminance output of the left and right channels — although high, with peak values of 808 and 1111 fL, respectively — did not provide the contrast values required by Comanche in all combinations of ambient luminance and protective visor. Of greatest concern was the modulation transfer function (MTF) — and the analogous Contrast Transfer Function (CTF) — exhibiting excessive rolloff at high spatial frequencies, and indicating a "soft" displayed image.

6.6.2 Improving RSD Image Quality

The second AIHS program phase concentrated on improving image quality. Microvision identified the sources of the luminance, contrast, and MTF/CTF deficiencies found by USAARL. A few relatively straightforward fixes such as better fiber coupling, stray light baffling, and scan-mirror edge treatment provided the luminance and low-spatial-frequency contrast improvements required to meet specification, but MTF/CTF performance at high spatial frequencies presented a more complex set of issues.

Each image-signal-handling component in the system contributes to the overall system MTF. Although the video electronics and AOM-controller frequency responses were inadequate, they were easily remedied through redesign and component selection. Inappropriate mounting of fixed fold mirrors in the projection path led to the accumulation of several wavelengths of wave-front error and resultant image blurring. This problem, too, is readily solved.

The second class of problems pertains to the figure of the scan mirrors. Interferometer analyses of the flying spot under dynamic horizontal scanning conditions indicated excessive mirror surface deformation ($\sim$2 peak-to-peak mechanical), resulting in irregular spot growth and reduced MTF/CTF performance (Figure 6.4).

Three fast-prototyping iterations brought the mirror surface under control ($\sim\lambda/4$) to achieve acceptable spot profiles at the raster edge. Thus, the component improvements described above are expected to result in MTF/CTF performance meeting U.S. Army specification.

6.7 Next Step

The next step in the evolution of the helicopter pilot's laser HMD is the introduction of daylight-readable color. Microvision first demonstrated a color VGA format RSD HMD in 1996, followed by SVGA in 1998. Development of a 1280 × 1024-color-pixel (SXGA) binocular HMD project is being made possible

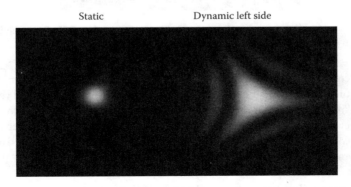

Static   Dynamic left side

FIGURE 6.4 The effect of improved mirror design is visible in these spot (pixel) images, normalized for size but not for intensity, for scanned spots at $\sim\lambda/4$ P-P mechanical mirror deformation (left image), and $\sim2\lambda$ P-P mechanical mirror deformation (right image).

by ACIS's Virtual Cockpit Optimization Program (VCOP), which began with software-reconfigurable virtual flight simulations in 2000 and proceeded to in-flight virtual cockpit demonstrations in 2001. For these demonstrations, the aircraft's traditional control-panel instrumentation served only as an emergency backup function. Figure 6.1, with which this chapter began, represents the VCOP RGB application concept.

One configuration of the VCOP simulation/operation HMD acknowledges the limited ability of the blue component to generate effective contrast against white clouds or blue sky. Because the helmet tracker used in any visually-coupled system will "know" when the pilot is "eyes out" or "head down", the HMD may employ graphics and imaging sensor formats in daylight readable greenscale, combined with red, for "eyes out" information display across established green/yellow/red caution advisory color codes, switching to full color formats at lower luminances for "head down" displays of maps, etc.

The fundamental capabilities of the human visual system, along with ever increasing imaging sensor and digital image generation bandwidths, require HMD spatial resolutions greater than SXGA. For this reason, the US Air Force Research Laboratory has contracted Microvision Inc. to build the first known HDTV HMD (1920 × 1080 pixels in a noninterlaced 60 Hz frame refresh digital video format). The initial system will be a monocular 100-fL monochrome green fighter pilot training HMD with growth-to-daylight readable binocular color operation.

An effort of 30 years has only scratched the surface of the HMD's pilot vehicle interfacing potential. It is expected that the RSD will open new avenues of pilot-in-the-loop research and enable safer, more effective air and ground operations.

Defining Terms

Optomechatronic: Application of integrated optical, mechanical, and electronic elements for imaging and display.

Helmet-Mounted Display (HMD): Head-Up Display (HUD); Head-Down Display (HDD).

ROSE: Raster Optical Scanning Engine.

Virtual Image Projection (VIP): An optical display image comprised of parallel or convergent light bundles.

Image Viewing Zone (IVZ): The range of locations from which an entire virtual image is visible while fixating any of the image's boundaries.

Optical Exit Pupil (OEP): The aerial image formed by all compound magnifiers, which defines the IVZ.

Retinal Scanning Display (RSD): A virtual image projection display which scans a beam of light to form a visible pattern on the retina. The typical 15-mm OEP of a helmet-mounted RSD OEP permits normal helmet shifting in operational helicopter environments without loss of image. Higher-*g* environments may require larger OEPs.

Virtual Retinal Display (VRD): A subcategory of RSD specifically characterized by an optical exit pupil less than 2 mm, for Low Vision Aiding (LVA), vision testing, narrow field of view, or "agile" eye-following OEP display systems. This is the most light-efficient form of RSD.

Acknowledgments

This work was partially funded by U.S. Army Contract No. DAAH23-99-C-0072, Program Manager, Aircrew Integrated Systems, Redstone Arsenal, AL. The author wishes to express appreciation for the outstanding efforts of the Microvision Inc. design and development team, and for the guidance and support provided by the U.S. Army Aviation community, whose vision and determination have made these advances in high-performance pilotage HMD systems possible.

References

1. Rash, C. E., Ed., Helmet-Mounted Displays: Design Issues for Rotary-Wing Aircraft, U.S. Army Medical Research and Materiel Command, Fort Detrick, MD, 1999.
2. Kollin, J., A retinal display for virtual environment applications, SID Int. Symp., Digest of Technical Papers, pp. 827–828, May 1993.
3. de Wit, G. C., A Virtual Retinal Display for Virtual Reality, Doctoral Dissertation, Ponsen & Looijen BV, Wageningen, Netherlands, 1997.
4. Gross, A., Lorenson, C., and Golich, D., Eye-safety analysis of scanning-beam displays, SID Int. Symp. Digest of Technical Papers, pp. 343–345, May 1999.
5. Urey, H., Nestorovic, N., Ng, B., and Gross, A., Optics designs and system MTF for laser scanning displays, Helmet and Head Mounted Displays IV, *Proc. SPIE,* 3689, 238–248, 1999.
6. Rash, C. E., Harding, T. H., Martin, J. S., and Beasley, H. H., Concept phase evaluation of the Microvision Inc., Aircrew Integrated Helmet System, HGU-56/P, virtual retinal display. Fort Rucker, AL: U.S. Army Aeromedical Research Laboratory, USAARL Report No. 99-18, 1999.

Further Information

Microvision Inc. Website: www.mvis.com.

7

Head-Up Displays

Robert B. Wood
Rockwell Collins Flight Dynamics

Peter J. Howells
Rockwell Collins Flight Dynamics

7.1   Introduction

During early military Head-Up Display (HUD) development, it was found that pilots using HUDs could operate their aircraft with greater precision and accuracy than they could with conventional flight instrument systems.[1,2] This realization eventually led to the development of the first HUD systems intended specifically to aid the pilot during commercial landing operations. This was first accomplished by Sextant Avionique for the Dassault Mercure aircraft in 1975, and then by Sundstrand and Douglas Aircraft Company for the MD80 series aircraft in the late 1970s (see Figure 7.1).

In the early 1980s, Flight Dynamics developed a holographic optical system to display an inertially derived aircraft flight path along with precision guidance, thus providing the first wide field-of-view (FOV) head-up guidance system. Subsequently, Alaska Airlines became the first airline to adopt this technology and perform routine fleet-wide manually flown CAT IIIa operations on B-727-100/200 aircraft using the Flight Dynamics system (see Figure 7.2). Once low-visibility operations were successfully demonstrated using a HUD in lieu of a fail passive autoland system, regional airlines opted for this technology to help maintain their schedules when the weather fell below CAT II minimums, and to help improve situational awareness.

By the end of the century, many airlines had installed head-up guidance systems, and thousands of pilots were fully trained in their use. HUD-equipped aircraft had logged more than 6,000,000 flight hours and completed over 30,000 low-visibility operations. HUDs are now well-established additions to aircraft cockpits, providing both additional operational capabilities and enhanced situational awareness, resulting in improved aircraft safety.

7.2   HUD Fundamentals

All head-up displays require an image source, generally a high-brightness cathode-ray tube, and an optical system to project the image source information at optical infinity. The HUD image is viewed by the pilot after reflecting from a semitransparent element referred to as the HUD combiner. The combiner is located

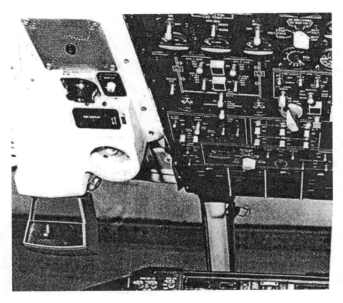

FIGURE 7.1   Early commercial HUD.[3]

FIGURE 7.2   Commercial manually flown CAT IIIa HUD installed in a B-737-800.

between the pilot's eyes and the aircraft windshield and is angled to reflect image-source light rays to the pilot for viewing. Special coatings on the combiner simultaneously reflect the HUD information and transmit the real-world scene, enabling the pilot to view both the outside world and the collimated display.

Head-up display systems are comprised of two major subsystems: the pilot display unit (PDU), and the HUD processor or HUD computer. The PDU interfaces electrically and mechanically with the aircraft structure and provides the optical interface to the pilot. The HUD processor interfaces electronically with aircraft sensors and systems, runs a variety of algorithms related to data verification and formatting, and generates the characters and symbols making up the display. Modern HUD processors are capable of generating high-integrity guidance commands and cues for precision low-visibility take-off, approach, landing (flare), and rollout. The interface between the HUD processor and the PDU can be either a serial digital display list or analog X and Y deflection and Z-axis video bright-up signals for controlling the display luminance.

The PDU is located within the cockpit to allow a pilot positioned at the cockpit Design Eye Position (DEP) to view HUD information which is precisely positioned with respect to the outside world. This allows, for example, the computer-generated and displayed horizon line to overlay the real-world horizon in all phases of flight.

The cockpit DEP is defined as the optimum cockpit location that meets the requirements of FAR 25.773[4] and 25.777.[5] From this location the pilot can easily view all relevant head-down instruments and the outside world scene through the aircraft windshield, while being able to access all required cockpit controls. The HUD "eyebox," is always positioned with respect to the cockpit DEP, allowing pilots to fly the aircraft using the HUD from the same physical location as a non-HUD-equipped aircraft would be flown.

7.2.1 Optical Configurations

The optics in head-up display systems are used to "collimate" the HUD image so that essential flight parameters, navigational information, and guidance are superimposed on the outside world scene.

The four distinct FOV characteristics used to fully describe the characteristics of the angular region over which the HUD image is visible to the pilot are illustrated in Figure 7.3, and summarized as follows:

Total FOV (TFOV) — The maximum angular extent over which symbology from the image source can be viewed by the pilot with either eye allowing vertical and horizontal head movement within the HUD eyebox.

Instantaneous FOV (IFOV) — The union of the two solid angles subtended at each eye by the clear apertures of the HUD optics from a fixed head position within the HUD eyebox. Thus, the instantaneous FOV is comprised of what the left eye sees plus what the right eye sees from a fixed head position within the HUD eyebox.

Binocular overlapping FOV — The binocular overlapping FOV is the intersection of the two solid angles subtended at each eye by the clear apertures of the HUD optics from a fixed head position within the HUD eyebox. The binocular overlapping FOV thus defines the maximum angular extent of the HUD display that is visible to both eyes simultaneously.

Monocular FOV — The solid angle subtended at the eye by the clear apertures of the HUD optics from a fixed eye position. Note that the monocular FOV size and shape may change as a function of eye position within the HUD eyebox.

The FOV characteristics are designed and optimized for a specific cockpit geometric configuration based on the intended function of the HUD. In some cases, the cockpit geometry may impact the maximum available FOV.

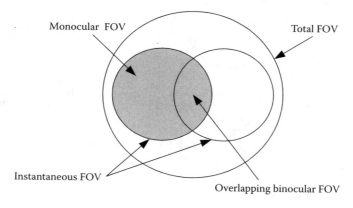

FIGURE 7.3 HUD fields-of-view defined.

One of the most significant advances in HUD optical design in the last 20 years is the change from optical systems that collimate by refraction to systems that collimate by reflection or, in some cases, by diffraction. The move towards more complex (and expensive) reflective collimation systems has resulted in larger display fields-of-view which expand the usefulness of HUDs as full-time primary flight references.

7.2.1.1 Refractive Optical Systems

Figure 7.4 illustrates the optical configuration of a refractive HUD system. This configuration is similar to the basic HUD optical systems in use since the 1950s.[6] In this optical configuration, the CRT image is collimated by a combination of refractive lens elements designed to provide a highly accurate display over a moderate display field of view. Note that an internal mirror is used to fold the optical system to reduce the physical size of the packaging envelope of the HUD. Also shown in Figure 7.4 is the HUD combiner glass, a flat semitransparent plate designed to reflect approximately 25% of the collimated light from the CRT, and transmit approximately 70% of the real-world luminance.

Note that the vertical instantaneous FOV can be increased by adding a second flat combiner glass, displaced vertically above and parallel with the first.

7.2.1.2 Reflective Optical Systems

In the late 1970s, HUD optical designers looked at ways to significantly increase the display total and instantaneous FOVs.[7,8] Figure 7.5 illustrates the first overhead-mounted reflective HUD optical system (using a holographically manufactured combiner) designed specifically for a commercial transport cockpit.[9] As in the classical refractive optical system, the displayed image is generated on a small CRT, about 3 in. in diameter. The reflective optics can be thought of as two distinct optical subsystems. The first is a relay lens assembly designed to re-image and pre-aberrate the CRT image source to an intermediate aerial image, located at one focal length from the optically powered combiner/collimator element.

The second optical subsystem is the combiner/collimator element that re-images and collimates the intermediate aerial image for viewing by the pilot. As in the refractive systems, the pilot's eyes focus at optical infinity, looking through the combiner to see the virtual image. To prevent the pilot's head from blocking rays from the relay lens to the combiner, the combiner is tilted off-axis with respect to the axial chief ray from the relay lens assembly. The combiner off-axis angle, although required for image viewing reasons, significantly increases the optical aberrations within the system, which must be compensated in the relay lens to have a well-correlated, accurate virtual display.

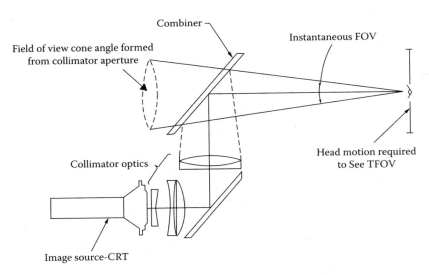

FIGURE 7.4 Refractive optical systems.

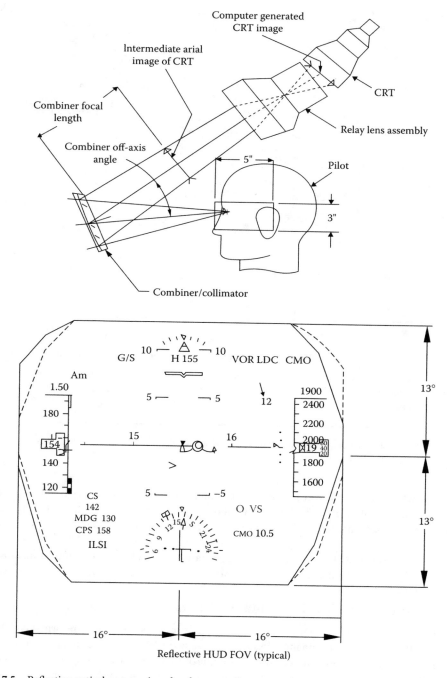

FIGURE 7.5 Reflective optical systems (overhead mounted).

Figure 7.6 illustrates the optical raytrace of a typical reflective HUD system showing the complexity of the relay lens assembly. (This is the optical system used on the first manually flown CAT IIIa HUD system ever certified.)

The complexity of the relay lens, shown in Figure 7.6, provides a large instantaneous FOV over a fairly large eyebox, while simultaneously providing low display parallax and high display accuracy.

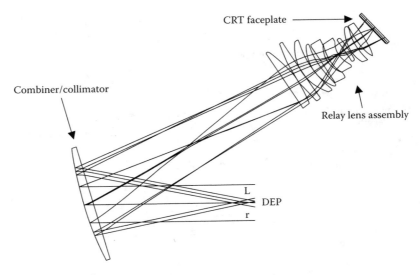

FIGURE 7.6   Reflective optical system raytrace.

TABLE 7.1   Typical HUD Fields-of-View

	Refractive HUD FOV Characteristics[a]		Reflective HUD Optics FOV Characteristics
	Single Combiner	Dual Combiners	
Total Field Of View	20°–25° Diameter	25°–30° Diameter	22–28° V × 28–34° H
Instantaneous FOV	12° V × 17.8° H	16° V × 17.8° H	22–28° V × 28–34° H
Overlapping	11° V × 6° H	16° V × 6° H	22–26° V × 25–30° H
Monocular FOV	12° Diameter	16° V × 12° H	22–28° V × 30° H

[a] Calculations assume a collimator exit aperture diameter of 5.0″, and a distance of 24″ between the pilot and the HUD collimator exit aperture.

The reflective optical system can provide an instantaneous and binocular overlapping FOV that is equal to the total FOV, allowing the pilot to view all of the information displayed on the CRT with each eye with no head movement. Table 7.1 summarizes typical field-of-view performance characteristics for HUD systems.

All commercially certified HUD systems in airline operation today use reflective optical systems because of the improved display FOV characteristics compared with refractive systems.

7.2.2   Significant Optical Performance Characteristics

This section summarizes other important optical characteristics associated with conformal HUD systems. It is clear that the HUD FOV, luminance, and display line width characteristics must meet basic performance requirements.[10] However, optical system complexity and cost are driven by HUD eyebox size, combiner off-axis angle, display accuracy requirements, and optical parallax errors. Without a well-corrected optical system, conformal symbology will not properly overlay the outside world view and symbology will not remain fixed with respect to the real-world view as the head is moved around within the HUD eyebox.

7.2.2.1   Display Luminance and Contrast Ratio

The HUD should be capable of providing a usable display under all foreseeable ambient lighting conditions, including a sun-lit cloud with a luminance of 10,000 foot-Lamberts (ft-L)(or 34,000 cd/m²), and a night approach to a sparsely lit runway. HUD contrast ratio is a measure of the relative luminance of the display with respect to the real-world background and is defined as follows:

$$\text{HUD Contrast Ratio} = \frac{\text{Display Luminance} + \text{Real World Luminance}}{\text{Real World Luminance}}$$

The display luminance is the photopically weighted CRT light output that reaches the pilot's eyes. Real-world luminance is the luminance of the real world as seen through the HUD combiner. (By convention, the transmission of the aircraft windshield is left out of the real-world luminance calculation.)

It is generally agreed that a contrast ratio (CR) of 1.2 is adequate for display viewing, but that a CR of 1.3 is preferable. A HUD contrast ratio of 1.3 against a 10,000-ft-L cloud seen through a combiner with an 80% photopic transmission requires a display luminance at the pilot's eye of 2400 ft-L, a luminance about 10 times higher than most head-down displays. (This luminance translates to a CRT faceplate brightness of about 9000 ft-L, a luminance easily met with high-brightness monochrome CRTs.)

7.2.2.2 Head Motion Box

The HUD head motion box, or "eyebox," is a three-dimensional region in space surrounding the cockpit DEP in which the HUD can be viewed with at least one eye. The center of the eyebox can be displayed forward or aft, or upward or downward, with respect to the cockpit DEP to better accommodate the actual sitting position of the pilot. The positioning of the cockpit eye reference point[11] or DEP is dependent on a number of ergonomically related cockpit issues such as head-down display visibility, the over-the-nose down-look angle, and the physical location of various controls such as the control yoke and the landing gear handle.

The HUD eyebox should be as large as possible to allow maximum head motion without losing display information. The relay lens exit aperture, the spacing between the relay lens and combiner and the combiner to DEP, and the combiner focal length all impact the eyebox size. Modern HUD eyebox dimensions are typically 5.2 in lateral, 3.0 in vertical, and 6.0 in longitudinal.

In all HUDs, the monocular instantaneous FOV is reduced (or vignettes) with lateral or vertical eye displacement, particularly near the edge of the eyebox. Establishing a minimum monocular FOV from the edge of the eyebox thus ensures that even when the pilot's head is de-centered so that one eye is at the edge of the eyebox, useful display FOV is still available. A 10° horizontal by 10° vertical monocular FOV generally can be used to define the eyebox limits. In reflective HUDs, relatively small head movements (>1.5 in laterally) will cause one eye to be outside of the eyebox and see no display. Under these conditions, the other eye will see the total FOV, so no information is lost to the pilot.

7.2.2.3 HUD Display Accuracy

Display accuracy is a measure of how precisely the projected HUD image overlays the real-world view seen through the combiner and windshield from any eye position within the eyebox. Display accuracy is a monocular measurement and, for a fixed display location, is numerically equal to the angular difference between a HUD-projected symbol element and the corresponding real-world feature as seen through the combiner and windshield. The total HUD system display accuracy error budget includes optical errors, electronic gain and offset errors, errors associated with the CRT and yoke, Overhead to Combiner misalignment errors, windshield variations, environmental conditions (including temperature), assembly tolerances, and installation errors. Optical errors are both head-position and field-angle dependent.

The following display accuracy values are achievable in commercial HUDs when all the error sources are accounted for:

Boresight	+/− 3.0 milliradians (mrad)
Total Display Accuracy	+/− 7.0 milliradians (mrad)

The boresight direction is used as the calibration direction for zeroing all electronic errors. Boresight errors include the mechanical installation of the HUD hardpoints to the airframe, electronic drift due to thermal variations, and manufacturing tolerances for positioning the combiner element. Refractive HUDs with integrated combiners (i.e., F-16) are capable of achieving display accuracies of about half of the errors above.

7.2.2.4 HUD Parallax Errors

Within the binocular overlapping portion of the FOV, the left and right eyes view the same location on the CRT faceplate. These slight angular errors between what the two eyes see are binocular parallax errors or collimation errors. The binocular parallax error for a fixed field point within the total FOV is the angular difference in rays entering two eyes separated horizontally by the interpupillary distance, assumed to be 2.5 in. If the projected virtual display image were perfectly collimated at infinity from all eyebox positions, the two ray directions would be identical, and the parallax errors would be zero. Parallax errors consist of both horizontal and vertical components.

Parallax errors in refractive HUDs are generally less than about 1.0 mrad due to the rotational symmetry of the optics, and because of the relatively small overlapping binocular FOV.

7.2.2.5 Display Line Width

The HUD line width is the angular dimension of displayed symbology elements. Acceptable HUD line widths are between 0.7 and 1.2 mrad when measured at the 50% intensity points. The displayed line width is dependent on the effective focal length of the optical system and the physical line width on the CRT faceplate. A typical wide FOV reflective HUD optical system with a focal length of 5 in. will provide a display line width of about 1 mrad given a CRT line width of 0.005 in. The display line width should be met over the full luminance range of the HUD, often requiring a high-voltage power supply with dynamic focus over the total useful screen area of the CRT.

HUD optical system aberrations will adversely affect apparent display line width. These aberrations include uncorrected chromatic aberrations (lateral color) and residual uncompensated coma and astigmatism. Minimizing these optical errors during the optimization of the HUD relay lens design will also help meet the parallax error requirements. Table 7.2 summarizes the optical performance characteristics of a commercial wide-angle reflective HUD optical system.

7.2.3 HUD Mechanical Installation

The intent of the HUD is to display symbolic information which overlays the real world as seen by the pilot. To accomplish this, the HUD PDU must be very accurately aligned with respect to the pitch, roll, and heading axis of the aircraft. For this reason, the angular relationship of the HUD PDU with respect to the cockpit coordinates is crucial. The process of installing and aligning the HUD attachment bushings or hardpoints into the aircraft is referred to as "boresighting" and occurs when the aircraft is first built. (Although the alignment of the HUD hardpoints may be checked occasionally, once installed, the hardpoints are permanent and rarely need adjustment.)

Some reflective HUDs utilize mating bushings for the PDU hardware which are installed directly to the aircraft structure. Once the bushings are aligned and boresighted to the aircraft axis, they are permanently fixed in place using a structural epoxy. Figure 7.7 illustrates this installation method for HUD boresighting. In this case, the longitudinal axis of the aircraft is used as the boresight reference direction. Using special tooling, the overhead unit and combiner bushings are aligned with a precisely positioned target board located near the aft end of the fuselage. This boresighting method does not require the aircraft to be jacked and leveled.

Other HUD designs utilize a tray that attaches to the aircraft structure and provides an interface to the HUD LRUs. The PDU tray must still be installed and boresighted to the aircraft axis.

TABLE 7.2 HUD Optical System Summary (Typical Reflective HUD)

1.	Combiner Design	Wide Field-of-View, Wavelength Selective, Stowable, Inertial Break-away (HIC[*]-Complaint)
2.	DEP to Combiner Distance	9.5 to 13.5 in. (Cockpit geometry dependent)
3.	Display Fields-of-View	
	Total Display FOV	24–28° Vertical × 30–34° Horizontal
	Instantaneous FOV	24–28° Vertical × 30–34° Horizontal
	Overlapping Binocular FOV	22–24° Vertical × 24–32° Horizontal
4.	Head Motion Box or Eyebox	Typical Dimensions (Configuration dependent)
	Horizontal	4.7 to 5.4 in.
	Vertical	2.5 to 3.0 in.
	Depth (fore/aft)	4.0 to 7.0 in.
5.	Head Motion Needed to View TFOV	None
6.	Display Parallax Errors (Typical)	
	Convergence	95% of data points <2.5 mrad
	Divergence	95% of data points <1.0 mrad
	Dipvergence	93% of data points <1.5 mrad
7.	Display Accuracy (2 sigma)	
	Boresight	<2.5–4.0 mrad
	Total Field-of-view	<5.0–9.0 mrad
8.	Combiner Transmission and Coloration	78–82% photopic (day-adapted eye)
		84% scotopic (night-adapted eye)
		<0.03 color shift u'v' coordinates
9.	Display Luminance and Contrast Ratio	
	Stroke Only	1,600–2,400 foot-Lambert (ft-L)
	Raster	600–1,000 ft-L
	Display Contrast Ratio	1.2 to 1.3:1 (10,000 ft-L ambient background)
10.	Display Line Width	0.7–1.2 mrads
11.	Secondary Display Image Intensity	<0.5% of the primary image from eyebox

[*]Head Injury Criteria

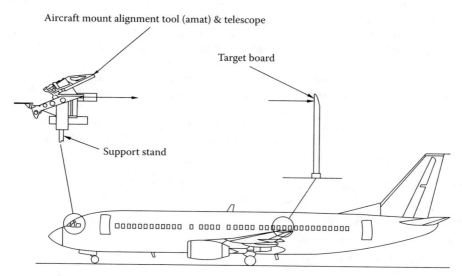

FIGURE 7.7 Boresighting the HUD hardpoints.

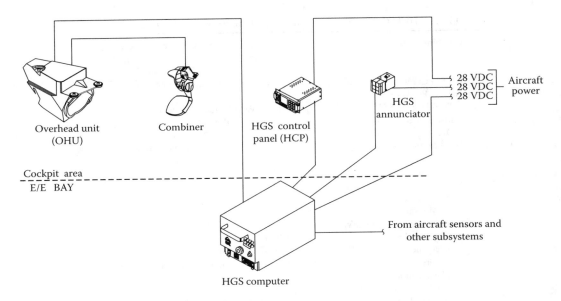

FIGURE 7.8 HUD interconnect diagram (RCFD HGS-4000).

7.2.4   HUD System Hardware Components

A typical commercial HUD system includes four principal line replaceable units (LRUs). (HUD LRUs can be interchanged on the flight deck without requiring any alignment or recalibration.) The cockpit-mounted LRUs include the overhead unit and combiner, the PDU, and the HUD control panel. The HUD computer is located in the electronics bay or another convenient location. A HUD interconnect diagram is shown in Figure 7.8.

7.2.4.1   HUD Overhead Unit

The Overhead Unit (OHU), positioned directly above the pilot's head, interfaces with the HUD computer receiving either analog X and Y deflection and Z-video data or a serial digital display list, as well as control data via a serial interface. The OHU electronics converts the deflection and video data to an image on a high-brightness cathode ray tube (CRT). The CRT is optically coupled to the relay lens assembly which re-images the CRT object to an intermediate aerial image one focal length away from the combiner LRU, as illustrated in the optical schematic in Figure 7.5. The combiner re-images the intermediate image at optical infinity for viewing by the pilot. The OHU includes all of the electronics necessary to drive the CRT and monitor the built-in-test (BIT) status of the LRU. The OHU also provides the electronic interfaces to the combiner LRU.

A typical Overhead Unit is illustrated in Figure 7.9. This LRU contains all electronic circuitry required to drive the high-brightness CRT, and all BIT-related functions. The following are the major OHU subsystems:

- Relay lens assembly
- Desiccant assembly (prevents condensation within the relay lens)
- Cathode ray tube assembly
- High-voltage power supplies
- Low-voltage power supplies and energy storage
- Deflection amplifiers (X and Y)
- Video amplifier
- Built-In-Test (BIT) and monitoring circuits

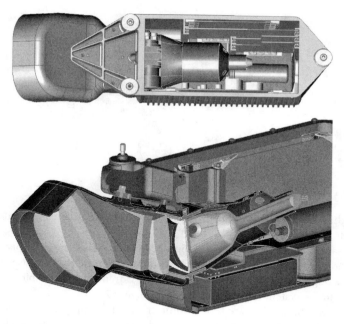

FIGURE 7.9 HUD overhead unit chassis (WFOV reflective optics).

- Motherboard assembly
- OHU chassis

In some HUD systems, the PDU may provide deflection data back to the HUD computer as part of the "wraparound" critical symbol monitor feature.[12] Real-time monitoring of certain critical symbol elements (i.e., horizon line) provides the high integrity levels required for certifying a HUD as a primary flight display. Other monitored critical data on the HUD may include ILS data, airspeed, flight path vector, and low-visibility guidance symbology.

7.2.4.2 HUD Combiner

The combiner is an optical-mechanical LRU consisting of a precision support structure for the wavelength-selective combiner element, and a mechanism allowing the combiner to be stowed and to breakaway. The combiner LRU interfaces with a precision pre-aligned mating interface permanently mounted to the aircraft structure. The combiner glass support structure positions the combiner with respect to the cockpit DEP and the overhead unit. The combiner mechanism allows the glass to be stowed upward when not in use, and to break away during a rapid aircraft deceleration, thus meeting the newly defined cockpit "head injury criteria" or HIC.[13] The combiner locks into both the stowed and breakaway positions and requires positive actions by the pilot to return it to the deployed position. Many HUD combiner assemblies include a built-in alignment detector that monitors the glass position in real time. Figure 7.10 shows a commercial HUD PDU and a wavelength-selective combiner. The combiner usually includes the HUD optical controls (brightness and contrast).

7.2.4.3 HUD Computer

The HUD computer interfaces with the aircraft sensors and systems, performs data conversions, validates data, computes command guidance (if applicable), positions and formats symbols, generates the display list, and converts the display list into X, Y, and Z waveforms for display by the PDU. In some commercial HUD systems, the HUD computer performs all computations associated with low-visibility take-off, approach, landing, and rollout guidance, and all safety-related performance and failure monitoring. Because of the critical functions performed by these systems, the displayed data

FIGURE 7.10   HUD PDU.

must meet the highest integrity requirements. The HUD computer architecture is designed specifically to meet these requirements.

One of the key safety requirements for a full flight regime HUD is that the display of unannunciated, hazardously misleading attitude on the HUD must be improbable, and that the display of unannunciated hazardously misleading low-visibility guidance must be extremely improbable. An analysis of these requirements leads to the system architecture shown in Figure 7.11.

In this architecture, primary data are brought into the HUD computer via dual independent input/ output (I/O) subsystems from the primary sensors and systems on the aircraft. (The avionics interface for a specific HUD computer depends on the avionics suite, and can include a combination of any of the following interfaces: ARINC 429, ARINC 629, ASCB-A, B, C, or D, or MIL STD 1553B.) Older aircraft will often include analog inputs as well as some synchro data. The I/O subsystem also includes the interfaces required for the overhead unit and combiner and will often include outputs to the flight data recorder and central maintenance computer.

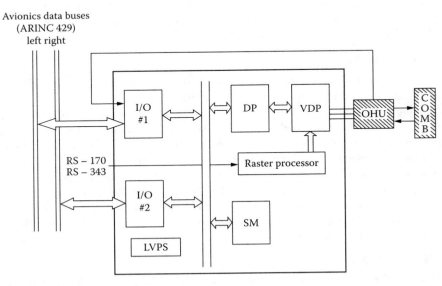

FIGURE 7.11   High integrity HUD computer architecture.

FIGURE 7.12   Commercial HUD symbology.

Figure 7.12 is a photograph of a typical commercial HUD symbology set. The aircraft sensor data needed to generate this display are in Table 7.3. In general two sources of the critical data are required to meet the safety and integrity requirements.

The display processor converts all input into engineering units, verifies the validity of the data, compares like data from the dual sensors, runs the control law algorithms, computes the display element locations, and generates a display list. The video display processor (VDP) converts the display list into X, Y, and Z signals that are output to the OHU.

The System Monitor processor (SM) verifies the display path by monitoring the displayed position of critical symbols using an inverse function algorithm,[12] independently computes the guidance algorithms using off-side data for comparison to the guidance solution from the display processor, and monitors the approach parameters to ensure a safe touchdown. The critical symbol monitor is a wraparound monitor that computes the state of the aircraft based on the actual display information on the CRT. The displayed state is compared to the actual aircraft state based on the latest I/O data. A difference between the actual state and the computed state causes the system monitor to blank the display through two independent channels, since any difference in states could indicate a display processor fault. All software in the HUD computer is generally developed to DO-178B Level A requirements due to the critical functions performed.

Also shown in Figure 7.11 is the raster processor subassembly, used in HUD systems that interface with Enhanced Vision sensors. This subsystem converts standard raster sensor video formats (RS-170 or RS-343) into a display format that is optimized for display on the HUD. In most raster-capable HUDs there is a trade-off between how much time is available for writing stroke information, and how much time is available for writing the raster image (the frame rate is fixed at 60 Hz, corresponding to 16.67 msec per frame). Some HUD systems "borrow" video lines from the raster image to provide adequate time to draw the stroke display (a technique called "line stealing"). The alternative is to limit the amount of stroke information that can be written on top of the raster image. Neither approach is optimal for a primary flight reference HUD required to display both stroke and raster images.

One solution is to convert the standard raster image format to a display format that is more optimized for HUD display. Specifically, the video input is digitized and scan-converted into a bi-directional display format, thus saving time from each horizontal line (line overscan, and flyback). This technique increases the time available for writing stroke information in the stroke-raster mode from about 1.6 msec to about 4.5 msec, adequate enough to write the entire worst-case stroke display. The bi-directional raster scan technique is illustrated in Figure 7.13, along with a photograph of a full-field raster image.

Figure 7.14 is a photograph of a HUD computer capable of computing take-off guidance, manual CAT IIIa landing guidance, rollout guidance, and raster image processing.

TABLE 7.3 Sensor Data Required for Full Flight Regime Operation

Input Data	Data Source
Attitude	Pitch and roll angles — 2 independent sources
Airspeed	Calibrated airspeed
	Low speed awareness speed(s) (e.g., Vstall)
	High speed awareness speed(s) (e.g., Vmo)
Altitude	Barometric altitude (pressure altitude corrected with altimeter setting)
	Radio altitude
Vertical Speed	Vertical speed (inertial if available, otherwise raw air data)
Slip/Skid	Lateral acceleration
Heading	Magnetic heading
	True heading or other heading (if selectable)
	Heading source selection (if other than magnetic selectable)
Navigation	Selected course
	VOR bearing/deviation
	DME distance
	Localizer deviation
	Glideslope deviation
	Marker beacons
	Bearings/deviations/distances for any other desired nav signals (e.g., ADF, TACAN, RNAV/FMS)
Reference	Selected airspeed
Information	Selected altitude
	Selected heading
	Other reference speed information (e.g., V_1, V_R, Vapch)
	Other reference altitude information (e.g., landing minimums [DH/MDA], altimeter setting)
Flight Path	Pitch angle
	Roll angle
	Heading (magnetic or true, same as track)
	Ground speed (inertial or equivalent)
	Track angle (magnetic or true, same as heading)
	Vertical speed (inertial or equivalent)
	Pitch rate, Yaw rate
Flight Path	Longitudinal acceleration
Acceleration	Lateral acceleration
	Normal acceleration
	Pitch angle
	Roll angle
	Heading (magnetic or true, same as track)
	Ground speed (inertial or equivalent)
	Track angle (magnetic or true, same as heading)
	Vertical speed (inertial or equivalent)
Automatic Flight	Flight director guidance commands
Control System	Autopilot/flight director modes
	Autothrottle modes
Miscellaneous	Wind speed
	Wind direction (and appropriate heading reference)
	Mach
	Windshear warning(s)
	Ground proximity warning(s)
	TCAS resolution advisory information

7.2.4.4   HUD Control Panel

Commercial HUD systems used for low-visibility operations often require some pilot-selectable data not available on any aircraft system bus as well as a means for the pilot to control the display mode. Some HUD operators prefer to use an existing flight deck control panel, e.g., an MCDU, for HUD data entry

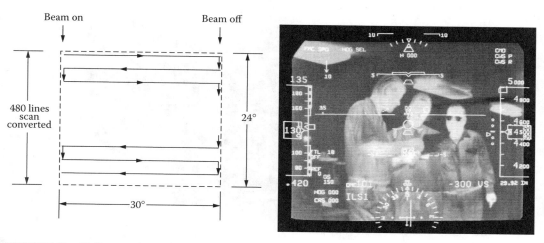

FIGURE 7.13 Bi-directional scan-converted raster image.

FIGURE 7.14 High integrity HUD computer.

and control. Other operators prefer a standalone control panel, dedicated to the HUD function. Figure 7.15 illustrates a standalone HUD control panel certified for use in CAT IIIa HUD systems.

7.2.5 Aspects of HUD Certification

Certification requirements for a HUD system depend on the functions performed. As the role of HUDs have expanded from CAT IIIa landing devices to full flight regime primary flight references including take-off and rollout guidance the certification requirements have become more complex. It is beyond the scope of this chapter to describe all the certification issues and requirements for a primary flight display HUD, however, the basic requirements are not significantly different from PFD head-down display certification requirements.

The FAA has documented the requirements for systems providing guidance in low-visibility conditions in Advisory Circular AC 120-28, "Criteria for Approval of Category III Weather Minima for Takeoff, Landing, and Rollout." The certification of the landing guidance aspects of the HUD are fundamentally different from automatic landing systems because the human pilot is in the active control loop during

FIGURE 7.15   HUD control and data entry panel.

the beam tracking and flare. The following summarizes the unique aspects of the certification process for a manual Category III system.

1. *Control Law Development* — The guidance control laws are developed and optimized based on the pilot's ability to react and respond. The control laws must be "pilot centered" and tailored for a pilot of average ability. The monitors must be designed and tuned to detect approaches that will be outside the footprint requirement, yet they cannot cause a go-around rate greater than about 4%.

2. *Motion-Based Simulator Campaign* — Historically, approximately 1400 manned approaches in an approved motion-based simulator, with at least 12 certification authority pilots, are required for performance verification for a FAA/EASA certification. The Monte Carlo test case ensemble is designed to verify the system performance throughout the envelope expected in field operation. Specifically, the full environment must be sampled (head winds, cross winds, tail winds, turbulence, etc.) along with variations in the airfield conditions (sloping runways, ILS beam offsets, beam bends, etc.). Finally, the sensor data used by the HUD must be varied according to the manufacturer's specified performance tolerances. Failure cases must also be simulated. Time history data for each approach, landing, and rollout is required to perform the required data reduction analysis. A detailed statistical analysis is required to demonstrate, among other characteristics, the longitudinal, lateral, and vertical touchdown footprint. Finally, the analysis must project out the landing footprint to a one-in-a-million (10^{-6}) probability.

3. *Aircraft Flight Test* — Following a successful simulator performance verification campaign, the HUD must be demonstrated in actual flight trials on a fully equipped aircraft. As in the simulator case, representative head winds, cross winds, and tail winds must be sampled for certification. Failure conditions are also run to demonstrate system performance and functionality.

This methodology has been used to certify head up display systems providing manual guidance for take-off, landing, and rollout on a variety of different aircraft types.

7.3   Applications and Examples

This section describes how the HUD is used on a typical aircraft using typical symbology sets that are displayed to a pilot in specific phases of flight. The symbology examples used in this section are taken from a Rockwell Collins Flight Dynamics Head-Up Guidance System (HGS®) installed on an in-service aircraft.

In addition to symbology, this section also discusses the pilot-in-the-loop optimized guidance algorithms that are provided as part of a HGS. Another feature of some HUDs is the display of video images on the HUD and the uses of this feature—where the HUD is only a display device—are discussed.

7.3.1   Symbology Sets and Modes

To optimize the presentation of information, the HUD has different symbology sets that present only the information needed by the pilot in that phase of flight. For example, the aircraft pitch information is not important when the aircraft is on the ground. These symbology sets are either selected as modes by the pilot or are displayed automatically when a certain condition is detected.

7.3.1.1   Primary Mode

The HGS primary (PRI) mode can be used during all phases of flight from take-off to landing. This mode supports low-visibility take-off operations, all en route operations, and approaches to CAT I or II minimums using FGS flight director guidance.

The HGS primary mode display is very similar to the Primary Flight Display (PFD) to enhance the pilot's transition from head down instruments to headup symbology. Figure 7.16 shows a typical in-flight primary mode display that includes the following symbolic information:

- Aircraft reference (boresight) symbol
- Pitch — scale and horizon relative to boresight
- Roll — scale and horizon relative to boresight
- Heading — horizon, horizontal situation indicator (HSI), and digital readouts
- Speeds — CAS (tape), vertical speed, ground speed, speed error tape
- Altitudes — barometric altitude (tape), digital radio altitude
- Flight path (inertial)
- Flight path acceleration
- Slip/skid indicators

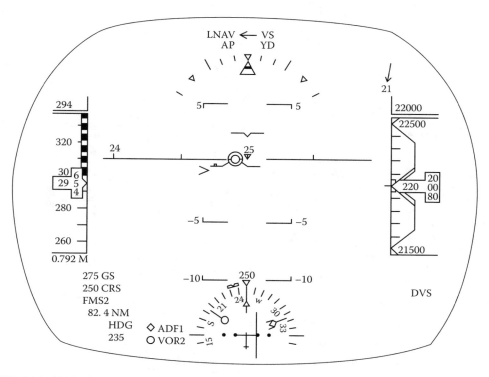

FIGURE 7.16   HGS primary mode symbology: in-flight.

- FGS flight director (F/D) guidance cue and modes
- Flight director armed and capture modes
- Navigation data — ILS, VOR, DME, FMS, marker beacons
- Wind — speed and direction
- Selected parameters — course, heading, airspeed, and altitude
- Attitude
- Altitude
- Airspeed
- Navigation data
- Warning and advisory

When the aircraft is on the ground, several symbols are removed or replaced as described in the following sections. After take-off rotation, the full, in-flight set of symbols is restored.

The primary mode is selectable at the HCP or by pressing the throttle go-around switch during any mode of operation.

7.3.1.1.1 *Primary Mode: Low-Visibility Take-off (HGS Guidance)*

The primary mode includes special symbology used for a low-visibility take-off as shown in Figure 7.17. The HGS guidance information supplements visual runway centerline tracking and enhances situational awareness.

For take-off operation, the HSI scale is removed from the primary display until the aircraft is airborne. Additional symbols presented during low-visibility take-off operation are

- Ground roll reference symbol (fixed position)
- Ground localizer scale and index

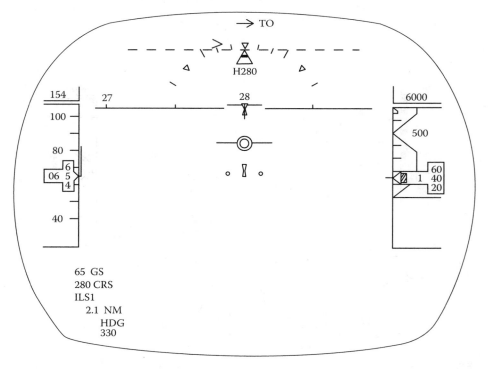

FIGURE 7.17   HGS primary mode: low-visibility take-off.

- Ground roll guidance cue (HGS-derived steering command)
- TOGA reference line

The ground localizer scale and index provide raw localizer information any time the aircraft is on the ground. For a low-visibility take-off, the general operating procedure is to taxi the aircraft into take-off position over the runway centerline. The selected course is adjusted as necessary to overlay the Selected Course symbol on the actual runway centerline at the furthest point of visibility. Take-off roll is started and the captain uses rudder control to center the ground roll guidance cue in the ground roll reference symbol (concentric circles). If the cue is to the right of the ground roll reference symbol then the pilot would need to apply right rudder to again center the two symbols. (At rotation, the ground roll reference and guidance cue symbols are replaced by the flight path symbol and the flight director guidance cue.)

7.3.1.1.2 Primary Mode: Climb

At rotation, a number of changes take place on the display (see Figure 7.16). Flight path acceleration, now positioned relative to flight path controlling the aircraft, is particularly useful in determining a positive climb gradient and in optimizing climb performance. With the appropriate airspeed achieved, to null flight path acceleration will maintain airspeed. Alternately, the flight director commands can be followed.

7.3.1.1.3 Primary Mode: Cruise

Figure 7.16 shows a typical HGS display for an aircraft in straight and level flight at 22,000 ft, 295 kn, and Mach .792. Ground speed is reduced to 275 kn as a result of a 21-kn, right-quartering headwind indicated by the wind arrow.

The aircraft is being flown by the autopilot with LNAV and VS modes selected. Holding the center of the flight path symbol level on the horizon, and the flight path acceleration symbol (>) on the flight path wing will maintain level flight.

7.3.2 AIII Approach Mode

The HGS AIII mode is designed for precision, manual ILS approach, and landing operations to CAT III minimums. Additionally, the AIII mode can be used for CAT II approaches at Type I airfields if operational authorization has been obtained (see Figure 7.18). The display has been de-cluttered to maximize visibility by removing the altitude and airspeed tape displays and replacing them with digital values. The HSI is also removed, with ILS raw data (localizer and glideslope deviation) now being displayed near the center of the display. (In the AIII mode, guidance information is shown as a circular cue whose position is calculated by the HGS.)

Tracking the HGS guidance cue, and ultimately the ILS, is achieved by centering and maintaining the flight path symbol over the cue. Monitoring localizer and glideslope lines relative to their null positions helps to minimize deviations and to anticipate corrections. Airspeed control is accomplished by nulling the speed error tape (left wing of flight path symbol) using the flight path acceleration caret to lead the airspeed correction. Any deviations in ILS tracking or airspeed error are easily identified by these symbolic relationships.

Following touchdown, the display changes to remove unnecessary symbology to assist with the landing rollout. The centerline is tracked while the aircraft is decelerated to exit the runway.

7.3.2.1 AIII Mode System Monitoring

The HGS computer contains an independent processor, the system monitor, which verifies that HGS symbology is positioned accurately and that the approach is flown within defined limits. If the system monitor detects a failure within the HGS or in any required input, it disables the AIII status, and an approach warning is annunciated to both crew members.

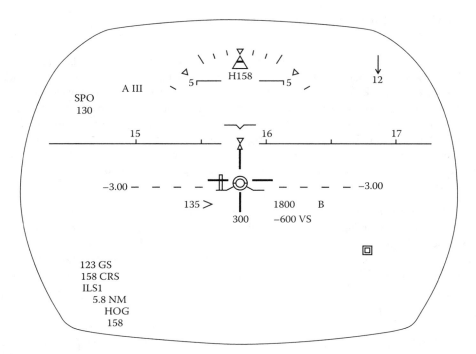

FIGURE 7.18   HGS AIII approach display.

7.3.2.2 Unusual Attitude

The HGS Unusual Attitude (UA) display is designed to aid the pilot in recognition of and recovery from unusual attitude situations. When activated, the UA display replaces the currently selected operational mode symbology, and the HCP continues to display the currently selected operational mode that will be reactivated once the aircraft achieves a normal attitude.

The UA symbology is automatically activated whenever the aircraft exceeds operational roll or pitch limits, and deactivated once the aircraft is restored to controlled flight, or if either pitch or roll data becomes invalid. When the UA symbology is deactivated, the HGS returns to displaying the symbology for the currently selected operational mode.

The UA symbology includes a large circle (UA Attitude Display Outline) centered on the combiner (see Figure 7.19). The circle is intended to display the UA attitude symbology in a manner similar to an Attitude Direction Indicator (ADI). The UA horizon line represents zero degrees pitch attitude and is parallel to the actual horizon. The UA horizon line always remains within the outline to provide a sufficient sky/ground indication, and always shows the closest direction to and the roll orientation of the actual horizon. The aircraft reference symbol is displayed on top of a portion of the UA horizon line and UA ground lines whenever the symbols coincide.

The three UA Ground Lines show the ground side of the UA horizon line corresponding to the brown side on an ADI ball or EFIS attitude display. The ground lines move with the horizon line and are angled to simulate a perspective view.

The UA pitch scale range is from −90° through +90° with a zenith symbol displayed at the +90° point, and a nadir symbol displayed at the −90° point.

The UA roll scale is positioned along the UA attitude display outline, with enhanced tic marks at ±90°. The UA roll scale pointer rotates about the UA aircraft reference symbol to always point straight up in the Earth frame.

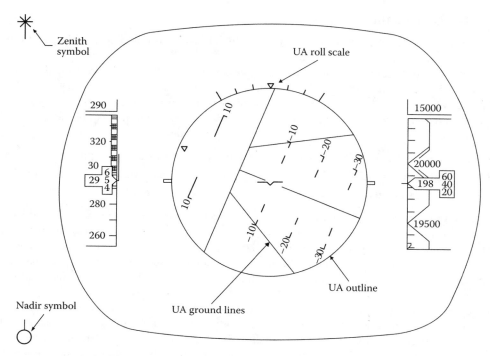

FIGURE 7.19 Unusual Attitude display.

7.3.3 Mode Selection and Data Entry

The data entry needs of the HUD are limited to mode selection and runway information for the guidance algorithms to work effectively. Data entry can be via a dedicated control panel or a Multipurpose Control Display Unit (MCDU), such as that defined by ARINC 739.

7.3.3.1 Mode Selection

On most aircraft the pilot has a number of ways to configure the HUD for an approach and landing based on the visibility conditions expected at the airport. In good weather, where the cloud "ceiling" is high and the runway visual range (RVR) is long, the pilot may leave the HUD in the primary mode or select a landing mode such as VMC, which removes some symbol groups, but has no guidance information. As the ceiling and/or RVR decreases the pilot may select the IMC mode to display FGS guidance (usually from the flight management system). If the visibility is at or near the Category III limit the pilot will select AIII mode, which requires an instrument landing system and special approach guidance. To reduce workload, the HUD can be configured to automatically select the appropriate landing mode when certain conditions are met, such as the landing system deviations become active.

Another mode that is available for selection, but only on the ground, is the test mode where the pilot, or more usually a maintenance person, can verify the health of the HUD and the sensors that are connected to the system.

7.3.3.2 Data Entry

To make use of the HUD-based guidance the pilot must enter the following information:

- Runway elevation — the altitude of the runway threshold
- Runway length — official length of the runway in feet or meters
- Reference glideslope — the published descent angle to the runway, e.g., 3°

On some aircraft, these data may be sent from the FMS and confirmed by the pilot.

7.3.4 HUD Guidance

On some aircraft the HUD can provide a pilot-in-the-loop low-visibility landing capability that is more cost-effective than that provided by an autoland system.

HUDs that compute guidance to touchdown use deviations from the ILS to direct the pilot back to the center of the optimum landing path. The method for guiding the pilot is the display of a guidance cue that is driven horizontally and vertically by the guidance algorithms. The goal of the pilot is to control the aircraft so that the flight path symbol overlays the guidance cue. The movement of the guidance cue is optimized for pilot-in-the-loop flying. This optimization includes:

- Limiting the movement of the cue to rates that are achievable by a normal pilot
- Anticipating the natural delay between the movement of the cue and reaction of the pilot/aircraft
- Filtering out short-term cue movements that may be seen in turbulent air

In addition to approach guidance where the goal is to keep the aircraft in the center of the ILS beam, guidance is also provided for other phases of the approach. During the flare phase — a pitch-up maneuver prior to touchdown — the guidance cue must emulate the normal rate and magnitude of pull-back that the pilot would use during a visual approach. During the rollout phase — where the goal is to guide the aircraft down the centerline of the runway — the pilot is given smooth horizontal commands that are easy to follow.

All these algorithms have to work for all normal wind and turbulence conditions. As following the guidance is critical to the safety of the aircraft, the algorithms include monitors to ensure that the information is not misleading and monitors to ensure that the pilot is following the commands. If the system detects the pilot is significantly deviating from the path or speed target the system will display an approach warning message that requires the pilot to abort the landing.

7.3.4.1 Annunciations

An important element of any system is the annunciations that inform or alert the pilots to problems that require their action. In a well-managed flight deck the role of each of the pilots is designed to be complementary. The pilot flying (PF) is responsible for control of the aircraft. The pilot not flying (PNF) is responsible for navigation and communication as well as monitoring the performance of the PF.

All the status information needed to safely fly the aircraft is displayed on the HUD for the pilot including:

- Mode status — modes of the HGS guidance or the guidance source.
- Cautions — approaching operating limitations or loss of a sensor.
- Warnings — loss of a critical sensor requiring immediate action.
- System failure — HUD has failed and the pilot should not use the system.

Because of the technology used in the HUD, the PNF can not directly monitor these annunciations. To support PNF monitoring the HUD outputs some or all of these annunciations to either a flight deck central warning system or to a dedicated annunciator panel in front of the other pilot.

7.3.5 Recent Developments

7.3.5.1 Color HUD

Due to the complexity of wide field-of-view reflective HUD optical systems, the optical designer must use all means available to meet display accuracy and parallax error requirements. All certified reflective HUDs today are monochromatic, generally using a narrow-band green emitting phosphor. The addition of a second color to the HUD is a desirable natural progression in HUD technology, however, one of the technical challenges associated with adding a second (or third) display color is maintaining the performance standards available in monochrome displays. One method for solving this problem uses a collimator with two independent embedded curvatures, one optimized for green symbology, the other optimized for red symbology, each with a wavelength-selective coating.[14]

One fundamental issue associated with color symbology on HUDs is the effects of the real-world background color "adding" to the display color (green), resulting in an unintended perceived display color.

7.3.5.2 Display of Enhanced Vision Sensor Images

Many modern HUD systems are capable of simultaneously displaying a real-time external video image and stroke symbology and guidance overlay. Given a sensor technology capable of imaging the real world through darkness, haze, or fog, the Enhanced Vision System (EVS) provides an image of the real world to the pilot while continuing to provide standard HUD symbology. This capability could provide benefit to the operator during taxi operations, low-visibility take-off, rollout, and perhaps during low-visibility approaches.

The interface between the EVS sensor and the HUD can be a standard video format (i.e., RS-170 or RS-343) or can be customized (i.e., serial digital). Sensor technologies that are candidates for EVS include:

- Forward-looking infrared, either cooled (InSb) or uncooled (InGaAs or microbolometer)
- MMW radar (mechanical or electronic scan)
- MMW radiometers (passive camera)
- UV sensors

Although the concept of interfacing a sensor with a HUD to achieve additional operational credit is straightforward, there are a number of technical and certification issues which must be overcome including pilot workload, combiner see-through with a raster image, sensor boresighting, integrity of the sensor, and potential failure modes. In addition, the location of the sensor on the aircraft can affect both parallax between the sensor image and the real world, and the aircraft aerodynamic characteristics.

Synthetic vision is an alternative approach to improving the pilot's situational awareness. In this concept, an onboard system generates a "real-world-like view" of the outside scene based on a terrain database using GPS position, track, and altitude. Some HUD systems today generate "artificial runway outlines" to improve the pilot's awareness of ground closure during low-visibility approach modes, a simple application of synthetic vision.

Defining Terms

Boresight: The aircraft longitudinal axis, used to position the HGS during installation and as a reference for symbol positioning. The process of aligning the HUD precisely with respect to the aircraft reference frame.

Collimation: The optical process of producing parallel rays of light, providing an image at infinity.

Eyebox: The HUD eyebox is a three-dimensional area around the flight deck eye reference point (ERP) where all of the data shown on the combiner can be seen.

References

1. Naish, J. Michael, Applications of the Head-Up Display (HUD) to a Commercial Jet Transport, *J. Aircraft*, August 1972, Vol. 9, No. 8, pp. 530–36.
2. Naish, J. Michael, Combination of Information in Superimposed Visual Fields, *Nature*, May 16, 1964, Vol. 202, No. 4933, pp. 641–46.
3. Sundstrand Data Control, Inc. (1979), *Head Up Display System*.
4. Part 25 — Airworthiness Standards: Transport Category Airplanes, Special Federal Aviation Regulation No. 13, Subpart A — General, Sec. 25.773 Pilot Compartment View.
5. Part 25 — Airworthiness Standards: Transport Category Airplanes, Special Federal Aviation Regulation No. 13, Subpart A — General, Sec. 25.775 Windshield and Windows.

6. Vallance, C.H. (1983). The approach to optical system design for aircraft head up display, *Proc. SPIE*, 399:15–25.

7. Hughes, U.S. Patent 3,940,204 (1976), Optical Display Systems Utilizing Holographic Lenses.

8. Marconi, U.S. Patent 4,261,647 (1981), Head Up Displays.

9. Wood, R. B. (1988), Holographic and classical head up display technology for commercial and fighter aircraft, *Proc. SPIE*, 883:36–52.

10. SAE (1998), AS8055 Minimum Performance Standard for Airborne Head Up Display (HUD).

11. Stone, G. (1987), The design eye reference point, SAE 6th Aerospace Behavioral Eng. Technol. Conf. Proc., *Human/Computer Technology: Who's in Charge?*, pp. 51–57.

12. Desmond, J., U.S. Patent 4,698,785 (1997), Method And Apparatus For Detecting Control System Data Processing Errors.

13. Part 25 — Airworthiness Standards: Transport Category Airplanes, Special Federal Aviation Regulation No. 13, Subpart A — General, Sec. 25.562 Emergency Landing Dynamic Conditions.

14. Gohman et al., U. S. Patent 5,710,668 (1988), Multi-Color Head-Up Display System.

8

Night Vision Goggles

Dennis L. Schmickley
The Boeing Company

8.1 Introduction

8.1.1 NVG as Part of the Avionics Suite

Night vision goggles (NVGs) are electronic devices that help observers see in the dark. NVGs made for aviation are generally termed night vision imaging systems (NVIS), although the term NVG is often used interchangeably with NVIS.

Visual reference to the aviator's outside world is essential for safe and effective flight. During the daylight hours and in visual meteorological conditions (VMC), the pilot relies heavily on the out-the-windshield view of the airspace and terrain for situational awareness. In addition, the pilot's visual system is augmented by the avionics, which provide communication, navigation, flight control, mission, and aircraft systems information. During nighttime VMC, the pilot can improve the out-the-windshield view with the use of NVIS. NVG lets the pilot see in the dark during VMC conditions.

This chapter deals with NVIS for aviation applications; it does not address the many nonaviation applications such as NVG for personnel on the ground or underwater and for ground and sea vehicles.

8.1.2 What Are NVIS?

NVIS are light image intensification I^2 devices that amplify the night-ambient-illuminated scenes by a factor of 10^4. For this application, "light" includes visual light and near infrared. The development of the microchannel plate (MCP) allowed miniature packaging of image intensifiers into a small, lightweight, helmet-mounted pair of goggles. With the NVIS, the pilot views the outside scene as a green phosphor image displayed in the eyepieces. Various terms are associated with NVG type equipment:

NVG — general term for any I² device, usually head-worn and binocular

I² — Image intensifier type of sensor device used in NVG

ANVIS — Aviator's night vision imaging system; a type of NVG designed for aviators

NVIS — Night vision imaging system; a general class of NVG for aviation including ANVIS

Gen II — Second-generation intensifier technology utilizing MCP and multi-alkali photocathode, which enabled construction of AN/PVS-5 NVG

Gen III — Third-generation intensifier technology utilizing improved MCP and gallium arsenide photocathode, which enabled construction of AN/AVS-6 ANVIS

NVG HUD — Night vision goggle with an attached head-up display

HMD — Helmet-mounted display; in this chapter, this includes NVG HUD

PNVG — Panoramic night vision goggle; usually about 100° field of view (FOV)

LPNVG — Low-profile night vision goggle that usually conforms to the face

AGC — Automatic gain control

8.1.3   History of NVIS in Aviation

8.1.3.1   1950s

In the 1950s, there was considerable diverse research on night image intensification as reported at the Image Intensifier Symposium.[4] The applications included devices for military sensing and for astronomy and scientific research, but they did not specifically address the needs of head-mounted pilotage devices. The U.S. Army first experimented with T-6A infrared (IR) driving binoculars in helicopters in the late 1950s.[2] The binocular device was a near-IR converter, which required an IR-filtered landing light for the radiant energy and was not satisfactory for aviation. In the late 1950s, the first continuous-channel electron multiplier research was being conducted at the Bendix Research Laboratories by George Goodrich, James Ignatowski, and William Wiley. The invention of the continuous-channel multiplier was the key step in the development of the MCP.[1]

8.1.3.2   1960s

In the early 1960s, first-generation tubes were developed that allowed operation as a passive system; however, the size of the three-stage I² tubes was too large for head-mounted applications. "Passive" means that there is no need for active projected illumination; the system can operate using the ambient starlight illumination, hence the name "starlight scope" from the Vietnam era foot soldier's sniper scope. In the late 1960s, the production of the MCPs used in the second-generation wafer technology I² tubes allowed night vision devices to be packaged small enough and light enough for head-mounted applications. Thus, in the late 1960s and early 1970s, the U.S. Army Night Vision and Electro-Optics Laboratory (NV&EOL) used Gen II I² tubes to develop NVG for foot soldiers, and some of these NVG were tested by aviators for night flight operations.

8.1.3.3   1970s

In 1971, the U.S. Air Force (USAF) began limited use of the SU-50 electronic binoculars. In 1973, the Army adopted the Gen II AN/PVS-5 as an interim NVG solution for aviators, although there were known deficiencies in low-light-level performance, weight, visual facemask obstruction, and refocusing (due to incompatibility with cockpit lighting systems). The ANVIS was the first NVG developed specifically to meet the visual needs of the aviator. The NV&EOL started ANVIS development in 1976, utilizing Gen III image intensifier technology and requiring high-performance, lightweight, and improved reliability and maintainability.

8.1.3.4   1980s

Two versions of the ANVIS were introduced into military aviation in the 1980s: AN/AVS-6(V)1, which was designed for most helicopters, fit onto the helmet with a centerline mount, and AN/AVS-6(V)2, which was for AH-1 Cobra only, fit onto the helmet with an offset mount.[13]

ANVIS operation would not have been feasible or safe in the aircraft if the cockpit lighting had remained the traditional red-lighted or white-lighted incandescent illumination. In 1981, the U.S. Army released an Aeronautical Design Standard (ADS-23 5)[5] to establish baseline requirements for development of cockpit lighting to be compatible with ANVIS. In 1986, the Joint Aeronautical Commanders Group (JACG) released a Tri-Service specification (MIL-L-85762 7)[7] that defined standards for designing and measuring ANVIS-compatible lighting. GEC-Marconi introduced a Gen III projected view NVG called the "Cat's Eye" for use in the AV-8 Harrier.

An updated MIL-L-85762A8[8] was released in 1988 that defined NVIS as a general term (replacing the specific ANVIS term) and expanded the lighting requirements to accommodate various types of NVIS. The controversial use of AN/PVS-5 continued in aviation pending full fielding of ANVIS. Based upon a series of nighttime accidents often involving NVGs, a Congressional hearing convened in 1989 to review the safety and appropriateness of NVGs in military helicopters deemed ANVIS necessary.

8.1.3.5 1990s

Within the NVG, head-up flight information symbology was desired, along with the out-the-window view. Integrating the symbology and imagery resulted in a new type of HMD referred to as the NVG HUD. Two types of NVG HUDs were placed in service: the AN/AVS-7 NVG HUD, which was installed on CH-47D and HH-60 aircraft, and the Optical Display Assembly (ODA) NVG HUD, which was installed on OH-58D.

NVIS-compatible cockpit lighting was incorporated into high-speed fixed-wing aircraft, but an additional requirement evolved for NVG to be safe during pilot ejection. The AN/AVS-9 (models F4949 and M949) was developed for the USAF for ejection capability. In an effort to provide a greater FOV than the normal 40° for NVIS the USAF developed a PNVG to provide about 100° FOV. Several other development programs attempted to reduce the size of the large protrusive goggle optics. Versions of the LPNVG folded the optics to fit conformly around the face. Several integrated helmet development programs incorporated integral I[2] devices and electronic projected display systems. In the early 1990s, several civilian helicopter operators expressed interest in using NVIS. Ongoing investigations into the use of NVIS in civil aviation delved into applications, safety, and Federal Aviation Administration (FAA) certification. The Society of Automotive Engineers (SAE) published several recommended lighting practices to accommodate NVIS-compatible lighting for nonmilitary aircraft. The Department of Defense created a lighting handbook (JSSG-2010-5) to be used as guidance for interior and exterior NVIS-compatible lighting. Although the requirements in MIL-L-85762A were still valid, the government sought to provide guidance documents rather than specifications.

8.1.3.6 2000s

MIL-STD-3009 was created to define performance requirements for lighting (both interior and exterior) and display equipment. The standard incorporated requirements from MIL-L-85762A and allowed extraction of applicable requirements from JSSG-2010-5. The SAE published ARP-5825 to define design requirements and test procedures for dual mode exterior lights — lights that can be used visually and with NVIS and covert lights that are only viewable with NVIS. Lighting manufacturers developed bright LED lights that could both provide exterior lighting for visual detection and allow for NVIS flying without sensor degradation.

8.2 Fundamentals

8.2.1 Theory of Operation

An image intensifier is an electronic device that amplifies light energy. Light energy (photons) enter into the device through the objective lens and are focused onto a photocathode detector that is receptive to both visible and near-IR radiation. Gen III devices use gallium arsenide as the detector. Due to the photoelectric effect, the photons striking the photocathode emit a current of electrons. Because the emitted electrons scatter in random directions, a myriad of parallel tubes (channels) are required to

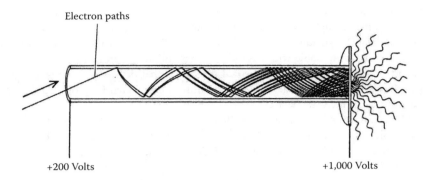

FIGURE 8.1 Electron amplification in a microchannel.[1]

provide separation and direction of the electron current to ensure that the final image will have sharp resolution. Each channel amplifier is microscopic, about 15 μm in diameter. A million or so microchannels are bundled in a wafer-shaped array called a MCP, which is about the diameter of a quarter. The thickness of the MCP, which is the length of the channels, is less than a quarter of an inch. Each channel is an electric amplifier. A bias potential of about 1000 V is established along the tube, and each electron produced by the photoelectric effect accelerates through the tube toward the anode (Figure 8.1). When an electron strikes other electrons in the coated channel they are knocked free and continue down the tube, hitting other electrons in a cascade effect. The result of this multiplication of electrons is a greatly amplified signal. The amplified stream of electrons finally hits a phosphor-type fluorescent screen which, in turn, emits a large number of photons and creates an image.

The MCP is a solid-state light amplifier. The intensity of the image is a product of the original signal strength (i.e., the number of photons in the night scene) and the amplification gain within the channel. The fine resolution of the total image is a product of the pixel size from the MCP array and the focusing optics.

The manufacture of MCPs requires complex processes that are dependent on a two-draw glass reduction technique (Figure 8.2). A concentric tube of an outer feed glass and an inner core glass is drawn into a fine fiber about 1 mm in diameter. Then a bundle of thousands of the fibers is drawn to form a multiple fiber about 50 mm in diameter. The core glass is etched out, leaving a matrix of hollow glass tubes. Wafer sections are sliced, and the wafers are plated with the metallic coatings necessary for the signal amplification.

The finished product is an NVG that contains an MCP packaged inside an optical housing containing an objective lens and eyepieces appropriate for the NVG's utilization. For aviators using the NVG for pilotage a one-to-one magnification is required. The pilot's perceived NVG image of the outside world must be equal to the actual size of the unaided-eye image of the outside real world to provide natural motion and depth perception. The image is displayed to the observer on an energized viewing screen at about 1 footlambert (fL). Screens may be the P20 or P25 phosphors. The light amplification may be 2000 or more and, to prevent phosphor damage, an AGC circuit limits the gain in high ambient conditions.

8.2.2   I² Amplification of the Night Scene

Gen II image intensifiers utilize multi-alkali photocathodes that are sensitive in the visible and near-IR bandwidth of 400 to 900 nm. Gen II utilization is generally limited to a minimum of quarter-moon or clear sky illumination (10^{-3} to 10^{-4} footcandle (fc)).

Gen III image intensifiers utilize gallium arsenide (GaAs) photocathodes, which are more sensitive than Gen II and have a bandwidth of 600 to 900 nm. Gen III NVIS are even effective in starlight and overcast conditions (10^{-4} to 10^{-5} fc).

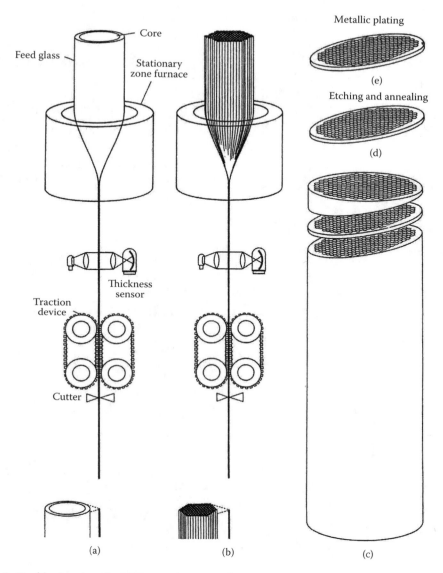

FIGURE 8.2 Double glass draw for MCP manufacture.[1]

8.2.3 NVIS Does Not Work without Compatible Lighting

NVIS lighting compatibility is required for effective NVIS use by pilots. If the cockpit lighting is not compatible and it emits energy with spectral wavelengths within the sensitivity range of the NVIS, the lighting will be amplified by the NVIS and will overpower the amplification of the lower illumination in the outside visual scene.

A lighting system is compatible with NVIS when it does not render the NVIS useless nor does it hamper the crew's visual tasks (with or without NVIS). NVIS compatibility permits a crew member to observe outside scenes through night vision goggles while maintaining necessary lighted information in the crew station. The Gen III NVIS are insensitive to blue-green light, so the cockpit lighting can be modified with blue cutoff filtering to reduce emitted energy in the red and near-IR regions to achieve compatibility. The complementary minus-blue coatings on the NVIS objective lens provide a sharp cutoff filter to block any red or near-IR light. Blue-green lighting allows external viewing through the NVIS

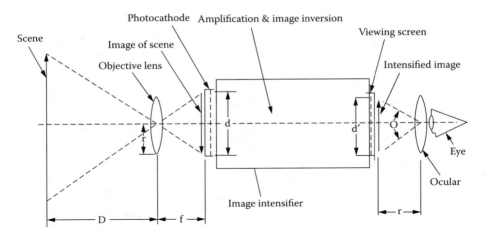

FIGURE 8.3 Typical NVIS image intensifier tube and optics.

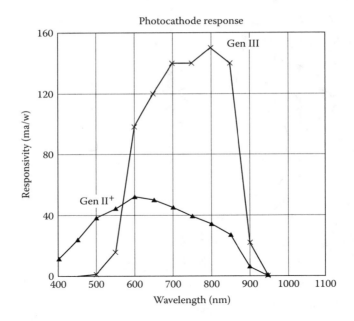

FIGURE 8.4 Photocathode sensitivity.

and internal viewing of the instruments by using the "look-around" technique. The NVIS look-around design allows the pilot visual access (with unaided eyes) into the blue-green lighted cockpit without head movement. NVIS compatibility requirements are defined by MIL-L-85762.

MIL-L-85762 lighting requirements, and by default the various NVIS, have been categorized into types and classes to match the appropriate cockpit lighting system depending on the type of NVIS being used in the aircraft. The original issue of MIL-L-85762 was based on recommendations for ANVIS compatibility[6] and addressed lighting only for ANVIS (Type I, Class A). MIL-L-86762A added Type II and Class B NVIS. A rationale was published to help manufacturers and evaluators interpret the requirements.[9] The USAF also defined NVIS compatible lighting,[10–12] and later defined a class C NVIS (JSSG-2010-5,[27] MIL-STD-3009[28]).

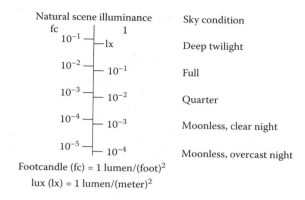

Natural scene illuminance

fc		Sky condition
10^{-1}	lx	Deep twilight
10^{-2}	10^{-1}	Full
10^{-3}	10^{-2}	Quarter
10^{-4}	10^{-3}	Moonless, clear night
10^{-5}	10^{-4}	Moonless, overcast night

Footcandle (fc) = 1 lumen/(foot)^2

lux (lx) = 1 lumen/(meter)^2

FIGURE 8.5 Illumination from the night sky.

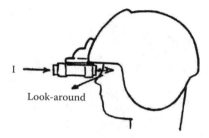

FIGURE 8.6 Type I (direct view) ANVIS with "look-around" vision into the cockpit.

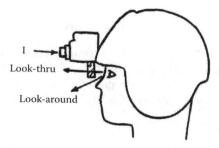

FIGURE 8.7 Type II (projected image). Cat's Eye with "look-through" outside viewing and "look-around" vision into the cockpit.

Type I — Type I lighting components are those compatible with Direct View Image NVIS. Direct View Image NVIS is defined as any NVIS using Gen III image intensifier tubes that displays the intensified image on a phosphor screen in the user's direct line of sight, such as the ANVIS.

Type II — Type II lighting components are those compatible with projected image NVIS. Projected image NVIS is defined as any NVIS using Gen III image intensifier tubes that projects the intensified image on a see-through medium that reflects the image into the user's direct line of sight, such as the Cat's Eyes (Figure 8.7).

Class A — Class A lighting components are those compatible with NVIS using a 625-nm minus-blue objective lens filter, which results in an NVIS sensitivity lens as shown in Figure 8.8. The standard AN/AVS-6 ANVIS are equipped with 625-nm minus-blue filters.

Class B — Class B lighting components are those compatible with NVIS using a 665-nm minus-blue objective lens as shown in Figure 8.9. Class B lighting allows red and yellow colors in cockpit displays,

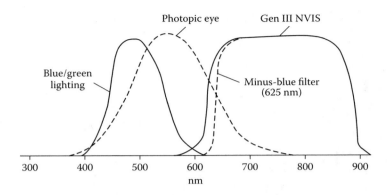

FIGURE 8.8 Typical Class A blue-green lighting and 625-nm minus-blue coating on NVIS.

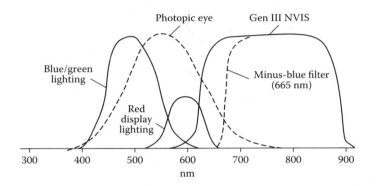

FIGURE 8.9 Typical Class B lighting allows blue-green, yellow, and red with 665-nm minus-blue coating on NVIS.

but the consequence is a reduced Gen III NVIS sensitivity to the outside visual scene. The 665-nm minus-blue filter reduces the NVIS sensitivity by 8 to 10% of the Class A NVIS in moonless conditions.

Class C — Class C NVIS have a special "leaky green" coating to allow selective green light near the 540 nm range into the night vision device (Figure 8.10). This allows direct view of the green HUD used in modern aircraft.

8.2.4   I² Integration into Aircraft

The integration of NVIS into an aircraft crew station usually requires very little modification with respect to the crew compartment space (volume). The primary aircraft requirements are:

1. Adequate helmet and NVIS motion envelope
2. Acceptable FOV and windshield transparency in the NVIS range
3. Compatible cockpit lighting and displays
4. Compatible interior (cabin) and exterior lighting

The alerting quality of warning, caution, and advisory color coding can be diminished with NVIS-compatible (Class A) cockpit lighting. Audio or voice warning messages may be considered to augment the alerting characteristics.

The NVIS is normally a self-contained stand-alone sensor powered by small batteries. The cost of a typical NVIS unit is $10,000, whereas the cost of an aircraft-mounted thermal-imaging IR sensor system is 10 to 20 times that amount. The integration of an NVG HUD requires more modification to the aircraft than the NVIS.

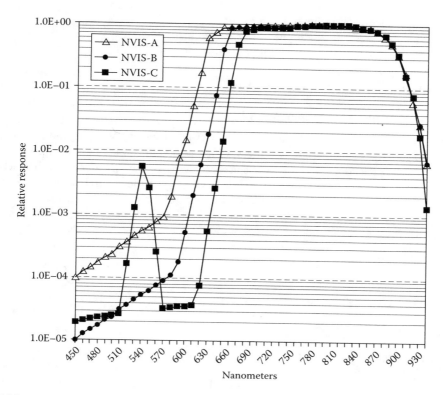

FIGURE 8.10

Incorporation of NVIS produces some advantages and some disadvantages for the aircraft and missions, but the advantages usually outweigh the disadvantages. The advantages are:

- Twenty-four-hour VFR operations; pilots say: "I'd rather fly with them"
- Enhanced situation awareness; pilots can see the terrain

The disadvantages are:

- Limited instantaneous FOV, which requires deliberate head movement
- Neck strain and fatigue (due to increased helmet weight and increased head movement)
- Cost of equipment (NVIS 1 compatible lighting)
- Pilot training, currency, and proficiency
- Not useful in Instrument Meteorological Conditions (IMC) such as fog
- Lack of safety (if there is inadequate training or overexpectations of system capability)

There are known limitations of the NVIS imposed by the limited FOV. Training is required to emphasize the required head motion scanning to compensate for the FOV. Depth perception is sometimes reported as a major deficiency, although it is most likely that inadequate motion perception cues due to limited peripheral vision are a contributor to this problem.

Military training programs have been implemented to exploit the capabilities of the NVIS sensor for various types of covert missions and to improve safety and situation awareness. Curricula have been developed "to assure that there is an appropriate balance of training realism and flight safety."[18] Training programs include academics, laboratory demonstrations with visual aids, and simulation to cover the following topics:

- Theory of I^2-operation
- FOV and field of regard (FOR)

- Focus, adjustment, and fitting procedures
- Moon, weather, ambient conditions
- Different visual scan patterns and head motion required with NVIS

8.3   Applications and Examples

8.3.1   Gen III and AN/AVS-6 ANVIS

To aid night flying, in the 1980s the Army developed ANVIS, a Gen III NVG. The ANVIS, designated as AN/AVS-6, is lightweight (total weight is 1.2 to 1.3 lb) and mounts on the pilot's helmet. The 25-mm eye relief allows the pilot to see around the eyepieces for viewing the instruments in the cockpit. The Gen III response characteristics are more sensitive than Gen II, and the spectral range, which covers 600 to 900 nm, takes advantage of the night sky illumination in the red and IR. Luminance gain is 2000 or greater. The FOV is 40° circular, and the resolution is about 1 cy/mr. Figure 8.11 shows several adjustment features on the ANVIS to accommodate each pilot's needs:

- Inter-ocular adjustment
- Tilt adjustment
- Vertical adjustment
- Eye relief (horizontal) adjustment
- Focus adjustment
- Diopter adjustment

The pilot can also flip up the ANVIS to a "helmet stow" position, and the mount has a break-away feature in case of a high g load or crash.

Early production models of ANVIS units produced system luminance gains of 2000. With improvements in manufacturing techniques and yields and increased photocathode sensitivities, newer units have system luminance gains of over 5000. The Army procured large lot quantities of AN/AVS-6 through "omnibus" purchase orders; Omni IV and Omni V AN/AVS-6 have system luminance gains of 5500. The luminance gains of the intensifiers may be 10,000 to 70,000, depending on the ambient illumination being amplified, but with optics and system throughput losses, the overall system gains are 50,001. Presently, the two major suppliers in the United States for AN/AVS-6 are ITT and Northrop Grumman

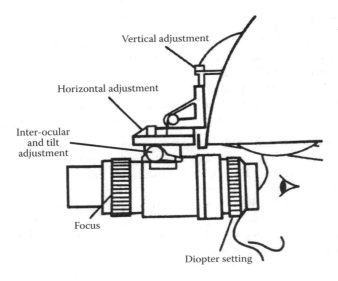

Vertical adjustment

Horizontal adjustment

Inter-ocular
and tilt
adjustment

Focus

Diopter setting

FIGURE 8.11   ANVIS adjustments.[17]

(formerly Litton), and the Army splits the procurement of the Omni lots. Adaptations and improved versions of the AN/AVS-6 include the AN/AVS-8 with a 45° FOV and the AN/AVS-9, which has a front-mounted battery to allow use in ejection seats. The AN/AVS-9 also can have a Class C "leaky green" sensitivity to allow viewing of the HUD symbology.

8.3.2 Gen II and AN/PVS-5 NVG

The Gen II AN/PVS-5 is outdated and is not now recommended for aviators. However, the AN/PVS-5 is discussed here because it was the most common device to allow night flying with NVG-aided vision. The AN/PVS-5A provided Army ground forces with enhanced night vision capability. Later, pilots used the NVG to fly helicopters. Tests indicated that pilots using NVG could fly lower and faster than pilots without NVG, and it was concluded that NVG provided considerable improvement over unaided, night-adapted vision.

The AN/PVS-5A weighs 2 lb and has a full face mask. Wearing these NVG requires the pilot to make all visual observations via the NVG, including cockpit instrument scanning. The pilot must move his head and refocus the lens to read the instruments. These restrictions result in annoyance, discomfort, and fatigue.

The spectral range of the Gen II NVG is from 350 to 900 nm, which includes the entire visual spectrum (380–760) plus some near-IR coverage. Most 1970s cockpits had red incandescent lamp lighting with large red and IR emissions. The NVG's AGC shuts down the NVG in the presence of large amounts of radiant energy in the goggles' range. Therefore, the use of Gen II NVG requires that all visible lighting must be reduced below the pilot's visual threshold that the lighting does not degrade the NVG operation. Commonly, this is accomplished by extinguishing the lights or using a "superdim" setting. Under these conditions, crew members without NVG cannot read the cockpit instruments. Crew members with NVG must refocus from outside viewing to read the instruments. U.S. and U.K. research on the shared-apertures and shared-lens attempted to provide viewing of the cockpit instruments with the NVG. Modifications to the face mask to provide peripheral and in-cockpit vision produced the "cut-away" mask.

The utilization of AN/PVS-5 NVG in aviation was controversial. The incorporation of NVG into aviation somewhat mirrored the development of aviation itself, with a period of trial and error, sometimes with inadequate or inappropriate equipment, producing some pioneering breakthroughs and some accidents. In the 1980s, there were nighttime accidents often involving NVGs. The *Orange County Register* published a lengthy investigative article because several of the helicopter crashes took place within the county,[14] and a Congressional hearing was convened to review the safety and appropriateness of NVGs in military helicopters.[15] The necessity of NVGs for night flight operations was confirmed, along with an emphasis on better equipment and training. A review of AN/PVS-5 and AN/AVS-6 testing concluded both were acceptable.[16] Since that time, AN/AVS-6 and AN/AVS-9 NVIS have become the preferred devices for aviators.

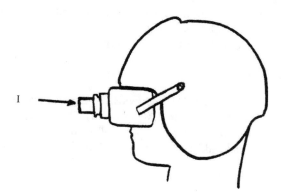

FIGURE 8.12 AN/PVS-5A NVG with full face mask.

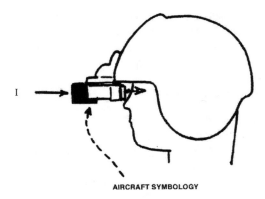

FIGURE 8.13 Aircraft symbology.

8.3.3 Cat's Eye

The Cat's Eye is a Type II (projected image) Gen III NVIS made by GEC-Marconi that is standard in the AV-8 series of Harrier aircraft. It weighs slightly over 1 lb, and its two optical combiner lenses display the image for out-of-the-cockpit viewing. The combiner has see-through capability to view the HUD. When the pilot is looking at the HUD, the I^2 imagery is automatically turned off to allow visibility of the HUD symbology. The combiner glass see-through transmission is <30%. The Cat's Eye has a 25-mm eye relief to allow look-under for cockpit instrument viewing.

8.3.4 NVG HUD

"NVG HUD" systems have been produced that add HUD symbology to the displayed night vision imagery provided by the NVG (Figure 8.13). Usually, the HUD portion is a cathode ray tube (CRT) image projected onto a combiner glass mounted in front of one of the NVG objective lens. The symbology displayed is aircraft information (attitude, altitude, airspeed, navigation data, etc.) that is generated in a processor box integrated to the aircraft systems.

8.3.5 ANVIS HUD

NVIS HUDs provide symbology, usually about 30° to 35° FOV, superimposed with the 40° the imagery of the NVIS. Honeywell produces the ODA that is integrated into the OH-58D. The OH-58D is one of the few aircraft that provides aircraft power for the NVIS (instead of a self-contained battery power). Elbit produces the AN/AVS-7 NVG HUD, which is used on the CH-47 and HH-60. Note that the AN/AVS-7 (NVG HUD) should not be confused with AN/PVS-7 single-tube NVG for ground troops. Rockwell Collins (Kaiser) has developed a day/night Eye-HUD™, which is used on the S-92; the nighttime module mounts onto the front of the NVIS. EFW has developed an NVIS symbology display unit (SDU) for the AH-64A/D as an NVIS HUD alternative to the helmet display unit (HDU).

8.3.6 PNVG

PNVG were originally developed for the USAF to provide an increased instantaneous horizontal FOV of the imagery to 100°. Night Vision Corporation developed the PNVG using four AN/PVS-7 image tubes, which produce a combined overlapping FOV of 100°. Kollsman developed the QuadEye™ 100° FOV PNVG with modular construction to add a 20° × 20° overlaid display and a 30° × 22.5° debriefing camera.

8.3.7 Low-Profile NVG

LPNVG have been developed for several reasons: to improve the head-borne center of gravity (CG), to allow visors, and to reduce possible injury caused by the protrusion of the longer I^2 tubes. The depth is 2 to 3 in. compared to 5 to 6 in. for other NVIS. ITT developed the Modular, Ejection-Rated, Low-profile Imaging for Night (MERLIN) Aviator Goggle for use by pilots in high-performance fixed-wing aircraft. Litton produces the AN/AVS-502 LPNVG for multirole missions (e.g., with parachute operations, weapons firing). Canadian Air Forces approved the AN/AVS-502 for flight engineers in the cabin, where head clearance and winds are issues. Systems Research Laboratories (SRL) developed the Eagle Eye™ for fixed-wing aircraft and multirole missions.

	Weight	I^2 FOV	Type
MERLIN	1.8 lb	35°	See-through optics
AN/AVS-502	1.5 lb	40°	See-through optics
Eagle Eye™	1.2 lb	40°	See-through optics

8.3.8 Integrated Systems

Sensors can be incorporated in avionics suites in several ways other than stand-alone NVG on a crew member's helmet. One method is to incorporate an I^2 on-board thermal imaging IR sensor pod to provide video imagery of either I^2 or IR to the crew. Several integrated helmet designs and future helmet concepts are integrating I^2 devices along with CRT, liquid crystal display (LCD), and LED helmet-mounted displays (Figure 8.14).

8.3.9 Testing and Maintaining the NVIS

NVIS manufacturers also supply testing and servicing equipment. Examples are the ANV-126 NVG Test Set, TS-6 Night Vision Device Test Set Kit, TS-4348/UV Night Vision Device Assessor, and the TS-10 Night Vision Leak Test and Purge Kit.

8.3.10 Lighting Design Considerations

NVIS-compatible aircraft interior lighting is essential to allow night flying with NVIS. Interior lighting consists of primary lighting (instrument and control panels), secondary lighting (task lights, area lights, floodlights), signals (warning, caution, advisory), and electronic displays. MIL-L-85762A, the key

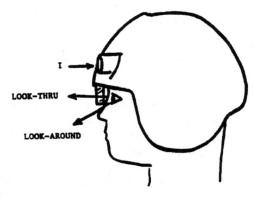

FIGURE 8.14 Integrated helmet.

specification that defines NVIS-compatible lighting, is unique in that it specifies two independent characteristics for the lighting system: luminance and chromaticity requirements for visual (unaided eye) viewing in a dark cockpit and radiance requirements for limiting any NVIS interference.

The luminance levels remain approximately the same as traditional red and white lighting systems in all previous aircraft. The chromaticity requirements generally produce a blue-green lighted cockpit. Four lighting colors for aviation have been defined in MIL-L-85762A (where u′ and v′ are 1976 UCS* chromaticity coordinates of the defined color):

- *NVIS Green A* — The color for primary, secondary, and advisory lighting. The chromaticity limits are within a circle of radius 0.037 with the center at u′ = 0.131, v′ = 0.623.
- *NVIS Green B* — The color for special lighting components needing saturated color (monochromatic) for contrast. The chromaticity limits are within a circle of radius 0.057 with the center at u′ = 0.131, v′ = 0.623.
- *NVIS Yellow* — The color for master caution and warning signals in Class A cockpits. The chromaticity limits are within a circle of radius 0.083 with the center at u′ =0.274, v′ = 0.622.
- *NVIS Red* — The color for warning signals in Class B cockpits. The chromaticity limits are within a circle of radius 0.060 with the center at u′ =0.450, v′ = 0.550.

Chromaticity and luminance requirements for various types of cockpit lighting and displays and cabin lighting are listed in Table VIII of MIL-L-85762A.

NVIS compatibility is not assured with proper chromaticity coordinates alone. Lights with different spectral compositions can appear visually as the same color; such similar visual colors are called metamers. All colored lights used in the NVIS-compatible cockpit must have filtering to block almost all the energy in the 600- to 900-nm range. The "NVIS-visible" portion of the lighting emission is to be limited per the NVIS radiance definition: NVIS radiance (NR) is the integral of the curve generated by multiplying the spectral radiance of the light source by the relative spectral response of the NVIS. "Formula 14a" of MIL-L-85762A is used to calculate the NVIS radiance of Class A lighting equipment, and "Formula 14b" is for the NVIS radiance of Class B equipment.

$$\text{NVIS radiance } (NR_A) = G(\lambda)_{max} \int_{450}^{930} G_A(\lambda)SN(\lambda)d\lambda \qquad \text{(Formula 14a)}$$

$$\text{NVIS radiance } (NR_B) = G(\lambda)_{max} \int_{450}^{930} G_B(\lambda)SN(\lambda)d\lambda \qquad \text{(Formula 14b)}$$

where:

$G_A(\lambda)$	= relative NVIS response of Class A equipment
$G_B(\lambda)$	= relative NVIS response of Class B equipment
$G(\lambda)_{max}$	= 1 ma/w
$N(\lambda)$	= spectral radiance of lighting component (w/cm² sr nm)
S	= scaling factor
dl	= 5 nm

For example, to be compatible a Class A lighting system requires that blue-green primary lighting not exceed 1.7×10^{-10} NR_A (NVIS radiance requirements for Class A equipment) when the lighting produces 0.1 fL luminance. If the lighting component is actually greater than 0.1 fL when it is measured, the scaling factor S scales the NR to 0.1 fL. For cockpits in which red or multicolor displays are desired, a similar equation for NR_B (NVIS radiance requirements for Class B equipment) applies to assure Class B compatibility. Note that Class B NVIS must be used with a Class B cockpit. NR requirements for various types of cockpit lighting and displays and cabin lighting are listed in the table below.

---

*Commission Internationale de l'Eclairage Uniform-Chromaticity-Scale.

Chromaticity Requirements (Table VIII, MIL-L-85762A)

Lighting Component(s)	TYPE 1 — Class A					Class B					TYPE II — Class A					Class B				
	u'_1	v'_1	r	Cd/m² (fL)	NVIS Color	u'_1	v'_1	r	Cd/m² (fL)	NVIS Color	u'_1	v'_1	r	Cd/m² (fL)	NVIS Color	u'_1	v'_1	r	Cd/m² (fL)	NVIS Color
Primary	.088	.543	.037	0.343 (0.1)	Green A						.088	.543	.037	0.343 (0.1)	Green A					
Secondary	.088	.543	.037	0.343 (0.1)	Green A						.088	.543	.037	0.343 (0.1)	Green A					
Illuminated Controls	.088	.543	.037	0.343 (0.1)	Green A						.088	.543	.037	0.343 (0.1)	Green A					
Compartment lighting	.088	.543	.037	0.343 (0.1)	Green A		Same				.088	.543	.037	0.343 (0.1)	Green A		Same			
Utility, work, and inspection	.088	.543	.037	0.343 (0.1)	Green A		as				.088	.543	.037	0.343 (0.1)	Green A		as			
Caution and advisory signals	.088	.543	.037	0.343 (0.1)	Green A		Class A				.088	.543	.037	0.343 (0.1)	Green A		Class A			
Jump lights	.088	.543	.037	17.2 (5.0)	Green A						.088	.543	.037	17.2 (5.0)	Green A					
	.274	.622	.083	51.5 (15.0)	Yellow						.274	.622	.083	51.5 (15.0)	Yellow					
Special lighting components where increased display emphasis by highly saturated (monochromatic) color is necessary, or adequate display light readability cannot be achieved with "GREEN A"	.131	.623	.057	0.343 (0.1)	Green B						.131	.623	.057	0.1	Green B					
Warning signal	.274	.622	.083	51.5 (15.0)	Yellow	.274	.622	.083	51.5 (15.0)	Yellow	.274	.622	.083	51.5 (15.0)	Yellow	.274	.622	.083	51.5 (15.0)	Yellow
	.450	.550	.060	51.5 (15.0)	Red	.450	.550	.060	51.5 (15.0)	Red						.450	.550	.060	51.5 (15.0)	Red
Master Caution signal	.274	.622	.083	51.5 (15.0)	Yellow		Same as Class A				.274	.622	.083	51.5 (15.0)	Yellow		Same as Class A			

Note: u'_1 and v'_1 = 1976 UCS chromaticity coordinates of the center point of the specified color area; r = radius of the allowable circular area on the 1976 UCS chromaticity diagram for the specified color; fl 5 footLamberts; Cd/m² = candela/(meter)².

NVIS Radiance Requirements (Table IX, MIL-L-85762A)

Lighting Components	TYPE 1 Class A Not Less than (NR_A)	Not Greater than: (NR_A)	fL	TYPE 1 Class B Not less than (NR_B)	Not Greater than: (NR_B)	fL	TYPE II Class A Not Less than (NR_A)	Not Greater than: (NR_A)	fL	TYPE II Class B Not Less than (NR_B)	Not Greater than: (NR_B)	fL
Primary	—	1.7×10^{-10}	0.1	Same as Class A (see Note)			—	1.7×10^{-10}	0.1	Same as Class A (see Note)		
Secondary	—	1.7×10^{-10}	0.1				—	1.7×10^{-10}	0.1			
Illuminated controls	—	1.7×10^{-10}	0.1				—	1.7×10^{-10}	0.1			
Compartment	—	1.7×10^{-10}	0.1				—	1.7×10^{-10}	0.1			
Utility, work and inspection lights	—	1.7×10^{-10}	0.1				—	$1.7 - 10^{-10}$	0.1			
Caution and advisory lights	—	1.7×10^{-10}	0.1				—	$1.7 - 10^{-10}$	0.1			
Jump lights	1.7×10^{-8}	5.0×10^{-8}	5.0	1.6×10^{-8}	4.7×10^{-8}	5.0	—	5.0×10^{-8}	5.0	—	4.7×10^{-8}	5.0
Warning signal	5.0×10^{-8}	1.5×10^{-7}	15.0	4.7×10^{-8}	1.4×10^{-7}	15.0	—	1.5×10^{-7}	15.0	—	1.4×10^{-7}	15.0
Master caution signal	5.0×10^{-8}	1.5×10^{-7}	15.0	4.7×10^{-8}	1.4×10^{-7}	15.0	—	1.5×10^{-7}	15.0	—	1.4×10^{-7}	15.0
Emergency exit lighting	5.0×10^{-8}	1.5×10^{-7}	15.0	4.7×10^{-8}	1.4×10^{-7}	15.0	—	1.5×10^{-7}	15.0	—	1.4×10^{-7}	15.0
Electronic and electro-optical displays (monochromatic)	—	1.7×10^{-10}	0.5	—	1.6×10^{-10}	0.5	—	1.7×10^{-10}	0.5	—	1.6×10^{-10}	0.5
Electronic and electro-optical displays (multicolor) { White	—	2.3×10^{-9}	0.5	—	2.2×10^{-9}	0.5	—	2.3×10^{-9}	0.5	—	2.2×10^{-9}	0.5
{ Max	—	1.2×10^{-8}	0.5	—	1.1×10^{-8}	0.5	—	1.2×10^{-8}	0.5	—	1.1×10^{-8}	0.5
HUD systems	1.7×10^{-9}	5.1×10^{-9}	5.0	1.6×10^{-9}	4.7×10^{-9}	0.5	—	1.7×10^{-9}	5.0	—	1.6×10^{-9}	5.0

NR_A = NVIS radiance requirements for Class A equipment.
NR_B = NVIS radiance requirements for Class B equipment.
fL = footLamberts.
Note: For these lighting components, Class B equipment shall meet all Class A requirements of this specification. The relative NVIS response data for Class A equipment, $G_A(\lambda)$, shall be substituted for GB(l) to calculate NVIS radiance.

All other aircraft lighting, not just the cockpit lighting, must be made compatible with NVIS. This includes stray light from the aircraft's interior cabin, the aircraft's exterior lighting system, and any external lights such as runway or shipboard lights. If exterior lights are required during NVG operations, they usually are in one of two categories:

- *Visible and NVIS compatible or NVIS friendly* — These include electroluminescent formation lights, which are green visible strips and do not degrade the pilot's NVIS or LED navigation lights, which are red, green and white position lights and do not degrade the pilot's NVIS.
- *Invisible and NVIS usable* — These include covert IR lights that provide illumination for the pilot using NVIS or allow signaling or alerting to other pilots operating with NVIS.

The cabin and cargo compartment interior lighting must be made NVG compatible if the aft crew uses NVG or if the cabin lighting is seen from the crew station. The cabin compartment in the HH-60Q "Medevac" helicopter requires white lighting for the medical personnel to attend to patients. The cabin has blackout curtains to protect the NVG compatibility of the crew station and to block any visual signature to the outside world.

8.3.11 Types of Filters/Lighting Sources

Aircraft lighting systems use various types of illuminating sources and lamps: incandescent, electroluminescent, fluorescent, LED, LCD, and CRT. Cockpit lighting can usually be modified to meet NVIS compatibility requirements by adding blue or blue-green glass filters from companies and suppliers such as Schott, Corning, Wamco, Hoffman Engineering, and Kopp. Usually, plastic filtering has not worked with incandescent sources since IR is transmitted freely, but Korry has developed a moldable plastic composition for NVIS-compatible products. Manufacturers of filters, measurement equipment, exterior lighting (Honeywell Grimes, Oxley, Luminescent Systems Inc., Goodrich, et al.) and interior lighting (Control Products Corp., Korry, Oppenheimer, IDD, Eaton, et al.) can be found through organizations involved in aircraft lighting, such as ALI and SAE.

8.3.12 Evaluating Aircraft Lighting

A field evaluation serves as a qualitative method of evaluating the NVIS compatibility of the overall cockpit. The evaluation should be conducted on a clear, moonless night with the aircraft parked in a secluded area away from disturbing light sources. A standard tri-bar resolution target board (e.g., USAF 1951, Figure 8.15), with patterns consisting of three horizontal and three vertical bar pairs arranged in decreasing size, is mounted in front of the aircraft. The resolution pattern is illuminated by the ambient starlight environment. The pilot (or observer) wears the NVIS and views the resolution pattern while looking through the windshield. With all the aircraft and cockpit lighting extinguished, the pilot first determines the smallest resolvable line pair that is observed. Then, as each lighting zone or display is turned on, the pilot continues to report the smallest resolvable line pair. Lighting zones and displays are activated individually and then simultaneously. If the lighting and displays have no effect on the minimum resolvable pattern observed, then the cockpit is considered to be compatible with the NVG because there is no impact on goggle performance. Visually observed reflections from the lighting on the canopy or windshield can also be evaluated for NVG compatibility. Compatibility usually is demonstrated if the reflections are not apparent when viewed through the NVIS.

8.3.13 Measurement Equipment

Laboratory measurements of the aircraft lighting components quantify photometric and radiometric characteristics of the light output, such as luminance, chromaticity, and NVIS radiance. These measurements use the guidelines of MIL-L-85762A to provide quantitative data to verify NVG compatibility of the lighting components. Units of radiometric measures are consistent with terms in other electromagnetic radiant energy applications. The measurements based upon the visual eye response of the average

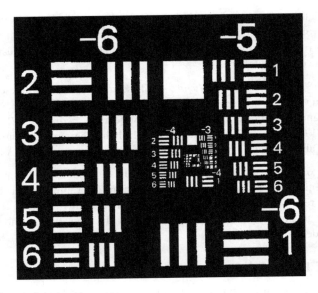

FIGURE 8.15 USAF 1951 Resolution Target.

human observer are termed photometric measurements. Luminance of lighted cockpit control panel and display presentation is frequently called "brightness." The color of the light is also necessary to define the visual characteristic of the lighted presentations, and spectroradiometric measurements determine the chromaticity to quantitatively define the color. The typical chromaticity coordinates used are from the 1976 CIE UCS system.

Radiometric Term	Unit	Photometric Term	SI Unit	English Unit
Radiant flux	Watt	Luminous flux	Lumen	Lumen
Radiant Intensity	Watt/steradian	Luminous Intensity	Candela	Candela
Radiance	Watt/steradian/m²	Luminance	Cd/m²	FootLambert
Irradiance	Watt/m²	Illuminance	Lux	Footcandle

Radiant energy for the NVIS-weighted response is measured with a radiometer with very low energy sensitivity, and the data is used to calculate the "NVIS Radiance" (as defined in MIL-L-85762) to determine the compatibility with the pilot's NVIS device. NVIS Radiant Intensity (NRI, as defined in ARP5825) can also be determined for intensity measurements for point source exterior lights. The following companies manufacture photometric, radiometric, and spectroradiometric measurement equipment that can determine visual and NVIS characteristics:

- Optronic Laboratories, Orlando, FL   www.olinet.com
- Photo Research, Chatsworth, CA   www.photoresearch.com
- Instrument Systems, Ottawa, Ontario   www.instrumentsystems.com
- Gamma Scientific, San Diego, CA   www.gamma-sci.com
- Hoffman Engineering, Stamford, CT   www.hoffmanengineering.com

8.3.14   Nighttime Illumination — Moon Phases

Flight planning requires knowledge of current weather conditions and the geography and topography along the route of the flight plan. Knowledge of the night sky illumination, including the moon's phase and position at various times, is very important in planning NVIS night flights. Astronomical data to determine times of sunrise, sunset, moonrise, moonset, and twilights, as well as data for positions of the sun and moon and moon phase and illumination can be obtained from the U.S. Naval Observatory (http://www.usno.navy.mil). The Observatory also offers a CD-Rom version of the Multi-year Interactive Computer Almanac (MICA) on their Astronomical Applications Web site (http://aa.usno.navy.mil/).

8.3.15   NVIS in Civil Aviation

In civil aviation applications, the NVIS enhances night VFR situation awareness and obstacle avoidance by allowing direct vision of the horizon, terrain, shadows, and other aircraft. While NVIS were primarily developed for military applications, NVIS are being used by civil authorities and private operators in a variety of civilian situations requiring increased night viewing and safe night flying conditions. The forestry service uses NVIS not only to increase the safety in night fire-fighting operations but also to find hot spots not readily seen by the unaided eye. Emergency Medical Services (EMS) helicopters use NVIS for navigating into remote rescue sites. Civilian and commercial use of NVIS in aircraft, land vehicles, and ships is growing.

The SAE G-10 Aerospace Behavioral Engineering Technology Committee's Vertical Flight Subcommittee assessed human factors issues associated with NVG for application to civil aviation. The SAE A-20 Aircraft Lighting Committee prepared the following Aerospace Recommended Practices (ARP) documents to allow general aviation design guidance similar to military specifications and standards that defined NVIS-compatible lighting:

- ARP4168 recommends considerations for light sources for designing NVIS-compatible lighting.[24]
- ARP4169 describes the functions and characteristics of NVIS filters used in NVIS-compatible lighting.[23]
- ARP4967 covers design considerations for NVIS-compatible panels (also known as "integrally illuminated information panels" or "lightplates"), which may use incandescent, electroluminescent (EL), or LED sources that are filtered to meet requirements specified in MIL-L-85762.[26]
- ARP4392 describes the recommended performance levels for NVIS-compatible aircraft exterior lighting equipment. Category I lights are compatible with NVIS, and category II lights allow NVIS viewing of the surroundings. The "lights" may not be in the visible spectrum.[25]
- ARP5825 describes the general requirements and test procedures for dual mode (NVIS-friendly visible and covert invisible) aircraft exterior lighting equipment. NVIS-friendly lights can be viewed with or without the use of NVIS and can meet FAA standards. The covert lights are for tactical use and can only be viewed with NVIS.

The FAA has conducted several studies and requested recommendations for civil application of NVG.[19–22] The primary emerging philosophy for the incorporation of NVG into civil aviation is that NVG do not enable flight.[29] The use of NVG will not enable any mode of flight that cannot be flown visually within the framework of the existing regulatory authority.

Because civil aviation does not have the regimented control of pilots and aircraft as in the military, there is a danger to the public if untrained operators fly in ill-equipped, unregulated, and incompatible aircraft. Therefore, minimum civil regulations and standards must be imposed. The future integration of NVG use in civil aviation will depend on the following key issues:

1. Limiting the I^2 device to Gen III
2. Modifying cockpit lighting
3. Modifying interior lighting
4. Modifying exterior lighting
5. Establishing training programs
6. Updating Federal Aviation Regulations (FARs) 61, 91, 135, et al.

Civil aviation should limit the Night Vision Device to Gen III NVIS. The military experience has demonstrated that an NVIS made for aviators is necessary. The third-generation sensor is preferred for starlight sensitivity. Gen II NVG with 625-nm minus-blue filters will work with MIL-L-85762A compatible lighting, but the filters reduce Gen II effectiveness. Without MIL-L-85762A lighting, the NVIS AGC can give a false sense of compatibility.

Cockpit lighting for civil aviation will have to be NVIS compatible. All nighttime lighting requires NVIS-compatible filtering, which normally includes control panel lightplates, numeric display read-outs, Warning/Caution/Advisory (W/C/A) legends, floodlights, flashlights, and electronic displays (CRTs, LCDs, LEDs). The MIL-L-85762A approach yields best compatibility results. An integral approach yields better lighting, although existing equipment can be modified with add-on bezels or filters. These add-ons can block viewing or reduce daylight readability. Color coding of W/C/A legends (if red warning lights are utilized) and use of multicolor electronic displays (e.g., weather radar) must be limited to the use of Class B NVIS with a 665-nm minus-blue filter.

Cabin and interior lighting for civil aviation also will need to be NVIS compatible. If the cabin and cargo compartment interior lighting is not NVIS compatible, the compartment and lighting must be shielded from the cockpit. If the compartment is not isolated from the cockpit, then the passengers and crew must not operate carry-on lighting sources that are not NVIS compatible. The carry-on equipment may include radios, television, computers, recorders, CD players, cellular phones, and flashlights. Also, smoking should be prohibited because smoking produces an incompatible glow.

Exterior lighting for civil aviation will have to be NVIS compatible. At present, most NVIS-compatible exterior lighting, including the ARP4392 exterior lighting, is not compliant with the FAR for "see and be seen" navigation and anticollision lights necessary for civil aviation VFR flight. Invisible (covert) lighting will not be allowed as the only lighting for civil aviation. New LED lights are being developed for exterior lighting that will be visible and useful with NVIS. Lights meeting the requirements of ARP5825 will have the following characteristics:

- Visible (red, green, white) to other aircraft VFR pilots not using NVIS
- Visible and NVIS-friendly (not degrading) to other aircraft VFR pilots who are using NVIS
- NVIS-friendly (not degrading) to allow the pilot of the aircraft to operate with NVIS

New training systems will have to be established to support NVIS use in civil aviation. For instance, civilian pilots utilizing NVIS will need minimum ground and flight training similar to that developed within the military. The basic ground training will include the theory of device, NVIS limitations, NVIS adjustments, nighttime moon and starlight illumination, FOV, and different visual scan and head motion techniques.

The FAA will have to establish certification and standards of NVIS use in civil aviation. In order to allow NVIS utilization in civil aviation, the FAA will have to modify regulations for pilot certification and ratings (FAR 61), equipment and flight rules (FAR 91), operating limitations (FAR 135), and airworthiness standards for various aircraft types (FAR 27, 29, etc.). Authority to operate with NVIS may be documented through FAR, Special Federal Aviation Regulation (SFAR), Advisory Circular (AC), Type Certificate (TC), Supplemental Type Certificate (STC), Technical Standards Orders (TSO), Kinds of Operations List (KOL), and Proposed Master Minimum Equipment List (PMMEL).

References

1. "The Microchannel Image Intensifier," Michael Lampton, Scientific American, Vol. 245, No. 5, November 1981, pp 62–71.
2. "Development of an Aviator's Night Vision Imaging System (ANVIS)," Albert Efkeman and Donald Jenkins, presented at SPIE Int. Tech. Symp. Exhibit, July 28–August 1, 1980, San Diego, CA.
3. TC 1-204 Night Flight Techniques and Procedures, U.S. Army.
4. Image Intensifier Symposium (Proceedings), U.S. Army Engineer Research and Development Laboratories, October 1958.
5. "Aircrew Station Lighting for Compatibility with Night Vision Goggle Use," ADS-23, Aeronautical Design Standard, U.S. Army Aviation Research and Development Command, May 1981.

6. "Aircraft Lighting Requirements for Aviator's Night Vision Imaging System (ANVIS) Compatibility," Dennis L. Schmickley, Rep. No. NADC-83032-60, Naval Air Development Center, April 1983.
7. "Lighting, Aircraft, Interior, Aviator's Night Vision Imaging System (ANVIS) Compatible," MIL-L-85762, Military Specification, January 1986.
8. "Lighting, Aircraft, Interior, Night Vision Imaging System (NVIS) Compatible", MIL-L-85762A, Military Specification, August 1988.
9. "Rationale Behind the Requirements Contained in Military Specifications MIL-L-85762 and MIL-L-85762A," Ferdinand Reetz, III, Rep. No. NADC-87060-20, Naval Air Development Center, September 1987.
10. "Lighting, Aircraft, Interior, Night Vision Imaging System (NVIS) Compatible," ASC/ENFC 96-01, Interface Document, March 1996.
11. "Rationale Behind the Requirements Contained in ASC/ENFC 96-01 Lighting, Aircraft, Interior, Night Vision Imaging System (NVIS) Compatible and Military Specification MIL-L-85762," James C. Byrd, Wright Patterson AFB, April 1996.
12. "Night Lighting and Night Vision Goggle Compatibility," Alan R. Pinkus, AGARD Lecture Series No. 156, Advisory Group for Aerospace Research and Development, North Atlantic Treaty Organization, April 1988.
13. "Aviator's Night Vision Imaging System AN/AVS-6(V)1, AN/AVS-6(V)2," MIL-A-49425(CR), Military Specification, November 1989.
14. "Death in the Dark," Edward Humes, The Orange County Register, CA, December 4, 1988.
15. "Night Vision Goggles," Hearing before the Investigations Subcommittee of the Committee on Armed Services, House of Representatives, held March 21, 1989, U.S. Government Printing Office, Washington, D.C.
16. "Review of Testing Performed on AN/PVS-5 and AN/AVS-6 Aviation Night Vision Goggles," Office of the Director, Operational Test and Evaluation, June 1989.
17. "Helicopter Flights with Night Vision Goggle — Human Factors Aspects," Michael S. Brickner, NASA Technical Memorandum 101039, March 1989.
18. "Review of the use of NVG in Flight Training," rep. for the Deputy Secretary of Defense, July 1989.
19. "Rotorcraft Night Vision Goggle Evaluation," David L. Green, Rep. DOT/FAA/RD-19/11.
20. "Civil Use of Night Vision Devices — Evaluation Pilot's Guide Part I," David L. Green, Rep. FAA/RD-94/18, July 1994.
21. "Civil Use of Night Vision Devices — Evaluation Pilot's Guide Part II," David L. Green, Rep. FAA/RD-94/19, July 1994.
22. "Assessment of Night Vision Goggle Workload — Flight Test Engineer's Guide," David L. Green, Rep. FAA/RD-94/20, July 1994.
23. "Night Vision Goggle (NVG) Filters," SAE Aerospace Recommended Practice ARP4169, reaffirmed September 1993.
24. "Night Vision Goggle (NVG) Compatible Light Sources," SAE Aerospace Recommended Practice ARP4168A, January 2004.
25. "Lighting, Aircraft Exterior, Night Vision Imaging System (NVIS) Compatible," SAE Aerospace Recommended Practice ARP4392, reaffirmed June 2002.
26. "Night Vision Imaging Systems (NVIS) Integrally Illuminated Information Panels," SAE Aerospace Recommended Practice ARP4967, reaffirmed January 2004.
27. "Crew Systems Aircraft Lighting Handbook," Department of Defense Joint Service Specification Guide JSSG-2010-5, 20 October 1998.
28. "Lighting, Aircraft, Night Vision Imaging System (NVIS) Compatible," MIL-STD-3009, 2 February 2001.
29. "Design Requirements and Test Procedures for Dual Mode Exterior Lights," SAE Aerospace Recommended Practice ARP5825, July 2005.

Further Information

"IESNA Lighting Handbook," published by The Illuminating Engineering Society of North America, 120 Wall Street, NYC, NY 10005. (httpwww.iesna.org/)

US Army Night Vision & Electronics Directorate (NVESD), Ft. Belvoir, VA

Aerospace Lighting Institute, Clearwater, FL (http://www.aligodfrey.com/)

Commission Internationale de l'Eclairage (International Commission on Illumination). (http://www.ping.at/cie/)

Society of Automotive Engineers [A-20 and G-10 committees]. (http://www.sae.org/)

9

Speech Recognition and Synthesis

Douglas W. Beeks

Beeks Engineering and Design

9.1 Introduction

The application of speech recognition (SR) in aviation is rapidly evolving and perhaps moving toward more common use on future flightdecks. The concept of using SR in aviation is not new. The use of speech recognition and voice control (VC) has been researched for more than 20 years, and many of the proposed benefits have been demonstrated in varied applications. Continuing advances in computer hardware and software are making the use of voice control applications on the flightdeck more practical, flexible, and reliable. There is little argument that the easiest and most natural and ideal way for a human to interact with a computer is by direct voice input (DVI).

While speech recognition has improved over the past several years, it has not reached the level of capability and reliability of one person talking to another. Using SR and DVI in a flightdeck atmosphere likely brings to mind thoughts of the computer on board the starship Enterprise from the science fiction classic *Star Trek*, or possibly of the HAL9000 computer from the movie *2001: A Space Odyssey*. The expectation of a voice control system like the computer on the Enterprise and the HAL9000 computer is that it be highly reliable, work in adverse and stressful conditions, be transparent to the user, and understand its users accurately without having to tailor their individual speech and vocabulary to suit the system. Current speech recognition and voice control systems are not able to achieve this level of performance expectations, although the ability and flexibility of speech recognition and its application to voice control has increased over the past few years. Whether or not a speech recognition system will ever be able to function to the level of one person speaking to another remains to be seen.

The current accuracy rate of speech recognition is in the lower to mid 90% range. Some speaker-dependent systems, and generally those with small vocabularies, have shown accuracy rates into the upper 90% range. While at first glance that might sound good, consider that with a 90% accuracy rate, 1 in 10 words will be incorrectly recognized. Also consider that this 90% and greater accuracy may be under ideal conditions; many times this high accuracy rate is achieved in a controlled and sterile lab environment. Under actual operating conditions, including cockpit noise, random noises, bumps and thumps, multiple people talking at once, etc., the accuracy rate of speech recognition systems can erode significantly.

Currently, a few military applications are planning on using SR to provide additional methods to support the Man-Machine Interface (MMI) to reduce the workload on the pilot in advanced aircraft. The Euro-fighter Typhoon has SR capabilities. Numerous aviation companies worldwide are conducting research and studies into how the available SR technology can be incorporated into current equipment designs and designs of the future for both the civilian and military marketplace.

9.2 How Speech Recognition Works: A Simplistic View

Speech recognition is based on statistical pattern matching. One of the more common methods of speech recognition based on pattern matching uses Hidden Markov Modeling (HMM) comprising two types of pattern models, the acoustical model and the language model. Which of the two models will be used, and in some cases both will be required, depends on the complexity of the application. Complex speech recognition applications, such as those supporting continuous or connected speech recognition, will use a combination of the acoustical and language models.

In a simple application using only the acoustical model, the application will process the uttered word into phonemes, which are the fundamental parts of speech. These phonemes are converted to a digital format. This digital format, or pattern, is then matched against stored patterns by the speech processor in search of a match from a stored database of word patterns. From the match, the phoneme and word can be identified.

In a more complex method, the speech processor will convert the utterance to a digital signal by sampling the voice input at some rate, commonly 16 kHz. The required acoustical signal processing can be accomplished using several techniques. Some commonly used techniques are Linear Predictive Coding (LPC) cochlea modeling, Mel Frequency Cepstral Coefficients (MFCC), and others. For this example, the sampled data is converted to the frequency domain using a fast-Fourier transformation. The transformation will analyze the stored data at $1/30^{th}$ to $1/100^{th}$ of a second (3.3 ms to 100 ms) intervals, and convert the value into the frequency domain. The resulting graph from the converted digital input will be compared against a database of known sounds. From these comparisons, a value known as a feature number will be determined.

The feature numbers will be used to reference a phoneme found using that feature number. This, ideally, would be all that is required to identify a particular phoneme, however, this will not work for a number of reasons. Background noises, the user not pronouncing a word the same way every time, and the sound of a phoneme will vary, depending on the surrounding phonemes that may add variance to the sound being processed. To overcome problems of variability of the different phonemes, the phonemes are assigned to more than one feature number. Since the speech input was analyzed at an interval of $1/30^{th}$ to $1/100^{th}$ of a second and a phoneme or sound may last from 500 ms to 2 s, many feature numbers may be assigned to a particular sound. By using statistical analysis of these feature numbers and the probability that any one sound may contain those feature numbers, the probability of that sound being a particular phoneme can be determined.

To be able to recognize words and complete utterances, the speech recognizer must also be able to determine the beginning and end of a phoneme. The most common method to determine the beginning and endpoint is by using the Hidden Markov Models (HMM) technique. The HMM is a state transition model and will use probabilities of feature numbers to determine the likelihood of transitioning from one state to another. Each phoneme is represented by a HMM. The English language is made up of 45 to 50 phonemes. A sequence of HMM will represent a word. This would be repeated for each word in the vocabulary. While the system can now recognize phonemes, phonemes do not always sound the same, depending on the phoneme preceding and following it. To address this problem, phonemes are placed in groups of three, called tri-phones, and as an aid in searching, similar sounding tri-phones are grouped together.

From the information obtained from the HMM state transitions, the recognizer is able to hypothesize and determine which phoneme likely was spoken, and then by referring this to a lexicon, the recognizer is able to determine the word that likely was spoken.

This is an overly simplified definition of the speech recognition process. There are numerous adaptations of the HMM technique and other modeling techniques. Some of these techniques are neural networks (NNs), dynamic time warping (DTW), and combinations of techniques.

9.2.1 Types of Speech Recognizers

There are two types of speech recognizers, speaker-dependent and speaker-independent.

9.2.1.1 Speaker-Dependent Systems

Speaker-dependent recognition is exactly that, speaker dependent. The system is designed to be used by one person. To operate accurately, the system will need to be "trained" to the user's individual speech patterns. This is sometimes referred to as "enrollment" of the speaker with the system. The speech patterns for the user will be recorded and patterned from which a template will be created for use by the speech recognizer. Because of the required training and storage of specific speech templates, the performance and accuracy of the speaker-dependent speech recognition engine will be tied to the voice patterns of a specific registered user. Speaker-dependent recognition, while being the most restrictive, is the most accurate, with accuracy rates in the mid to upper 90% range. For this reason, past research and applications for cockpit applications have opted to use speaker-dependent recognition.

The major drawback of this system is that it is dedicated to a single user, and that it must be trained prior to its use. Many applications will allow the speech template to be created elsewhere prior to use on the hosting system. This can be done at separate training stations prior to using the target system by transferring the created user voice template to the target system. If more than one user is anticipated, or if the training of the system is not desirable, a speaker-independent system might be an option.

9.2.1.2 Speaker-Independent Recognizers

Speaker-independent recognition systems are independent of the user. This type of system is intended to allow multiple users to access a system using voice input. Examples of speaker-independent systems are directory assistance programs and an airline reservation system with a voice input driven menu system. Major drawbacks with a speaker-independent system, in addition to increased complexity and difficult implementation, are its lower overall accuracy rate, higher system overhead, and slower response time. The impact of these drawbacks continues to lessen with increased processor speeds, faster hardware, and increased data storage capabilities.

A variation of the speaker-independent system is the speaker-adaptive system. The speaker-adaptive system will adapt to the speech pattern, vocabulary, and style of the user. Over time, as the system adapts to the users' speech characteristics, the error rate of the system will improve, exceeding that of the independent recognizer.

9.2.2 Vocabularies

A vocabulary is a list of words that are valid for the recognizer. The size of a vocabulary for a given speech recognition system affects the complexity, processing requirements, and the accuracy of that system. There are no established definitions for how large a vocabulary should be, but systems using smaller vocabularies can result in better recognizer accuracy. As a general rule, a small vocabulary may contain up to 100 words, a medium vocabulary may contain up to 1000 words, a large vocabulary may contain up to 10,000 words, and a very large vocabulary may contain up to 64,000 words, and above that the vocabulary is considered unlimited. Again, this is a general rule and may not be true in all cases.

The size of a vocabulary will be dependent upon the purpose and intended function of the application. A very specific application may require only a few words and make use of a small vocabulary, while an application that would allow dictation or setting up airline reservations would require a very large vocabulary.

How can the size and contents of a vocabulary be determined? The words used by pilots are generally specific enough to require a small to medium vocabulary. Words that can or should be in the vocabulary could be determined in a number of ways. Drawing from the knowledge of how pilots would engage a desired function or task is one way. This could be done using a questionnaire or some similar survey method.

Another way to gather words for the vocabulary is to set up a lab situation and use the "Wizard of Oz" technique. This technique would have a test evaluator behind the scenes acting upon the commands given by a test subject. The test subject would have various tasks and scenarios to complete. While the test subject runs through the tasks, the words and phrases used by the subject are collected for evaluation. After running this process numerous times, the recorded spoken words and phrases will be used to construct a vocabulary list and command syntax, commonly referred to as a grammar. The vocabulary could be refined in further tests by only allowing those contained words and phrases to be valid, and have test subjects again run through a suite of tasks. Observations would be made as to how well the test subjects were able to complete the tasks using the defined vocabulary and syntax. Based on these tests, and the evaluation results, the vocabulary is modified as required.

A paper version of the evaluation process could be administered by giving the pilot a list of tasks, and then asking them to write out what commands they would use to perform the task. Following this data collection step, a second test could be generated having the pilot choose from a selected list of words and commands what he would likely say to complete the task. As a rule, pilots will tend to operate in a predictable manner, and this lends itself to a reduced vocabulary size and structured grammar.

9.2.3 Modes of Operation for Speech Recognizers

There are two modes of operation for a speech recognizer: continuous recognition, and discrete or isolated word recognition.

9.2.3.1 Continuous Recognition

Continuous speech recognition systems are able to operate on a continuous spoken stream of input in which the words are connected together. This type of recognition is more difficult to implement due to several inherent problems such as determining start and stop points in the stream and the rate of the spoken input.

The system must be able to determine the start and endpoint of a spoken stream of continuous speech. Words will have varied starting and ending phonemes depending on the surrounding phonemes. This is called "co-articulation." The rate of the spoken speech has a significant impact on the accuracy of the recognition system. The accuracy will degrade with rapid speech.

9.2.3.2 Discrete Word Recognition

Discrete or isolated word recognition systems operate on single words at a time. The system requires a pause between saying each word. The pause length will vary, and on some systems the pause length can be set to determined lengths. This type of recognition system is the simplest to perform because the endpoints are easier for the system to locate, and the pronunciation of a word is less likely to affect the pronunciation of other words (co-articulation effects are reduced). A user of this type of system will speak in a broken fashion. This system is the type most people think of in terms of a voice recognition system.

9.2.4 Methods of Error Reduction

There are no real standards by which error rates of various speech recognizers are measured and defined. Many systems claim accuracy rates in the high 90% range, but under actual usage with surrounding noise conditions, the real accuracy level may be much less. Many factors can impact the accuracy of SR systems.

Some of these factors include the individual speech characteristics of the user, the operating environment, and the design of the SR system itself.

There are four general error types impacting the performance of a SR system; these are substitution errors, insertion errors, rejection errors, and operator errors:

- Substitution errors occur when the SR system incorrectly identifies a word from the vocabulary. An example might be the pilot calling out "Tune COM one to one two four point seven" and the SR system incorrectly recognizes that the pilot spoke "Tune NAV one to one two four point seven." The SR system substituted NAV in place of COM. Both words may be defined and valid in the vocabulary, but the system selected the wrong word.

- Insertion errors may occur when some source of sound other than a spoken word is interpreted by the system as valid speech. Random cockpit noise might at some time be identified as a valid word to the SR system. The use of noise-canceling microphones and PTT can help to reduce this type of error.

- Rejection errors occur when the SR system fails to respond to the user's speech, even if the word or phrase was valid.

- Operator errors occur when the user is attempting to use words or phrases that are not identifiable to the SR system. A simple example might be calling out "change the radio frequency to one one eight point six" instead of "Tune COM one to one one point eight six," which is recognized by the vocabulary.

When designing a speech recognition application, several design goals and objectives should be kept in mind:

- **Limitations of the hardware and the software** — Keep in mind the limitations of the hardware and the software being used for the application. Will the system need to have continuous recognition and discrete word recognition? Will the system need to be speaker independent, or will the reduced accuracy in using a speaker-independent recognizer be acceptable. Will the system be able to handle the required processing in an acceptable period of time? Will the system operate acceptably in the target environment?

- **Safety** — Will using SR to interface with a piece of equipment compromise safety? Will an error in recognition have a serious impact on the safety of flight? If the SR system should fail, is there an alternate method of control for that application?

- **Train the system in the environment in which it is intended to be used** — As discussed earlier, a SR system that has a 99% accuracy in the lab, may be frustrating and unusable in actual cockpit conditions. The speech templates or the training of the SR system needs to be done in the actual environment, or in as similar an environment as possible.

- **Don't try to use SR for tasks that don't really fit** — The problem with a new tool, like a new hammer, is that everything becomes a nail to try out that new hammer. Some tasks are natural candidates for using SR, many are not. Do not force SR onto a task if it is not appropriate for use of SR. Doing so will add significant risk and liability. Good target applications for SR include radio tuning functions, navigation functions, FMS functions, and display mode changes. Bad target applications for SR would be things that can affect the safety of flight, in short, anything that will kill you.

- **Incorporate error correction mechanisms** — Have the system repeat, using either voice synthesis or through a visual display, what it interprets, and allow the pilot to accept or reject this recognition. Allow the system to be able to recognize invalid recognition. If the recognizer interprets that it heard the pilot call out an invalid frequency, it should recognize it as invalid and possibly query the pilot to repeat, or prompt the pilot by saying or displaying that the frequency is invalid.

- **Provide feedback of the SR system's activities** — Allow the user to interact with the SR system. Have the system speak, using voice synthesis, or display what it is doing. This will allow the user to either accept or reject the recognizer interpretation. This may also serve as a way to prompt a user for more data that may have been left out of the utterance. "Tune COM 1 to...." After a delay, the system might query the user for a frequency: "Please select frequency for COM1." If the user selects some repeated command, the system may repeat back the command as it is executed: "Tuning COM 1 to ...."

9.2.4.1 Reduced Vocabulary

One way to dramatically increase the accuracy of a SR system is to reduce the number of words in a vocabulary. In addition to the reduction in words, the words should be carefully chosen to weed out words that sound similar.

Use a trigger phrase to gain the attention of the recognizer. The trigger phrase might be as simple as "computer... " followed by some command. In this example, "computer" is the trigger phrase and alerts the recognizer that a command is likely to follow. This can be used with a system that is always on-line and listening.

Speech recognition errors can be reduced using a noise-canceling microphone. The flightdeck is not the quiet, sterile place a lab or a desktop might be. There are any number of noises and chatter that could interfere with the operation of speech recognition. Like humans, a recognizer can have increased difficulty in understanding commands in a noisy environment. In addition to the use of noise-canceling microphones, the use of high-quality omnidirectional microphones will offer further reduction in recognition errors. Using push-to-talk (PTT) microphones will help to reduce the occurrence of insertion errors as well as recognition errors.

9.2.4.2 Grammar

Grammar definition plays an important role in how accurate a SR application may be. It is used to not only define which words are valid to the system, but what the command syntax will be. A grammar notation frequently used in speech recognition is Context Free Grammar (CFG). A sample of a valid command in CFG is

$$\langle \text{start} \rangle \ = \ \text{tune(COM} | \text{NAV) radio}$$

This definition would allow valid commands of "tune COM radio," and "tune NAV radio." Word order is required, and words cannot be omitted. However, the grammar can be defined to allow for word order and omitted words.

9.3   Recent Applications

Though speech recognition has been applied to various flightdeck applications over the past 20 years, limitations in both hardware and software capability have kept the use of speech recognition from serious contention as a flightdeck tool. Even though there have been several notable applications of speech recognition in the recent past, and there are several current applications of speech recognition in the cockpit of military aircraft, it will likely be several more years before the civilian market will see such applications reach the level of reliability and pilot acceptance to see them commonly available.

In the mid 1990s, NASA performed experiments using speech recognition and voice control on an OV-10A aircraft. The experiment involved 12 pilots. The speech recognizer used for this study was an ITT VRS-1290 speaker-dependent system. The vocabulary used in this study was small, containing 54 words. The SR system was tested under three separate conditions: on the ground, 1g conditions, and 3g conditions. There was no significant difference in SR system performance found between the three conditions. The accuracy rates for the SR system under these three test conditions was 97.27% in hangar conditions, 97.72 under 1g conditions, and 97.11% under 3g conditions.[1]

A recent installation that is now in production is a military fighter, the Eurofighter Typhoon. This aircraft will be the first production aircraft with voice interaction as a standard OEM configuration with speech recognition modules (SRMs). The speech recognizer is speaker dependent, and sophisticated enough to recognize continuous speech. The supplier of the voice recognition system for this aircraft is Smiths Industries. In addition, the system has received general pilot acceptance. Since the system is speaker dependent, the pilot must train the speech recognizer to his unique voice patterns prior to its use. This is done at ground-based, personal computer (PC) support stations. The PC is used to create a voice template for a specific pilot. The created voice template is then transferred to the aircraft prior to flight, via a data loader. Specifications for the recognizer include a 250-word vocabulary, a 200-ms response time, continuous speech recognition, and an accuracy rate of 95–98%.[2]

9.4   Flightdeck Applications

The use of speech recognition, the enabling technology for voice control, should not be relied on as the sole means of control or entering data and commands. Speech recognition is more correctly defined as an assisted method of control and should have reversionary controls in place if the operation and performance of the SR system is no longer acceptable. It is not a question of whether voice control will find its way into mainstream aviation cockpits, but a question of when and to what degree. As the technology of SR continues to evolve, care must be exercised so that SR does not become a solution looking for a problem to solve. Not all situations will be good choices for the application of SR. In a high workload atmosphere, such as the flightdeck, the use of SR could be a logical choice for use in many operations, leading to a reduction in workload and head-down time.

Current speech recognition systems are best assigned to tasks that are not in themselves critical to the safety of flight. In time, this will change as the technology evolves. The thought of allowing the speech recognition system to gain the ability to directly impact flight safety brings to mind an example that occurred at a speech recognition conference several years ago. While a speech recognition interface on a PC was being discussed and demonstrated before an audience, a member of the audience spoke out "format C: return," or something to that effect. The result was the main drive on the computer was formatted, erasing its contents. Normally an event such as this impacts no one's safety, however, if such unrestricted control were allowed on an aircraft, there would be serious results.

Some likely applications for voice control on the flightdeck are navigation functions, communications functions such as frequency selection, toggling of display modes, checklist functions, etc.

9.4.1   Navigation Functions

For navigation functions, SR could be used as a method of entering waypoints and inputting FMS data. Generally, most tasks requiring the keyboard to be used to enter data into the FMS would make good use of a SR system. This would allow time and labor savings in what is a repetitive and time consuming task. Another advantage of using SR is that the system is able to reduce confusion and guide the user by requesting required data. The use of SR with FMS systems is being evaluated and studied by both military and civilian aviation.

9.4.2   Communication Functions

For communication functions, voice control could be used to tune radio frequencies by calling out that frequency. For example, "Tune COM1 to one one eight point seven." The SR system would interpret this utterance, and would place the frequency into stand-by. The system may be designed to have the SR system repeat the recognized frequency back through a voice synthesizer to the pilot for confirmation prior to the frequency being placed into standby. The pilot would then accept the frequency and make it active or reject it. This would be done with a button press to activate the frequency. Another possible method of making a frequency active would be to do this by voice alone. This does bring about some added risk, as the pilot will no longer be physically making the selection. This could be done by a simple, "COM one Accept" to accept the frequency, but leave it in pre-select. Reject the frequency by saying, "COM one Reject," and to activate the frequency by saying, "COM one activate."

The use of SR would also allow a pilot to query systems, such as by requesting a current frequency setting; "What is COM one?" The ASR system could then respond with the current active frequency and make possible pre-select. This response could be by voice or by display. Other possible options would be to have the SR respond to ATC commands by moving the command frequency change to the pre-select automatically. Having done this, the pilot would only have to command "Accept," "Activate," or "Reject." The radio would never on its own, place a frequency from standby to active mode.

With the use of a GPS position-referenced database, a pilot might only have to call out "Tune COM one Phoenix Sky Harbor Approach." By referencing the current aircraft location to a database, the SR systems could look up the appropriate frequency and place it into pre-select. The system might respond back with, "COM one Phoenix Sky Harbor Approach at one two oh point seven." The pilot would then be able to accept and activate the frequency without having to know the correct frequency numbers or having to dial the frequency into the radio. Clearly a time-saving operation. Possible drawbacks are out-of-date radio frequencies in the database or no frequency listing. This can be overcome by being able to call out specific frequencies if required. "Tune COM one to one two oh point seven."

9.4.3   Checklist

The use of speech recognition is almost a natural for checklist operations. The pilot may be able to command the system with "configure for take-off." This could lead to the system bringing up an appropriate checklist for take-off configuration. The speech system could call out the checklist items as they occur and the pilot, having completed and verified the task, could press a button to accept and move on to the next task. It may be possible to allow a pilot to verbally check-off a task vs. a button selection; however, that does bring about an opportunity for a recognition error.

Defining Terms

Accuracy: Generally, accuracy refers to the percentage of times that a speech recognizer will correctly recognize a word. This accuracy value is determined by dividing the number of times that the recognizer correctly identifies a word by the number of words input into the SR system.

Continuous speech recognition: The ability of the speech recognition system to accept a continuous, unbroken stream of words and recognize it as a valid phrase.

Discrete word recognition: This refers to the ability of a speech recognizer to recognize a discrete word. The words must be separated by a gap or pause between the previous word and successive words. The pause will typically be 150 ms or longer. The use of such a system is characterized by "choppy" speech to ensure the required break between words.

Grammar: This is a set of syntax rules determining valid commands and vocabulary for the SR system. The grammar will define how words may be ordered and what commands are valid. The grammar definition structure most commonly used is known as "context free grammar" or CFG.

Isolated word recognition: The ability of the SR system to recognize a specific word in a stream of words. Isolated word recognition can be used as a "trigger" to place the SR system into an active standby mode, ready to accept input.

Phonemes: Phonemes are the fundamental parts of speech. The English language is made up from 45 to 50 individual phonemes.

Speaker Dependent: This type of system is dependent upon the speaker for operation. The system will be trained to recognize one person's speech patterns and acoustical properties. This type of system will have a higher accuracy rate than a speaker-independent system, but is limited to one user.

Speaker Independent: A speaker-independent system will operate regardless of the speaker. This type of system is the most desirable for a general use application, however the accuracy rate and response rate will be lower than the speaker-dependent system.

Speech Synthesis: The use of an artificial means to create speech-like sounds.

Text to Speech: A mechanism or process in which text is transformed into digital audio form and output as "spoken" text. Speech synthesis can be used to allow a system to respond to a user verbally.

Tri-Phones: These are groupings of three phonemes. The sound a phoneme makes can vary depending on the phoneme ahead of it and after it. Speech recognizers use tri-phones to better determine which phoneme has been spoken based upon the sounds preceding and following it.

Verbal Artifacts: These are words or phrases, spoken with the intended command that have no value content to the command. This is sometimes referred to simply as garbage when defining a specific grammar. Grammars may be written to allow for this by disregarding and ignoring these utterances, for example, the pilot utterance, "uhhhhhmmmmmmmm, select north up mode." The "uhhhh-hmmmmmmmm" would be ignored as garbage.

Vocabulary: The vocabulary a speech recognition system is made up of the words or phrases that the system is to recognize. Vocabulary size is generally broken into four sizes; small, with tens of words, medium with a few hundred words, large with a few thousand words, very large with up to 64,000 words, and unlimited. When a vocabulary is defined, it will contain words that are relative, and specific to the application.

References

1. Williamson, David T., Barry, Timothy P., and Liggett, Kristen K., Flight test results of ITT VRS-1290 in NASA OV10A. Pilot-Vehicle Interface Branch (WL/FIGP), WPAFB, OH.
2. The Eurofighter Typhoon Speech Recognition Module, [On-Line]. Available: www.smithsind-aerospace.com/PRODS/CIS/Voice.htm

Bibliography

Anderson, Timothy R., Applications of speech-based control, in *Proc. Alternative Control Technologies: Human Factors Issues,* 14-15 Oct., 1998, Wright-Patterson AFB, OH, (ISBN 92-837-1003-7).

Anderson, Timothy R., The technology of speech-based control, in *Proc. Alternative Control Technologies: Human Factors Issues,* 14-15 Oct., 1998, Wright-Patterson AFB, OH, (ISBN 92-837-1003-7).

Bekker, M. M., "A comparison of mouse and speech input control of a text-annotation system," Faculty of Industrial Design Engineering, Delft University of Technology, Jaffalaan 9, 2628 BX Delft, The Netherlands.

Eurofighter Typhoon Speech Recognition Module, Available: www.smithsind-aerospace.com/PRODS/CIS/Voice.htm.

Hart, Sandra G., Helicopter human factors, in *Human Factors in Aviation,* Wiener, Earl L. and Nagel, David C., Eds., Academic Press, San Diego, 1988, chap. 18.

Hopkin, V. David, Air traffic control, in *Human Factors in Aviation,* Wiener, Earl L. and Nagel, David C., Eds., Academic Press, San Diego, 1988, chap. 19.

Jones, Dylan M., Frankish, Clive R., and Hapeshi, K., Automatic Speech Recognition in Practice, Behav. Inf. Technol., 2, 109–22, 1992.

Leger, Alain, Synthesis and expected benefits analysis, in *Proc. Alternative Control Technologies: Human Factors Issues,* 14-15 Oct., 1998, Wright-Patterson AFB, OH, (ISBN 92-837-1003-7).

Rood, G. M., Operational rationale and related issues for alternative control technologies, in *Proc. Alternative Control Technologies: Human Factors Issues,* 14-15 Oct., 1998, Wright-Patterson AFB, OH, (ISBN 92-837-1003-7).

Rudnicky, Alexander I. and Hauptmann, Alexander G., Models for evaluating interaction protocols in speech recognition, School of Computer Science, Carnegie Mellon University, Pittsburgh, PA, 1991.

Wickens, Christopher D. and Flach, John M., Information processing, in *Human Factors in Aviation,* Wiener, Earl L. and Nagel, David C., Eds., Academic Press, San Diego, 1988, chap. 5.

Williamson, David T., Barry, Timothy P., and Liggett, Kristen K., Flight test results of ITT VRS-1290 in NASA OV10A. Pilot-Vehicle Interface Branch (WL/FIGP), WPAFB, OH.

Williges, Robert C., Williges, Beverly H., and Fainter, Robert G., Software interfaces for aviation systems, in *Human Factors in Aviation,* Wiener, Earl L. and Nagel, David C., Eds., Academic Press, San Diego, 1988, chap. 14.

Further Information

There are numerous sources for additional information on speech recognition. A search of the Internet on "speech recognition" will yield many links and information sources. The list will likely contain companies and corporations that deal primarily in speech recognition products. Some of these companies include:

• Analog Devices	(800) 262-5643	www.analog.com
• AT&T Adv Speech Products Group	(800) 592-8766	www.att.com/aspg
• Brooktrout Technology	(617) 449-4100	www.techspk.com
• Dialogic	(201) 993-3000	www.dialogic.com
• Dragon Systems	(800) 825-5897	www.dragonsys.com
• Entropic Cambridge Research Labs	(202) 547-1420	www.entropic.com
• IBM Speech Products	(800) 825-5263	www.software.ibm.com/is/voicetype
• Kurzweil Applied Intelligence	(617) 883-5151	www.kurzweil.com
• Lernout & Hauspie	(617) 238-0960	www.lhs.com
• Nuance Communications	(415) 462-8200	www.nuance.com
• Oki Semiconductor	(408) 720-1900	www.oki.com
• Philips Speech Processing	(516) 921-9310	www.speech.be.philips.com
• PureSpeech	(617) 441-0000	www.speech.com
• Sensory	(408) 744-1299	www.SensoryInc.com
• Smith Industries	(610) 296-5000	www.smithsind-aerospace.com/
• Speech Solutions	(800) 773-3247	www.speechsolutions.com
• Texas Instruments	(800) 477-8924 x 4500	www.ti.com

10

Human Factors Engineering and Flight Deck Design

Kathy H. Abbott
Federal Aviation Administration

10.1 Introduction

This chapter briefly describes human factors engineering and considerations for civil aircraft flight deck design. The motivation for providing the emphasis on the human factor is that the operation of future aviation systems will continue to rely on humans in the system for effective, efficient, and safe operation. Pilots, mechanics, air traffic service personnel, designers, dispatchers, and many others are the basis for successful operations now and for the foreseeable future. There is ample evidence that failing to adequately consider humans in the design and operations of these systems is at best inefficient and at worst unsafe.

This becomes especially important with the continuing advance of technology. Technology advances have provided a basis for past improvements in operations and safety and will continue to do so in the future. New alerting systems for terrain and traffic avoidance, data link communication systems to augment voice-based radiotelephony, and new navigation systems based on Required Navigation Performance are just a few of the new technologies being introduced into flight decks.

Often such new technology is developed and introduced to address known problems or to provide some operational benefit. While introduction of new technology may solve some problems, it often introduces others. This has been true, for example, with the introduction of advanced automation.[1,2] Thus, while new technology can be part of a solution, it is important to remember that it will bring issues that may not have been anticipated and must be considered in the larger context (equipment design, training, integration into existing flight deck systems, procedures, operations, etc.). These issues are especially important to address with respect to the human operator.

The chapter is intended to help avoid vulnerabilities in the introduction of new technology and concepts through the appropriate application of human factors engineering in the design of flight decks. The chapter first introduces the fundamentals of human factors engineering, then discusses the flight deck design process. Different aspects of the design process are presented, with an emphasis on the incorporation of human factors in flight deck design and evaluation. To conclude the chapter, some additional considerations are raised.

10.2   Fundamentals

This section provides an overview of several topics that are fundamental to the application of Human Factors Engineering (HFE) in the design of flight decks. It begins with a brief overview of human factors, then discusses the design process. Following that discussion, several topics that are important to the application of HFE are presented: the design philosophy, the interfaces and interaction between pilots and flight decks, and the evaluation of the pilot/machine system.

10.2.1   Human Factors Engineering

It is not the purpose of this section to provide a complete tutorial on human factors. The area is quite broad and the scientific and engineering knowledge about human behavior and human performance, and the application of that knowledge to equipment design (among other areas), is much more extensive than could possibly be cited here.[3–8] Nonetheless, a brief discussion of certain aspects of human factors is desirable to provide the context for this chapter.

For the purposes of this chapter, human factors and its engineering aspects involve the application of knowledge about human capabilities and limitations to the design of technological systems.[9] Human factors engineering also applies to training, personnel selection, procedures, and other topics, but those topics will not be expanded here.

Human capabilities and limitations can be categorized in many ways, with one example being the SHEL model.[6] This conceptual model describes the components *Software, Hardware, Environment, and Liveware*. The SHEL model, as described in Reference 6, is summarized below.

The center of the model is the human, or *Liveware*. This is the hub of human factors. It is the most valuable and most flexible component of the system. However, the human is subject to many limitations, which are now predictable in general terms. The "edges" of this component are not simple or straight, and it may be said that the other components must be carefully matched to them to avoid stress in the system and suboptimal performance. To achieve this matching, it is important to understand the characteristics of this component:

- **Physical size and shape** — In the design of most equipment, body measurements and movement are important to consider at an early stage. There are significant differences among individuals, and the population to be considered must be defined. Data to make design decisions in this area can be found in anthropometry and biomechanics.

- **Fuel requirements** — The human needs fuel (e.g., food, water, and oxygen) to function properly. Deficiencies can affect performance and well-being. This type of data is available from physiology and biology.

- **Input characteristics** — The human has a variety of means for gathering input about the world around him or her. Light, sound, smell, taste, heat, movement, and touch are different forms of information perceived by the human operator; for effective communication between a system and the human operator this information must be understood to be adequately considered in design. This knowledge is available from biology and physiology.

- **Information processing** — Understanding how the human operator processes the information received is another key aspect of successful design. Poor human-machine interface or system design that does not adequately consider the capabilities and limitations of the human information processing system can strongly affect the effectiveness of the system. Short- and long-term memory limitations are factors, as are the cognitive processing and decision-making processes used. Many human errors can be traced to this area. Psychology, especially cognitive psychology, is a major source of data for this area.

- **Output characteristics** — Once information is sensed and processed, messages are sent to the muscles and a feedback system helps to control their actions. Information about the kinds of forces that can be applied and the acceptable direction of controls are important in design decisions. As another example, speech characteristics are important in the design of voice communication systems. Biomechanics and physiology provide this type of information.
- **Environmental tolerances** — People, like equipment, are designed to function effectively only within a narrow range of environmental conditions such as temperature, pressure, noise, humidity, time of day, light, and darkness. Variations in these conditions can all be reflected in performance. A boring or stressful working environment can also affect performance. Physiology, biology, and psychology all provide relevant information on these environmental effects.

It must be remembered that humans can vary significantly in these characteristics. Once the effects of these differences are identified, some of them can be controlled in practice through selection, training, and standardized procedures. Others may be beyond practical control and the overall system must be designed to accommodate them safely. This *Liveware* is the hub of the conceptual model. For successful and effective design, the remaining components must be adapted and matched to this central component.

The first of the components that requires matching to the characteristics of the human is *Hardware*. This interface is the one most generally thought of when considering human-machine systems. An example is designing seats to fit the sitting characteristics of the human. More complex is the design of displays to match the human's information processing characteristics. Controls, too, must be designed to match the human's characteristics, or problems can arise from, for example, inappropriate movement or poor location. The user is often unaware of mismatches in this liveware-hardware interface. The natural human characteristic of adapting to such mismatches masks but does not remove their existence. Thus this mismatch represents a potential hazard to which designers should be alerted.

The second interface with which human factors engineering is concerned is that between *Liveware* and *Software*. This encompasses the nonphysical aspects of the systems such as procedures, manual and checklist layout, symbology, and computer programs. The problems are often less tangible than in the *Liveware-Hardware* interface and more difficult to resolve.

One of the earliest interfaces recognized in flying was between the human and the environment. Pilots were fitted with helmets against the noise, goggles against the airstream, and oxygen masks against the altitude. As aviation matured, the environment became more adapted to the human (e.g., through pressurized aircraft). Other aspects that have become more of an issue are disturbed biological rhythms and related sleep disturbances because of the increased economic need to keep aircraft, and the humans that operate them, flying 24 hours a day. The growth in air traffic and the resulting complexities in operations are other aspects of the environment that are becoming increasingly significant now and in the future.

The last major interface described by the SHEL model is the human-human interface. Traditionally, questions of performance in flight have focused on individual performance. Increasingly, attention is being paid to the performance of the team or group. Pilots fly as a crew; flight attendants work as a team; maintainers, dispatchers, and others operate as groups; therefore, group dynamics and influences are important to consider in design.

The SHEL model is a useful conceptual model, but other perspectives are important in design as well. The reader is referred to the references cited for in-depth discussion of basic human behavioral considerations, but a few other topics are especially relevant to this chapter and are discussed here: usability, workload, and situation awareness.

10.2.1.1 Usability

The usability of a system is very pertinent to its acceptability by users; therefore, it is a key element to the success of a design. Nielsen[10] defines usability as having multiple components:

- Learnability — the system should be easy to learn
- Efficiency — the system should be efficient to use
- Memorability — the system should be easy to remember

- Error — the system should be designed so that users make few errors during use of the system, and can easily recover from those they do make
- Satisfaction — the system should be pleasant to use so users are subjectively satisfied when using it.

This last component is indicated by subjective opinion and preference by the user. This is important for acceptability, but it is critical to understand that there is a difference between subjective preference and performance of the human-machine system. In some cases, the design that was preferred by the user was not the design that resulted in the best performance. This illustrates the importance of both subjective input from representative end users and objective performance evaluation.

10.2.1.2 Workload

In the context of the commercial flight deck, workload is a multidimensional concept consisting of: (1) the duties, amount of work, or number of tasks that a flight crew member must accomplish; (2) the duties of the flight crew member with respect to a particular time interval during which those duties must be accomplished; and/or (3) the subjective experience of the flight crew member while performing those duties in a particular mission context. Workload may be either physical or mental.[11]

Both overload (high workload, potentially resulting in actions being skipped or executed incorrectly or incompletely) and underload (low workload, leading to inattention and complacency) are worthy of attention when considering the effect of design on human-machine performance.

10.2.1.3 Situation Awareness

This can be viewed as the perception on the part of a flight crew member of all the relevant pieces of information in both the flight deck and the external environment, the comprehension of their effects on the current mission status, and the projection of the values of these pieces of information (and their effect on the mission) into the near future.[11]

Situation awareness has been cited as an issue in many incidents and accidents, and can be considered as important as workload. As part of the design process, the pilot's information requirements must be identified, and the information display must be designed to ensure adequate situation awareness. Although the information is available in the flight deck, it may not be in a form that is directly usable by the pilot, and therefore of little value.

Another area that is being increasingly recognized as important is the topic of organizational processes, policies and practices.[12] It has become apparent that the influence of these organizational aspects is a significant, if latent, contributor to potential vulnerabilities in design and operations.

10.2.2 Flight Deck Design

The process by which commercial flight decks are designed is complex, largely unwritten, variable, and nonstandard.[11] That said, Figure 10.1 is an attempt to describe this process in a generic manner. It represents a composite flight deck design process based on various design process materials. The figure is not intended to exactly represent the accepted design process within any particular organization or program; however, it is meant to be descriptive of generally accepted design practice. (For more detailed discussion of design processes for pilot-system integration and integration of new systems into existing flight decks, see References 13 and 14.)

The figure is purposely oversimplified. For example, the box labeled "Final Integrated Design" encompasses an enormous number of design and evaluation tasks, and can take years to accomplish. It could be expanded into a figure of its own that includes not only the conceptual and actual integration of flight deck components, but also analyses, simulations, flight tests, certification and integration based on these evaluations.

Flight deck design necessarily requires the application of several disciplines, and often requires trade-offs among those disciplines. Human factors engineering is only one of the disciplines that should be part of the process, but it is a key part of ensuring that the flight crew's capabilities and limitations are considered. Historically, this process tends to be very reliant on the knowledge and experiences of individuals involved in each program.

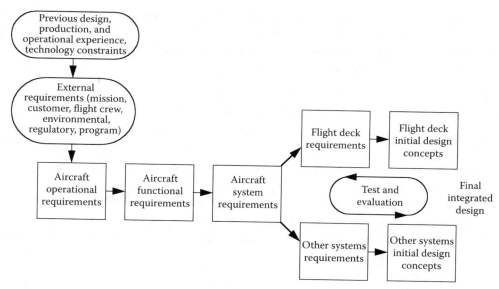

FIGURE 10.1 Simplified representation of the flight deck design process (from NASA TM 109171).

Human-centered or user-centered design has been cited as a desirable goal. That is, design should be focused on supporting the human operator of the system, much as discussed above on the importance of matching the hardware, software, and environment to the human component. A cornerstone of human-centered design is the design philosophy.

10.2.2.1 Flight Deck Design Philosophy

The design philosophy, as embodied in the top-level philosophy statements, guiding principles, and design guidelines, provides a core set of beliefs used to guide decisions concerning the interaction of the flight crew with the aircraft systems. It typically deals with issues such as allocation of functions between the flight crew and the automated systems, levels of automation, authority, responsibility, information access and formatting, and feedback, in the context of human use of complex, automated systems.[1,11]

The way pilots operate airplanes has changed as the amount of automation and its capabilities have increased. Automation has both provided alternate ways of accomplishing pilot tasks performed on previous generations of airplanes and created new tasks. The increased use of and flight crew reliance on flight deck automation makes it essential that the automation act predictably with actions that are well understood by the flight crew. The pilot has become, in some circumstances, a supervisor or manager of the automation.

Moreover, the automation must be designed to function in a manner that directly supports flight crews in performing their tasks. If these human-centered design objectives are not met, the flight crew's ability to properly control or supervise system operation is limited, leading to confusion, automation surprises, and unintended airplane responses.

Each airplane manufacturer has a different philosophy regarding the implementation and use of automation. Airbus and Boeing are probably the best-known for having different flight deck design philosophies. However, there is general agreement that the flight crew is and will remain ultimately responsible for the safety of the airplane they are operating.

Airbus has described its automation philosophy as:

- Automation must not reduce overall aircraft reliability, it should enhance aircraft and systems safety, efficiency, and economy
- Automation must not lead the aircraft out of the safe flight envelope and it should maintain the aircraft within the normal flight envelope

- Automation should allow the operator to use the safe flight envelope to its full extent, should this be necessary due to extraordinary circumstances
- Within the normal flight envelope, the automation must not work against operator inputs, except when absolutely necessary for safety

Boeing has described its philosophy as follows:

- The pilot is the final authority for the operation of the airplane
- Both crew members are ultimately responsible for the safe conduct of the flight
- Flight crew tasks, in order of priority, are safety, passenger comfort, and efficiency
- Design for crew operations based on pilot's past training and operational experience
- Design systems to be error tolerant
- The hierarchy of design alternatives is simplicity, redundancy, and automation
- Apply automation as a tool to aid, not replace, the pilot
- Address fundamental human strengths, limitations, and individual differences — for both normal and nonnormal operations
- Use new technologies and functional capabilities only when:
  - They result in clear and distinct operational or efficiency advantages, and
  - There is no adverse effect to the human-machine interface

One of the significant differences between the design philosophies of the two manufacturers is in the area of envelope protection. Airbus' philosophy has led to the implementation of what has been described as "hard" limits, where the pilot can provide whatever control inputs he or she desires, but the airplane will not exceed the flight envelope. In contrast, Boeing has "soft" limits, where the pilot will meet increasing resistance to control inputs that will take the airplane beyond the normal flight envelope, but can do so if he or she chooses. In either case, it is important for the pilot to understand what the design philosophy is for the airplane being flown.

Other manufacturers may have philosophies that differ from Boeing and Airbus. Different philosophies can be effective if each is consistently applied in design, training, and operations, and if each supports flight crew members in flying their aircraft safely. To ensure this effectiveness, it is critical that the design philosophy be documented explicitly and provided to the pilots who will be operating the aircraft, the trainers, and the procedure developers.

10.2.2.2 Pilot/Flight Deck Interfaces

The layout, controls, displays and amount of automation in flight decks have evolved tremendously in commercial aviation.[15,16] What is sometimes termed the "classic" flight deck, which includes the B-727, the DC-10, and early series B-747, is typically characterized by dedicated displays, where one piece of data is generally shown on a dedicated gage or dial as the form of display. These aircraft are relatively lacking in automation. A representative "classic" flight deck is shown in Figure 10.2. All of these aircrafts are further characterized by the relative simplicity of their autopilot, which offers one or a few simple modes in each axis. In general, a single instrument indicates the parameter of a single sensor. In a few cases, such as the horizontal situation indicator, a single instrument indicates the "raw" output of multiple sensors. Regardless, the crew is generally responsible for monitoring the various instruments and realizing when a parameter is out of range. A simple caution and warning system exists, but it covers only the most critical system failures.

The first generation of "glass cockpit" flight decks, which include the B-757/767, A-310, and MD-88, receive their nickname due to their use of cathode ray tubes (CRTs). A representative first-generation "glass cockpit" flight deck is shown in Figure 10.3. A mix of CRTs and instruments was used in this generation of flight deck, with instruments used for primary flight information such as airspeed and altitude. A key innovation in this flight deck was the "map display" and its coupling to the Flight Management System (FMS). This enabled the crew to program their flight plan into a computer and see their planned track along the ground, with associated waypoints, on the map display. Accompanying the introduction of the map

FIGURE 10.2   Representative "classic" flight deck (DC-10).

FIGURE 10.3   Representative first-generation "glass cockpit" (B-757) flight deck.

display and FMS were more complex autopilots (added modes from the FMS and other requirements). This generation of aircraft also featured the introduction of an integrated Caution and Warning System, usually displayed in a center CRT with engine information. A major feature of this Caution and Warning System was that it prioritized alerts according to a strict hierarchy of "warnings" (immediate crew action required), "cautions" (immediate crew awareness and future action required), and "advisories" (crew awareness and possible action required).[17]

The second generation of "glass cockpit" flight decks, which include the B-747-400, A-320/330/340, F-70/100, MD-11, and B-777, are characterized by the prevalence of CRTs (or LCDs in the case of the B-777) on the primary instrument panel. A representative second-generation "glass cockpit" flight deck is shown in Figure 10.4. CRT/LCDs are used for all primary flight information, which is integrated on a few displays. In this generation of flight deck, there is some integration of the FMS and autopilot — certain pilot commands can be input into either the FMS or autopilot and automatically routed to the other.

There are varying levels of aircraft systems automation in this generation of flight deck. For example, the MD-11 fuel system can suffer certain failures and take corrective action — the crew is only notified if they must take some action or if the failure affects aircraft performance. The caution and warning systems in this generation of flight decks are sometimes accompanied by synoptic displays that graphically

FIGURE 10.4 Representative second-generation "glass cockpit" (Airbus A320) flight deck.

indicate problems. Some of these flight decks feature fly-by-wire control systems — in the case of the A-320/330/340, this capability has allowed the manufacturer to tailor the control laws such that the flying qualities of these various size aircraft appear similar to pilots. The latest addition to this generation of flight deck, the B-777, has incorporated "cursor control" for certain displays, allowing the flight crew to use a touchpad to interact with "soft buttons" programmed on these displays.

Of note is the way that this flight deck design evolution affects the manner in which pilots access and manage information. Figure 10.2 illustrates the flight deck with dedicated gages and dials, with one display per piece of information. In contrast, the flight deck shown in Figure 10.4 has even more information available, and the pilot must access it in an entirely different manner. Some of the information is integrated in a form that the pilot can more readily interpret (e.g., moving map displays). Other information must be accessed through pages of menus. The point is that there has been a fundamental change in information management in the flight deck, not through intentional design but through introduction of technology, often for other purposes.

An example is shown in Figure 10.5 from the business aircraft community illustrating that the advanced technology discussed here is not restricted to large transport aircraft. In fact, new technology is quite likely to be more quickly introduced into these smaller, sophisticated aircraft.

Major changes in the flight crew interface with future flight decks are expected. While it is not known exactly what the flight decks of the future will contain or how they will function, some possible elements may include:

- Sidestick control inceptors, interconnected and with tailorable force/feel, preferably "backdriven" during autopilot engagement.
- Cursor control devices, which the military has used for many years, but the civil community is just starting to use (e.g., in the Boeing 777).
- Multifunction displays.
- Management of subsystems through displays and control-display units.
- "Mode-less" flight path management functions.
- Large, high-resolution displays having multiple signal sources (computer-generated and video).
- Graphical interfaces for managing certain flight deck systems.

FIGURE 10.5 Gulfstream GV flight deck.

- High-bandwidth, two-way datalink communication capability embedded in appropriate flight deck systems
- Replacement of paper with "electronic flight bags."
- Voice interfaces for certain flight deck systems.

These changes will continue to modify the manner in which pilots manage information within the flight deck, and the effect of such changes should be explicitly considered in the flight deck design process.

10.2.2.3 Pilot/Flight Deck Interaction

Although it is common to consider the pilot interfaces to be the only or primary consideration in human factors in flight deck design, the interaction between the pilot(s) and the flight deck must also be considered. Some of the most visible examples of the importance of this topic, and the consequences of vulnerabilities in this area, are in the implementation of advanced automation.

Advanced automation (sophisticated autopilots, autothrust, flight management systems, and associated displays and controls) has provided large improvements in safety (e.g., through reduced pilot workload in critical or long-range phases of flight) and efficiency (improved precision of flying certain flight paths). However, vulnerabilities have been identified in the interaction between the flight crews and modern systems.[2]

For example, on April 26, 1994, an Airbus A300–600 operated by China Airlines crashed at Nagoya, Japan killing 264 passengers and flight crew members. Contributing to the accident were conflicting actions taken by the flight crew and the autopilot. During complex circumstance, the flight crew attempted to stay on glide slope by commanding nose-down elevator. The autopilot was then engaged, and because it was still in go-around mode, commanded nose-up trim. A combination of an out-of-trim condition, high engine thrust, and retracting the flaps too far led to a stall. The crash provided a stark example of how a breakdown in the flight crew/automation interaction can affect flight safety. Although this particular accident involved an A300–600, other accidents, incidents, and safety indicators demonstrate that this problem is not confined to any one airplane type, airplane manufacturer, operator, or geographical region.

A lesson to be learned here is that design of the interaction between the pilot and the systems must consider human capabilities and limitations. A good human-machine interface is necessary but may not be sufficient to ensure that the system is usable and effective. The interaction between the pilot and the system, as well as the function of the system itself, must be carefully "human engineered."

10.2.3   Evaluation

Figure 10.1 showed test and evaluation (or just evaluation, for the remainder of the discussion) as an integral part of the design process. Because evaluation is (or should be) such an important part of design, some clarifying discussion is appropriate here. (See Reference 18 for a more detailed discussion of the evaluation issues that are summarized below.)

Evaluation often is divided into verification (the process of demonstrating that the system works as designed) and validation (the process of assessing the degree to which the design achieves the system objectives of interest). Thus, validation goes beyond asking whether the system was built according to the plan or specifications; it determines whether the plan or specifications were correct for achieving the system objectives.

One common use of the term "evaluation" is as a synonym of "demonstration." That is, evaluation involves turning on the system and seeing if it basically resembles what the designer intended. This does not, however, provide definitive information on safety, economy, reliability, maintainability, or other concerns that are generally the motivation for evaluation.

It is not unusual for evaluation to be confused with demonstration, but they are not the same. In addition, there are several different types and levels of evaluation that are useful to understand. For example, **formative** evaluation is performed during the design process. It tends to be informal and subjective, and its results should be viewed as hypotheses, not definitive results. It is often used to evaluate requirements. In contrast, **formal** evaluation is planned during the design but performed with a prototype to assess the performance of the human/machine system. Both types of evaluations are required, but the rest of this discussion focuses on formal evaluation.

Another distinction of interest in understanding types of evaluation is the difference between **absolute** vs. **comparative** evaluations. **Absolute** evaluation is used when assessing against a standard of some kind. An example would be evaluating whether the pilot's response time using a particular system is less than some prespecified number. **Comparative** evaluation compares one design to another, typically an old design to a new one. Evaluating whether the workload for particular tasks in a new flight deck is equal to or less than in an older model is an example comparative evaluation. This type of evaluation is often used in the airworthiness certification of a new flight deck, to show its acceptability relative to an older, already certified flight deck. It may be advantageous for developers to expand an absolute evaluation into a comparative evaluation (through options within the new system) to assess system sensitivities.

Yet another important distinction is between **objective** vs. **subjective** evaluation. **Objective** evaluation measures the degree to which the objective criteria (based on system objectives) have been met. **Subjective** evaluation focuses on users' opinions and preferences. Subjective data are important but should be used to support the objective results, not replace them.

Planning for the evaluation should proceed in parallel with design rather than after the design is substantially completed. Evaluation should lead to design modification, and this is most effectively done in an iterative fashion.

Three basic issues, or levels of evaluation, are worth considering. The first is **compatibility**. That is, the physical presentation of the system must be compatible with human input and output characteristics. The pilot has to be able to read the displays, reach the controls, etc. Otherwise, it doesn't matter how good the system design is; it will not be usable.

Compatibility is important but not sufficient. A second issue is **understandability**. That is, just because the system is compatible with human input-output capabilities and limitations does not necessarily mean that it is understandable. The structure, format, and content of the pilot-machine dialogue must result in meaningful communication. The pilot must be able to interpret the information provided, and be able to "express" to the system what he or she wishes to communicate. For example, if the pilot can read the menu, but the options available are meaningless, that design is not satisfactory.

A designer must ensure that the design is both compatible and understandable. Only then should the third level of evaluation be addressed: that of **effectiveness**. A system is effective to the extent that it supports a pilot or crew in a manner that leads to improved performance, results in a difficult task being

TABLE 10.1 Methods of Evaluation[18]

Method	Levels of Evaluation		
	Compatibility	Understandability	Effectiveness
Paper Evaluation: Static	Useful and efficient	Somewhat useful but inefficient	Not useful
Paper Evaluation: Dynamic	Useful and efficient	Somewhat useful but inefficient	Not useful[a]
Part-Task Simulator: "Canned" Scenarios	Useful but inefficient	Useful and efficient	Marginally useful but efficient[a]
Part-Task Simulator: Model Driven	Useful but inefficient	Useful and efficient	Somewhat useful and efficient
Full-Task Simulator	Useful but very inefficient	Useful but inefficient	Useful but somewhat inefficient
In-Service Evaluation	Useful but extremely inefficient	Useful but very inefficient	Useful but inefficient

[a] Can be effective for formative evaluation.

made less difficult, or enables accomplishing a task that otherwise could not have been accomplished. Assessing effectiveness depends on defining measures of performance based on the design objectives. Regardless of these measures, there is no use in attempting to evaluate effectiveness until compatibility and understandability are ensured.

Several different methods of evaluation can be used, ranging from static paper-based evaluations to in-service experience. The usefulness and efficiency of a particular method of evaluation naturally depends on what is being evaluated. Table 10.1 shows the usefulness and efficiency of several methods for each of the levels of evaluation.

As can be seen from this discussion, evaluation is an important and integral part of successful design.

10.3 Additional Considerations

10.3.1 Standardization

Generally, across manufacturers, there is a great deal of variation in existing flight deck systems design, training, and operation. Because pilots often operate different aircraft types, or similar aircraft with different equipage, at different points in time, another way to avoid or reduce errors is standardization of equipment, actions, and other areas.[19]

It is not realistic (or even desirable) to think that complete standardization of existing aircraft will occur. However, for the sake of the flight crews who fly these aircraft, appropriate standardization of new systems/technology/operational concepts should be pursued, as discussed below.

Appropriate standardization of procedures/actions, system layout, displays, color philosophy, etc. is generally desirable, because it has several potential advantages, including:

- Reducing potential for crew error/confusion due to negative transfer of learning from one aircraft to another;
- Reducing training costs, because you only need to train once; and
- Reducing equipment costs because of reduced part numbers, inventory, etc.

A clear example of standardization in design and operation is the Airbus A320/330/340 commonality of flight deck and handling qualities. This has advantages of reduced training and enabling pilots to easily fly more than one airplane type.

If standardization is so desirable, why is standardization not more prevalent? There are concerns that inappropriate standardization, rigidly applied, can be a barrier to innovation, product improvement, and product differentiation. In encouraging standardization, known issues should be recognized and addressed.

One potential pitfall of standardization that should be avoided is to standardize on the lowest common denominator. Another question is to what level of design prescription should standardization be done, and when does it take place? From a human performance perspective, consistency is a key factor. The actions and equipment may not be exactly the same, but should be consistent. An example where this has been successfully applied is in the standardization of alerting systems,[16] brought about by the use of industry-developed design guidelines. Several manufacturers have implemented those guidelines into designs that are very different in some ways, but are generally consistent from the pilot's perspective.

There are several other issues with standardization. One of them is related to the introduction of new systems into existing flight decks. The concern here is that the new system should have a consistent design/operating philosophy with the flight deck into which it is being installed. This point can be illustrated by the recent introduction of a warning system into modern flight decks. In introducing this new system, the question arose whether the display should automatically be brought up if an alert occurs (replacing the current display selected by the pilot). One manufacturer's philosophy is to bring the display up automatically when an alert occurs; another manufacturer's philosophy is to alert the pilot, then have the pilot select the display when desired. This is consistent with the philosophy of that flight deck of providing the pilot control over the management of displays. The trade-off between standardization across aircraft types (and manufacturers) and internal consistency with flight deck philosophy is very important to consider and should probably be done on a case-by-case basis.

The timing of standardization, especially with respect to introduction of new technology, is also critical.[4] It is desirable to deploy new technology early, because some problems are only found in the actual operating environment. However, if we standardize too early, then there is a risk of standardizing on a design that has not accounted for that critical early in-service experience. We may even unintentionally standardize a design that is error inducing. However, attempt to standardize too late and there may already be so many variations that no standard can be agreed upon. It is clear that standardization must be done carefully and wisely.

10.3.2 Error Management

Human error, especially flight crew error, is a recurring theme and continues to be cited as a primary factor in a majority of aviation accidents.[2,20] It is becoming increasingly recognized that this issue must be taken on in a systematic way or it may prove difficult to make advances in operations and safety improvements. However, it is also important to recognize that human error is also a normal by-product of human behavior, and most errors in aviation do not have safety consequences. Therefore, it is important for the aviation community to recognize that error cannot be completely prevented and that the focus should be on error management.

In many accidents where human error is cited, the human operator is blamed for making the error; in some countries the human operator is assigned criminal responsibility, and even some U.S. prosecutors seem willing to take similar views. While the issue of personal responsibility for the consequences of one's actions is important and relevant, it also is important to understand why the individual or crew made the error(s). In aviation, with very rare exceptions, flight crews (and other humans in the system) do not intend to make errors, especially errors with safety consequences. To improve safety through understanding of human error, it may be more useful to address errors as *symptoms* rather than *causes* of accidents. The next section discusses understanding of error and its management, then suggests some actions that might be constructive.

Human error can be distinguished into two basic categories: (a) those which presume the intention is correct, but the action is incorrect, (including *slips* and *lapses*), and (b) those in which the intention is wrong (including *mistakes* and *violations*).[21–23]

> *Slips* are where one or more incorrect actions are performed, such as in a substitution or insertion of an inappropriate action into a sequence that was otherwise good. An example would be setting the wrong altitude into the mode selector panel, even when the pilot knew the correct altitude and intended to enter it.

Lapses are the omission of one or more steps of a sequence. For example, missing one or more items in a checklist that has been interrupted by a radio call.

Mistakes are errors where the human did what he or she intended, but the planned action was incorrect. Usually mistakes are the result of an incorrect diagnosis of a problem or a failure to understand the exact nature of the current situation. The plan of action thus derived may contain very inappropriate behaviors and may also totally fail to rectify a problem. For example, a mistake would be shutting down the wrong engine as a result of an incorrect diagnosis of a set of symptoms.

Violations are the failure to follow established procedures or performance of actions that are generally forbidden. Violations are generally deliberate (and often well-meaning), though an argument can be made that some violation cases can be inadvertent. An example of a violation is continuing on with a landing even when weather minima have not been met before final approach. It should be mentioned that a "violation" error may not necessarily be in violation of a regulation or other legal requirement.

Understanding differences in the types of errors is valuable because management of different types may require different strategies. For example, training is often proposed as a strategy for preventing errors. However, errors are a normal by-product of human behavior. While training can help reduce some types of errors, they cannot be completely trained out. For that reason, errors should also be addressed by other means, and considering other factors, such as the consequences of the error or whether the effect of the error can be reversed. As an example of using design to address known potential errors, certain switches in the flight deck have guards on them to prevent inadvertent activation.

Error management can be viewed as involving the tasks of error avoidance, error detection, and error recovery.[23] Error avoidance is important, because it is certainly desirable to prevent as many errors as possible. Error detection and recovery are important, and in fact it is the safety consequences of errors that are most critical.

It seems clear that experienced pilots have developed skills for performing error management tasks. Therefore, it is possible that design, training, and procedures can directly support these tasks if we get a better understanding of those skills and tasks. However, the understanding of those skills and tasks is far from complete.

There are a number of actions that should be taken with respect to dealing with error, some of them in the design process:

Stop the blame that inhibits in-depth addressing of human error, while appropriately acknowledging the need for individual and organizational responsibility for safety consequences. The issue of blaming the pilot for errors has many consequences, and provides a disincentive to report errors.

Evaluate errors in accident and incident analyses. In many accident analyses, the reason an error is made is not addressed. This typically happens because the data are not available. However, to the extent possible with the data available, the types of errors and reasons for them should be addressed as part of the accident investigation.

Develop a better understanding of error management tasks and skills that can support better performance of those tasks. This includes:

- Preventing as many errors as possible through design, training, procedures, proficiency, and any other intervention mechanism;
- Recognizing that it is impossible to prevent all errors, although it is certainly important to prevent as many as possible; and
- Addressing the need for error management, with a goal of error tolerance in design, training, and procedures.

System design and associated flight crew interfaces can and should support the tasks of error avoidance, detection, and recovery. There are a number of ways of accomplishing this, some of which are mentioned here. One of these ways is through user-centered design processes that ensure that the design supports the human performing the desired task. An example commonly cited is the navigation display in modern flight decks, which integrates information into a display that provides information in a manner directly usable by the flight crew. This is also an example of a system that helps make certain errors more detectable, such as entering an incorrect waypoint. Another way of contributing to error resistance is designing systems that cannot be used or operated in an unintended way. An example of this is designing connectors between a cable and a computer such that the only place the cable connector fits is the correct place for it on the computer; it will not fit into any other connector on the computer.

10.3.3 Integration with Training/Qualification and Procedures

To conclude, it is important to point out that flight deck design should not occur in isolation. It is common to discuss the flight deck design separately from the flight crew qualification (training and recency of experience), considerations, and procedures. And yet, flight deck designs make many assumptions about the knowledge and skills of the pilots who are the intended operators of the vehicles. These assumptions should be explicitly identified as part of the design process, as should the assumptions about the procedures that will be used to operate the designed systems. Design should be conducted as part of an integrated, overall systems approach to ensuring safe, efficient, and effective operations.

References

1. Billings, Charles E., *Aviation Automation: The Search for a Human-Centered Approach,* Lawrence Erlbaum Associates, 1997.
2. Federal Aviation Administration, The Human Factors Team Report on: The Interfaces Between Flightcrews and Modern Flight Deck Systems, July 1996.
3. Sanders, M. S. and McCormick, E. J., *Human Factors in Engineering and Design,* 7th ed., New York: McGraw-Hill, 1993.
4. Norman, Donald A., *The Psychology of Everyday Things,* also published as *The Design of Everyday Things,* Doubleday, 1988.
5. Wickens, C. D., *Engineering Psychology and Human Performance,* 2nd ed., New York: Harper Collins College, 1991.
6. Hawkins, F., *Human Factors in Flight,* 2nd ed., Avebury Aviation, 1987.
7. Bailey, R. W., *Human Performance Engineering: A Guide for System Designers,* Englewood Cliffs, NJ: Prentice-Hall,1982.
8. Chapanis, A., *Human Factors in Systems Engineering,* New York: John Wiley & Sons, 1996.
9. Cardosi, K. and Murphy, E. Eds., Human Factors in the Design and Evaluation of Air Traffic Control Systems, DOT/FAA/RD-95/3, 1995.
10. Nielsen, Jakob, *Usability Engineering,* New York: Academic Press, 1993.
11. Palmer, M. T., Roger, W. H., Press, H. N., Latorella, K. A., and Abbott, T. S., NASA Tech. Memo., 109171, January 1995.
12. Reason, J., *Managing the Risks of Organizational Accidents,* Ashgate Publishing, 1997.
13. Society of Automotive Engineers, *Pilot-System Integration,* Aerospace Recommended Practice (ARP) 4033, 1995.
14. Society of Automotive Engineers, *Integration Procedures for the Introduction of New Systems to the Cockpit,* ARP 4927, 1995
15. Sexton, G., Cockpit: Crew Systems Design and Integration, in Wiener, E. and Nagel, D., Eds., *Human Factors in Aviation,* San Diego, CA: Academic Press, 1988.
16. Arbuckle, P. D., Abbott, K. H., Abbott, T. S., and Schutte, P. C., Future Flight Decks, 21st Congr. Int. Council Aeronautical Sci., Paper Number 98-6.9.3, September, 1998.

17. Federal Aviation Administration, Aircraft Alerting Systems Standardization Study, Volume II: Aircraft Alerting Systems Design Guidelines, FAA Rep. No. DOT/FAA/RD/81–38, II, 1981.
18. Electric Power Research Institute, Rep. NP-3701: Computer-Generated Display System Guidelines, Vol. 2: Developing an Evaluation Plan, September 1984.
19. Abbott, K., Human Error and Aviation Safety Management, *Proc. Flight Saf. Found. 52ⁿᵈ Int. Air Saf. Semin.*, November 8–11, 1999.
20. Boeing Commercial Airplane Group, *Statistical Summary of Commercial Jet Aircraft Accidents, World Wide Operations 1959–1995,* April 1996.
21. Reason, J. T., *Human Error,* New York: Cambridge University Press, 1990.
22. Hudson, P.T.W., van der Graaf, G.C., and Verschuur, W.L.G., Perceptions of Procedures by Operators and Supervisors, Paper SPE 46760, *HSE Conf. Soc. Pet. Eng.,* Caracas, 1998.
23. Hudson, P.T.W., Bending the Rules. II. Why do people break rules or fail to follow procedures? and, What can you do about it?
24. Wiener, Earl L., Intervention Strategies for the Management of Human Error, *Flight Safety Digest,* February 1995.

11

Synthetic Vision

Russell V. Parrish
MVP Technologies, Inc.

Michael S. Lewis
The Boeing Company

Daniel G. Baize
NASA Kennedy Space Center

11.1 Introduction

A large majority of the avionics systems introduced since the early days of flight (attitude indicators, radio navigation, instrument landing systems, etc.) have sought to overcome the issues resulting from limited visibility. Limited visibility is the single most critical factor affecting both the safety and capacity of worldwide aviation operations. In commercial aviation, over 30% of all fatal accidents worldwide are categorized as controlled flight into terrain (CFIT)—accidents in which a functioning aircraft impacts terrain or obstacles that the flight crew could not see. In general aviation the largest accident category is continued flight into instrument meteorological conditions, in which pilots with little experience continue to fly into deteriorating weather and visibility conditions and either collide with unexpected terrain or lose control of the vehicle because of the lack of familiar external cues. Finally, the single largest factor causing airport flight delays is the limited runway capacity and increased air traffic separation distances resulting when visibility conditions fall below visual flight rule operations. Now, synthetic vision technology will provide a visibility solution, making every flight the equivalent of a clear daylight operation.

Initial attempts to solve the visibility problem with a visibility solution have used imaging sensors to enhance the pilot's view of the outside world. Such systems, termed "enhanced vision systems," attempt to improve visual acquisition by enhancing significant components of the real-world scene. Enhanced vision systems typically use advanced sensors to penetrate weather phenomena such as darkness, fog, haze, rain, and snow, and the resulting enhanced scene is presented on a head-up display (HUD) through which the outside real world may be visible. The sensor technologies involved include either active or passive radar or infrared systems of varying frequencies.

These systems have been the subject of experiments for over three decades and the military has successfully deployed various implementations; however, few sensor-based applications have seen commercial aircraft success for a variety of reasons, including cost, complexity, and technical performance. Though technological advances are making radar and infrared sensors more affordable, they still suffer from deficiencies for commercial applications, particularly when combined with the pragmatic difficulties of obtaining operational credit for equipage. High-frequency radars (e.g., 94 GHz) and infrared sensors have degraded range performance in heavy rain and certain fog types. Low-frequency (e.g., 9.6 GHz) and mid-frequency (e.g., 35 GHz) radars have improved range but have poor resolution displays. Active radar sensors also suffer from mutual interference issues with multiple users in close proximity. All such sensors yield only monochrome displays with potentially misleading visual artifacts in certain temperature or radar reflective environments.

A "synthetic vision system" is a display system in which the view of the outside world is provided by melding computer-generated airport scenes from on-board databases and flight display symbologies with information derived from a weather-penetrating sensor (e.g., information from runway edge detection or object detection algorithms) or with actual imagery from such a sensor. These systems are characterized by their ability to represent, in an intuitive manner, the visual information and cues that a flight crew would have in daylight — Visual Meteorological Conditions (VMC). The visual information and cues are depicted based on precise positioning information relative to an onboard terrain database and possibly includes traffic information from surveillance sources (such as the Traffic Alert and Collision Avoidance System [TCAS], Airport Surface Detection Equipment [ASDE], etc.) and other hazard information (such as wind shear).

Synthetic vision displays are unlimited in range, unaffected by atmospheric conditions, and require only precise on-ship location and attitude and readily available display media, computer memory, and processing to function. The rapid emergence of reliable Global Positioning System (GPS) information and precise digital terrain maps, including data from the Space Shuttle Radar Topography Mission (SRTM), make this approach capable of both true all-weather performance as well as extremely low cost, low maintenance operations, although it too faces significant difficulties in obtaining operational credit for equipage. When fully implemented, successful synthetic vision technologies will be a revolutionary improvement in aviation safety and utility.

11.2   Background

Synthetic vision systems are intended to reduce accidents by improving a pilot's situation and spatial awareness during low-visibility conditions, including night and Instrument Meteorological Conditions (IMC). Synthetic vision technologies are most likely to help reduce the following types of accidents: CFIT, Loss of Control (LOC), and Runway Incursion (RI). CFIT is the number one cause of fatalities in revenue service flights, and the majority of CFIT accidents, runway incursion accidents, and general aviation (GA) loss-of-control accidents can be considered to be visibility-induced crew error, in which better pilot vision would have been a substantial mitigating factor. Better pilot vision is provided by synthetic and enhanced vision display systems. Such display systems will substantially reduce the following accident precursors:

- Loss of vertical and lateral spatial awareness
- Loss of terrain and traffic awareness on approach
- Unclear escape or go-around path even after recognition of problem
- Loss of attitude awareness
- Loss of situation awareness relating to the runway environment
- Unclear path guidance on the surface

Many laboratory research efforts have investigated replacing the conventional attitude direction indicator or primary flight display for transport airplanes with a pictorial display to increase situation awareness and operational capability for landing in low-visibility weather conditions. These research efforts have consistently demonstrated the advantages of pictorial displays over conventional display formats, and the technologies involved in implementing such concepts appear to be available in the near term. Over the past ten years, a number of organizations have demonstrated synthetic vision-based flight, landings, and taxi operations in research aircraft, as well as digital data links and displays of the positions and paths of airborne and ground traffic.

The practical implementation tasks remaining are to define requirements for display configurations and associated human performance criteria and to resolve human performance and technology issues relating to the development of synthetic vision concepts. These same tasks also remain for the necessary enabling technologies and the supporting infrastructure and certification strategies. The move toward implementation requires aggressive, active participation by synthetic vision advocates with appropriate standards and regulatory groups.

11.3 Applications

All aircraft categories are expected to benefit from synthetic vision applications, including GA aircraft, rotorcraft, business jets, and both cargo and passenger commercial transports. The concepts will emphasize the cost-effective use of synthetic and enhanced vision displays, worldwide navigation, terrain, obstruction, and airport databases, and GPS-derived navigation to eliminate "visibility-induced" (lack of visibility) errors for all aircraft categories.

Application of synthetic vision to commercial transports will prevent CFIT and RI accidents in high-end GA aircraft (business jets) as well by improving the pilot's situation awareness of terrain, obstacle, and airport surface operations during all phases of flight, with particular emphasis on the approach and landing phases, airport surface navigation, and missed approaches. Current accident data indicate that the majority of CFIT accidents involving transports occur during nonprecision approaches. This application will require the examination of technology issues related to implementation of an infrastructure for autonomous precision guidance systems. The standards committees (RTCA SC-193 and EUROCAE WG-44) that developed the requirements for terrain, obstacles, and airport surface databases and maintained coordination with the Federal Aviation Administration's (FAA) Local Area Augmentation System (LAAS) and Wide-Area Augmentation System (WAAS) programs, were well aware of synthetic vision applications.

In the United States runway incursions have increased substantially over the last decade. A runway incursion occurs any time a plane, vehicle, person, or object on the ground creates a collision hazard with an airplane that is taking off or landing at an airport under the supervision of air traffic controllers. The FAA established the Runway Incursion Reduction Program (RIRP) to develop surface technology at major airports to help reduce runway incursions. NASA and the FAA jointly conducted additional activities to integrate the RIRP surface infrastructure with the flight deck to enhance situation awareness of the airport surface and further reduce the possibility of runway incursions. Runway incursion reduction efforts continue to target surface surveillance, GPS-based navigation, and Cockpit Display of Traffic Information (CDTI) to improve situational awareness on the surface. Also to be considered are surface route clearance methodologies and onboard alerting strategies during surface operations (runway incursion, route deviation, and hazard detection alerting).

Advanced synthetic vision technologies are already being applied to low-end GA aircraft to prevent GA CFIT and loss-of-control accidents by improving the pilot's situation awareness during up and away flight. Current accident data indicate a leading cause for GA loss-of-control accidents is pilot disorientation after inadvertent flight into low-visibility weather conditions. Low-cost synthetic vision display systems for the low-end GA aircraft, which often operate in marginal VMC, enable safe landing or transit to VMC in the event of the unplanned, inadvertent encounter of IMC, including low-ceiling and low-visibility weather conditions. These systems also address loss of spatial situation awareness and unusual attitude issues. Successful synthetic vision concepts will also lower the workload and increase the safety of the demanding single-pilot GA Instrument Flight Rules (IFR) operations.

Synthetic vision applications in rotorcraft will be forced to supplement the database view of the outside world with a heavier dependence on imaging sensors because the rotorcraft environment has requirements that exceed the current expectations for the content of available databases. For example, Emergency Medical Services (EMS) vehicles operate to and from innumerable ball fields and hospitals at very low altitudes among power and telephone wires; hence rotorcraft applications will employ more of the features of enhanced vision, although low-cost imaging sensors for civilian applications will present significant challenges.

Block diagram: Synthetic vision concept

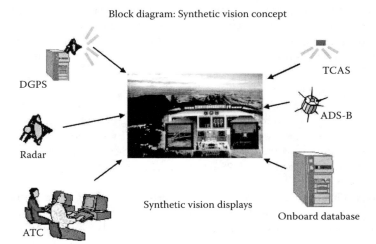

FIGURE 11.1   Possible synthetic vision system.

11.4   Concepts

Synthetic vision systems are based on precise on-ship positioning information relative to an onboard terrain database and traffic information from surveillance sources (such as TCAS, ADS-B, or TIS-B, air-to-air modes of the weather radar, ASDE, Airport Movement Area Safety System [AMASS], etc.). Enhanced vision systems are based on display presentations of onboard weather-penetrating sensor data combined with some synthetic vision elements. Figure 11.1 illustrates a top-level view of a potential high-end synthetic vision system. The specific architecture and use of specific technology is for illustration only. In this flight operations concept, the traffic surveillance sources of information are represented, as are the other enabling technologies of terrain, obstacle, and airport databases (including curvilinear approach waypoint data), data link, and Differential Global Positioning System (DGPS), LAAS, and WAAS. Surface operations sources of surveillance information could be ASDE and AMASS (via TIS-B) or ADS-B. Controller Pilot Datalink Communications (CPDLC) may also be considered.

These system concepts address information conveyance for two separate functional levels: tactical and strategic. Tactical concepts generally provide alerting information and may include escape guidance methodologies, whereas strategic concepts are more concerned with incident prevention and avoidance. These display concepts allow for presentation of three-dimensional perspective scenes with necessary and sufficient information and realism to be nearly equivalent to daylight VMC, thus increasing spatial awareness of terrain, attitude, and traffic. Symbolic information can be overlaid on these scenes to enhance situational awareness through, for example, presentation of an artificial horizon, heading, attitude indications, and pitch and/or velocity vector references. Tactical guidance capability can also be provided by highway or pathway-in-the-sky symbology for approach and landing, departure, and go-around or other Air Traffic Control (ATC) cleared routings.

11.5   Challenges

Of the technologies and issues involved in the cost-effective use of synthetic and enhanced vision displays, worldwide navigation, terrain, obstacle, and airport databases, and GPS-derived navigation to eliminate visibility-induced errors, challenges exist in the areas of human factors, avionics displays, DGPS, certification, and databases.

As with any new avionics concept, the important principles of human-centered design and other human factors considerations must be applied because human error is a principal cause of aircraft accidents of any type. A majority of CFIT and approach, landing, takeoff, and runway incursion accidents have been attributed to visibility-induced crew error. Pilot situation and spatial awareness should be dramatically improved by synthetic vision display systems, and the maturity of the human factors discipline is such that effective designs can be confidently expected. For avionics applications, the following are the most significant display technology issues that have the potential for limiting the implementation of synthetic vision:

- Limited cockpit space available for display media
- Limited display capability (graphics, size, field of view, brightness, color) of current commercial aircraft
- Retrofit cost

The tactical (Primary Flight Display [PFD]) and strategic (Navigation [Nav] Display) pictorial concepts may be presented on existing form factor A, B, or D displays in a retrofit application. However, as larger displays may be more effective in presenting these concepts, other form factors will be considered for new cockpits. Other display mechanisms, such as HUDs or head-mounted systems, are not illustrated in Figure 11.1 but have been investigated. The following are expectations for display solutions:

- Panel-mounted displays for near-term applications in modern "glass" flight decks
- HUDs for retrofit to older "steam gauge" flight decks
- Head-mounted (and head-tracked for unlimited field-of-view) or direct in-window displays for future applications
- Guidance and other flight symbology superimposed on the terrain image for all display types

Another important enabling technology for synthetic vision systems is the DGPS infrastructure. It is anticipated that unaugmented civil-code GPS will be suitable for en route operations, and that the FAA's WAAS and/or LAAS is required for approach and landing (meter/submeter accuracy). In this arena, the intent for synthetic vision system enthusiasts is to support, supplement, and complement research and modernization work currently underway by the FAA, NASA, and other governmental and private entities around the world to implement LAAS, WAAS, and other precision positioning systems.

Implementation issues such as cost-effectiveness and certification, however, provide perhaps the greatest challenges to full realization of commercially viable synthetic vision systems. The successful development of a compelling business case to serve as an economic incentive over and above the safety benefits of a synthetic vision system is a significant hurdle. The leading candidate for that business case is increased operational capability in low-visibility conditions. Tactical use of synthetic vision as a replacement for today's PFD requires the certification of a flight-critical system. While certification to that level is a lengthy and expensive process, that effort is beginning. Solid, certifiable processes for database development assurance, quality assurance, and integrity assurance, and standards for the database content sources and maintenance are also needed, and efforts are underway in this area.

Database implementation issues are equally challenging. However, the efforts of the joint RTCA/European Organisation for Civil Aviation Equipment (RTCA/EuroCAE) committee to develop the industry requirements for terrain, obstacle, and airport databases, indicated a desire by the world aeronautical community to solve the database issues. In line with this activity was the FAA's Notice of Proposed Rulemaking (NPRM) requiring all airplanes with turbine engines and six or more passenger seats to carry a Terrain Awareness Warning System (TAWS) using a computer database for providing terrain displays and warnings. To carry such technology beyond the terrain warning stage to applications of strategic planning and tactical navigation and guidance, however, compounds the database implementation issues. The most significant are cost and validation of accurate high-resolution worldwide terrain, obstacle, and airport databases ($50 million by some estimates) and liability, maintenance, and ownership of the data. It seems clear the database implementation issues will require not only involvement of the appropriate American government agencies (FAA, NASA, NOAA [National Oceanic and Atmospheric Administration], NIMA [National Imagery and Mapping Agency]) but also International Civil Aviation Organization (ICAO) and member governments' funding and sponsorship.

The capability of synthetic vision systems is limited only by the resolution and accuracy of the terrain database. Two key potential concerns with a synthetic vision approach and their mitigation strategies follows.

11.5.1   How Can You Trust the Database Is Correct?

The terrain database will be certified to necessary standards at the start. Aircraft operations (cruise flight, approach, landing, taxi) will only be allowable to the certified fidelity of the database. It will constantly improve over time; every clear daytime approach will be confirmation of the basic presentation. If necessary, processing of radar altimeter or existing weather radar signals may be used to "confirm" the actual terrain surface with the displayed database in real time. Flight guidance cues used by the flight crew will come from the same GPS positioning as will be certifiably acceptable for instrument-only (no synthetic vision) approaches.

11.5.2   What about Obstacles or Traffic That Are Not in the Database?

Airborne and ground traffic position information data-linked to the aircraft would be readily displayed. The proper flight path — always clear of buildings and towers — would be clearly displayed and obvious to follow. In addition, the database would be updated on a regular cycle, much like today's paper approach charts. Such obstacles and traffic are, of course, not able to be seen in today's nonsynthetic vision, low-visibility operations. If necessary, additional modes may be added to the onboard weather radar to detect obstacles and traffic not in the database, and imaging sensors may be added to augment the synthetic scene.

The approximate requirements for the database, "nested" in four layers of resolution, for example, are as follows:

Location from Airport Spacing	Resolution (m)	Accuracy (m)	Grid (m)
30 miles—Enroute	~150	~50	~500
30 – 5 miles — Approach/Departure	~50	~30	~200
5 – 0.5 miles — Landing/Takeoff	~10	~5	~50
0 miles — Surface Ops	~0.5	~1	NA
For comparison:			
Shuttle SRTM	20	8–16	30

The realization of such a database and its supporting infrastructure is somewhat dependent on the following:

- The shuttle radar topography mapping mission provided more than adequate terrain data for the enroute and approach/departure databases for approximately 80% of the earth's surface.
- NIMA support is considered critical for developing and releasing the worldwide enroute and approach/departure level data. International defense/security issues may limit the release of higher-resolution data.
- Local airport authorities and/or other providers (not the government) are expected to develop the landing and takeoff and surface ops databases to specified certification standards for individual airports. The safety and operational benefits to be gained by adding an airport to the evolving database are expected to be a significant motivation.

11.6   Conclusion

Synthetic vision display concepts allow for presentation of three-dimensional perspective scenes with necessary and sufficient information and realism to be equivalent to a bright, clear, sunny day, regardless of the outside weather condition, for increased spatial awareness of terrain, attitude, and traffic. Symbolic information can be overlaid on these scenes to enhance situational awareness and to provide tactical guidance capability through, for example, presentation of pathway-in-the-sky symbology. The technologies are available in the near term and the safety and operational benefits seem obvious in spite of the numerous challenges and hurdles that must be faced and overcome to prove synthetic vision applications is practical, not just as a research demonstration but as a viable, implementable capability. Solving a visibility problem with a visibility solution just plain makes sense. There is little doubt that synthetic vision-based flight will be the standard method for low-visibility operations in the future.

Defining Terms

ADS-B: Automated Dependent Surveillance — Broadcast
AMASS: Airport Movement Area Safety System
ASDE: Airport Surface Detection Equipment
ATC: Air Traffic Control
CDTI: Cockpit Display of Traffic Information
CFIT: Controlled Flight into Terrain
CPDLC: Controller Pilot Datalink Communications
DGPS: Differential Global Positioning System
EUROCAE: European Organisation for Civil Aviation Equipment
FAA: Federal Aviation Administration
GA: General Aviation
GHz: Gigahertz
GPS: Global Positioning System
HUD: Head-Up Display
IFR: Instrument Flight Rules
IMC: Instrument Meteorological Conditions
LAAS: Local Area Augmentation System
NASA: National Aeronautics and Space Administration
Nav: Navigation
NIMA: National Imagery and Mapping Agency
NOAA: National Oceanic and Atmospheric Administration
NPRM: Notice of Proposed Rulemaking
PFD: Primary Flight Display
RI: Runway Incursion
RIRP: Runway Incursion Reduction Program
SC: Special Committee
SRTM: Space Shuttle Radar Topography Mission
TAWS: Terrain Awareness Warning System
TCAS: Traffic Alert and Collision Avoidance System
TIS-B: Traffic Information Services — Broadcast
VMC: Visual Meteorological Conditions
WAAS: Wide-Area Augmentation System
WG: Working Group

Further Information

NASA's Aviation Safety Program, Synthetic Vision Project http://avsp.larc.nasa.gov

12

Terrain Awareness

Barry C. Breen
Honeywell

12.1 Enhanced Ground Proximity Warning System

The Enhanced Ground Proximity Warning System (EGPWS)* is one of the newest systems becoming standard on all military and civilian aircraft. Its purpose is to help provide situational awareness to terrain and to provide predictive alerts for flight into terrain. This system has a long history of development and its various modes of operation and warning/advisory functionality reflect that history:

- Controlled Flight Into Terrain (CFIT) is the act of flying a perfectly operating aircraft into the ground, water, or a man-made obstruction. Historically, CFIT is the most common type of fatal accident in worldwide flying operations.

- Analysis of the conditions surrounding CFIT accidents, as evidenced by flight recorder data, Air Traffic Control (ATC) records, and experiences of pilots in Controlled Flight Towards Terrain (CFTT) incidents, have identified common conditions which tend to precede this type of accident.

- Utilizing various onboard sensor determinations of the aircraft current state, and projecting that state dynamically into the near future, the EGPWS makes comparisons to the hazardous conditions known to precede a CFIT accident. If the conditions exceed the boundaries of safe operation, an aural and/or visual warning/advisory is given to alert the flight crew to take corrective action.

---

*There are other synonymous terms used by various government/industry facets to describe basically the same equipment. The military (historically at least) and at least one non-U.S. manufacturer refer to GPWS and EGPWS as Ground Collision Avoidance Systems (GCAS), although the military is starting to use the term EGPWS more frequently. The FAA, in its latest regulations concerning EGPWS functionality, have adopted the term Terrain Awareness Warning System (TAWS).

12.2   Fundamentals of Terrain Avoidance Warning

The current state of the aircraft is indicated by its position relative to the ground and surrounding terrain, attitude, motion vector, accelerations vector, configuration, current navigation data, and phase of flight. Depending upon operating modes (see next section) required or desired, and EGPWS model and complexity, the input set can be as simple as GPS position and pressure altitude or fairly large including altimeters, air data, flight management data, instrument navigation data, accelerometers, inertial references, etc. (see Figure 12.1.)

The primary input to the "classic" GPWS (nonenhanced) is the Low Range Radio (or Radar) Altimeter (LRRA), which calculates the height of the aircraft above the ground level (AGL) by measuring the time it takes a radio or radar beam directed at the ground to be reflected back to the aircraft. Imminent danger of ground collision is inferred by the relationship of other aircraft performance data relative to a safe height above the ground. With this type of system, level flight toward terrain can only be implied by detecting rising terrain under the aircraft; for flight towards steeply rising terrain, this may not allow enough time for corrective action by the flight crew.

The EGPWS augments the classic GPWS modes by including in its computer memory a model of the earth's terrain and man-made objects, including airport locations and runway details. With this digital terrain elevation and airports database, the computer can continuously compare the aircraft state vector to a virtual three-dimensional map of the real world, thus predicting an evolving hazardous situation much in advance of the LRRA-based GPWS algorithms.

The EGPWS usually features a colored or monochrome display of terrain safely below the aircraft (shades of *green* for terrain and *blue* for water is standard). When a potentially hazardous situation exists,

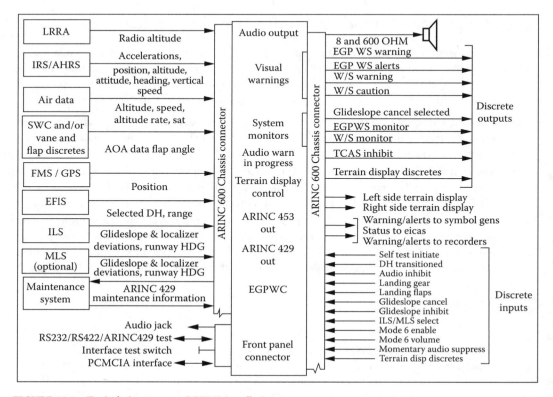

FIGURE 12.1   Typical air transport EGPWS installation.

the EGPWS alerts the flight crew with aural and/or visual warnings. Advisory (informative) situational awareness information may consist simply of an aural statement, for example, "One Thousand," as the aircraft height AGL passes from above to below 1000 ft. Cautionary alerts combine an informative aural warning, e.g., "Too Low Flaps" for flying too low and slow without yet deploying landing flaps, with a *yellow* visual alert. The cautionary visual alert can be an individual lamp with a legend such as "GPWS" or "TERRAIN;" it can be a yellow text message displayed on an Electronic Flight Instrument System (EFIS) display; or, in the case of the enhanced alert with a display of surrounding terrain, the aural "CAUTION TERRAIN" or "TERRAIN AHEAD," accompanied by both a *yellow* lamp and a rendering of the hazardous terrain on the display in *bright yellow*.

When collision with terrain is imminent and immediate drastic recovery action must be taken by the flight crew to avert disaster, the standard aural alert is a loud, commanding "PULL UP" accompanied by a *red* visual alert. Older aircraft with no terrain display utilize a single red "pull up" lamp; modern EFIS-equipped aircraft put up the words PULL UP in bright red on the Primary Flight Display (PFD). On the display of surrounding terrain, usually integrated on the EFIS Horizontal Situation Indicator, the location of hazardous terrain turns *bright red*.*

12.3 Operating Modes

The various sets of hazardous conditions that the EGPWS monitors and provides alerts for are commonly referred to as **Modes**.** These are described in detail in the following paragraphs.

Modes 1 through 4 are the original classic GPWS modes, first developed to alert the pilots to unsafe trajectory with respect to the terrain. The original analogue computer model had a single red visual lamp and a continuous siren tone as an aural alert for all modes. Aircraft manufacturer requirements caused refinement to the original modes, and added the voice "Pull Up" for Modes 1 through 4 and a new Mode 5 "Glideslope". Mode 6 was added with the first digital computer models about the time of Boeing 757/767 aircraft introduction; and Mode 7 was added when windshear detection became a requirement in about 1985.***

The latest addition to the EGPWS are the Enhanced Modes: Terrain Proximity Display, Terrain Ahead Detection, and Terrain Clearance Floor. For many years, pilot advocates of GPWS requested that Mode 2 be augmented with a display of approaching terrain. Advances in memory density, lower costs, increased computing power, and the availability of high-resolution maps and Digital Terrain Elevation Databases (DTED) enabled this advancement. Once displayable terrain elevation database became a technical and economic reality, the obvious next step was to use the data to *look ahead* of the aircraft path and predict terrain conflict well before it happened, rather than waiting for the downward-looking sensors.

Combining the DTED with a database of airport runway locations, heights, and headings allows the final improvement — warnings for normal landing attempts where there is no runway.

---

*Note that all of the EGPWS visual indication examples in this overview discussion are consistent with the requirements of FAR 25.1322.

**The EGPWS modes described here are the most common for commercial and military transport applications. Not discussed here are more specialised warning algorithms, closely related to terrain-following technology, that have been developed for military high-speed low-altitude operations. These are more related to advanced terrain database guidance, which is outside the scope of enhanced situation awareness function.

***Though not considered CFIT, analysis of windshear-related accidents has resulted in the development of reactive windshear detection algorithms. At the request of Boeing, their specific reactive windshear detection algorithm was hosted in the standard commercial GPWS, about the same time the 737-3/4/500 series aircraft was developed. By convention this became Mode 7 in the GPWS. The most common commercially available EGPWS computer contains a Mode 7 consisting of both Boeing and non-Boeing reactive windshear detection algorithms, although not all aircraft installations will use Mode 7. There also exist "standalone" reactive windshear detection computers; and some aircraft use only predictve wind shear detection, which is a function of weather radar.

12.3.1   Mode 1 — Excessive Descent Rate

The first ground proximity mode warns of an excessive descending barometric altitude rate near the ground, regardless of terrain profile. The original warning was a straight line at 4000 ft/min barometric sinkrate, enabled at 2400 ft AGL, just below the altitude at which the standard commercial radio altimeters came into track (2500 ft AGL). This has been refined over the years to a current standard for Mode 1 consisting of two curves, an outer cautionary alert and a more stringent inner warning boundary. Exceeding the limits of the outer curve results in the voice alert "Sinkrate;" exceeding the inner curve results in the voice alert "Pull Up."

Figure 12.2 illustrates the various Mode 1 curves, including the current standard air transport warnings, the DO-161A minimum warning requirement, the original curve and the Class B TSO C151 curves for 6 to 9 passenger aircraft and general aviation. Note that the Class B curves, which use GPS height above the terrain database instead of radio altitude, are not limited to the standard commercial radio altimeter range of 2500 ft AGL.

12.3.2   Mode 2 — Excessive Closure Rate

Rate-of-change of radio altitude is termed the closure rate, with the positive sense meaning that the aircraft and the ground are coming closer together. When the closure rate initially exceeds the Mode 2 warning boundary, the alert "Terrain Terrain" is given. If the warning condition persists, the voice is changed to a "Pull Up" alert.

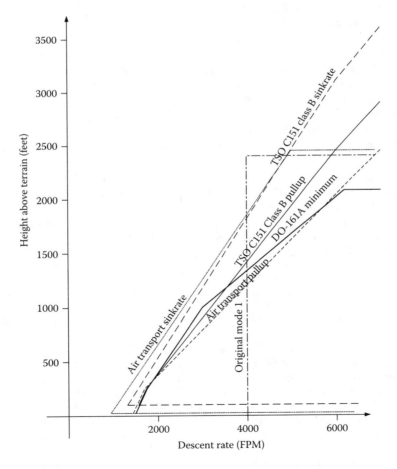

FIGURE 12.2   Mode 1 warning curves.

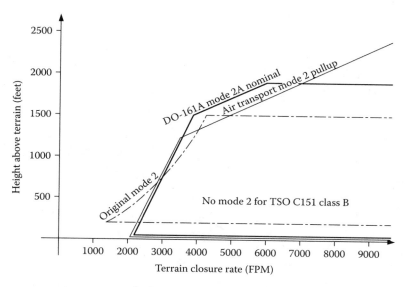

FIGURE 12.3 Mode 2 curves.

Closure rate detection curves are the most difficult of the classic GPWS algorithms to design. Tall buildings, towers, trees, and rock escarpments in the area of final approach can cause sharp spikes in the computed closure rate. Modern Mode 2 algorithms employ complex filtering of the computed rate, with varying dynamic response dependent upon phase of flight and aircraft configuration. The Mode 2 detection algorithm is also modified by specific problem areas by using latitude, longitude, heading, and selected runway course — a technique in the EGPWS termed "Envelope Modulation."

Landing configuration closure rate warnings are termed Mode 2B; cruise and approach configurations are termed Mode 2A. Figure 12.3 illustrates some of the various Mode 2A curves, including the original first Mode 2 curve, the current standard air transport Mode 2A "Terrain-Terrain-Pull-Up" warning curve, and the DO-161A nominal Mode 2A warning requirement. Note that the Class B TSO C151 EGPWS does not use a radio altimeter and therefore has no Mode 2.

12.3.3 Mode 3 — Accelerating Flight Path Back into the Terrain after Take-off

Mode 3 (Figure 12.4) is active from liftoff until a safe altitude is reached. This mode warns for failure to continue to gain altitude. The original Mode 3, still specified in DO-161A as Mode 3A, produced warnings for any negative sinkrate after take-off until 700 ft of ground clearance was reached. The mode has since been redesigned (designated 3B in DO-161A) to allow short-term sink after take-off but detect a trend to a lack of climb situation. The voice callout for Mode 3 is "Don't Sink." This take-off mode now remains active until a time-integrated ground clearance value is exceeded; thus allowing for a longer protection time with low-altitude noise abatement maneuvering before climb-out.

Altitude loss is computed by either sampling and differentiating altitude MSL or integrating altitude rate during loss of altitude. Because a loss is being measured, the altitude can be a corrected or uncorrected pressure altitude, or an inertial or GPS height. Typical Mode 3 curves are linear, with warnings for an 8-ft loss at 30 ft AGL, increasing to a 143-ft loss at 1500 ft AGL.

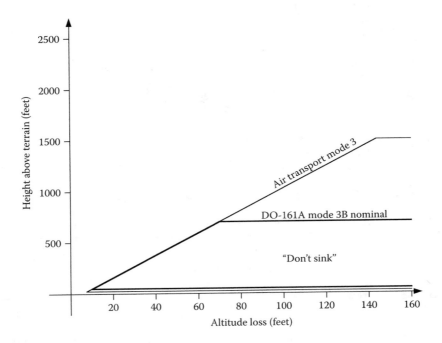

FIGURE 12.4 Mode 3 curves.

12.3.4 Mode 4 — Unsafe Terrain Clearance Based on Aircraft Configuration

The earliest version of Mode 4 was a simple alert for descent below 500 ft with the landing gear up. Second generations of Mode 4 added additional alerting at lower altitudes for flaps not in landing position. The warning altitude for flaps was raised to the 500-ft level for higher descent rates. There are three of these types of Mode 4 curves still specified as alternate minimum performance requirements in DO-161A (see Figure 12.5).

Modern Mode 4 curves are airspeed-enhanced, rather than descent rate alone, and for high airspeeds will give alerts at altitudes up to 1000 ft AGL.

Currently, EGPWS Mode 4 has three types of alerts based upon height AGL, mach/airspeed, and aircraft configuration, termed Modes 4A, 4B, and 4C (Figure 12.6). Two of the curves (4A, 4B) are active during cruise until full landing configuration is achieved with a descent "close to the ground" — typically 700 ft for a transport aircraft. Mode 4C is active on take-off in conjunction with the previously described Mode 3. All three alerts are designed with the intent to warn of flight "too close to the ground" for the current speed/configuration combination. At higher speeds, the alert commences at higher AGL and the voice alert is always "Too Low Terrain." At lower speeds, Mode 4A warning is "Too Low Gear" and the Mode 4B warning is "Too Low Flaps."

Mode 4C compliments Mode 3, which warns on an absolute loss of altitude on climb-out, by requiring a continuous gain in height above the terrain. If the aircraft is rising, but the terrain under is also rising, Mode 4C will alert "Too Low Terrain" on take-off if sufficient terrain clearance is not achieved prior to Mode 3 switching out.

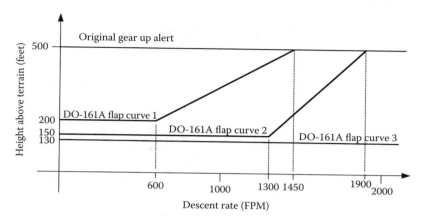

FIGURE 12. 5 Old GPWS Mode 4 curves.

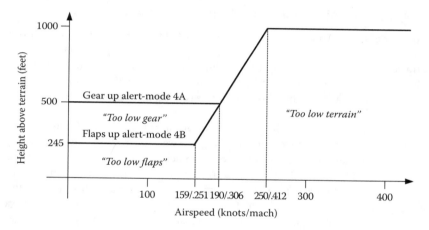

FIGURE 12.6 EGPWS Mode 4.

12.3.5 Mode 5 — Significant Descent below the ILS Landing Glide Path Approach Aid

This mode warns for failure to remain on an instrument glide path on approach. Typical warning curves alert for 1.5 to 2.0 dots below the beam, with a wider divergence allowed at lower altitudes. The alerts and warnings are only enabled when the crew is flying an ILS approach, as determined by radio frequency selections and switch selection. Most installations also include separate enable switch and a warning cancel for crew use when flying some combination of visual and or other landing aids and deviation from the ILS glide path is intentional. Although the mode is typically active from 1000 ft AGL down to about 30 ft, allowance in the design of the alerts must also be made for beam capture from below, and level maneuvering between 500 and 1000 ft without nuisance alerting.

Figure 12.7 illustrates the Mode 5 warnings for a typical jet air transport. When the outer curve is penetrated, the voice message "Glideslope" is repeated at a low volume. If the deviation below the beam increases or altitude decreases, the repetition rate of the voice is increased. If the altitude/deviation combination falls within the inner curve, the voice volume increases to the equivalent of a warning message and the repetition rate is at maximum.

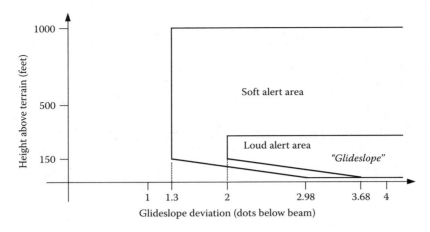

FIGURE 12.7   EGPWS Mode 5.

12.3.6   Mode 6 — Miscellaneous Callouts and Advisories

The first application of this mode consisted of a voice alert added to the activation of the decision height discrete on older analog radio altimeters. This voice alert is "Minimums" or "Decision Height," which adds an extra level of awareness during the landing decision point in the final approach procedure. Traditionally, this callout would be made by the pilot not flying (PNF). Automating the callout frees the PNF from one small task enabling him to more easily monitor other parameters during the final approach.

This mode has since been expanded as a "catch all" of miscellaneous aural callouts requested by air transport manufacturers and operators, many of which also were normally an operational duty of the PNF (see Figure 12.8). In addition to the radio altitude decision height, callouts are now available at barometric minimums, at an altitude approaching the decision height or barometric minimums, or at various combinations of specific altitudes. There are also "smart callouts" available that only call the altitude for nonprecision approaches (ILS not tuned). The EGPWS model used on Boeing aircraft will also callout for V_1 on take-off and give aural "engine out" warnings. Finally, included in the set of Mode 6 callouts are warnings of overbanking (excessive roll angle).

Callout voice	Description
Radio altimeter	Activates at 2500 feet as radio altimeter comes into track
Twenty five hundred	(Alternate to radio altimeter)
One thousand	Activates at 1000 feet AGL
Five hundred (smart)	Activates at 500 feet AGL for non-precision approaches only
One hundred	Activates at 100 feet AGL
Fifty	50 feet AGL
Forty	40 feet AGL
Thirty	30 feet AGL
Twenty	20 feet AGL
Ten	10 feet AGL
Approaching minimums	100 feet above the selected decision height
Minimums	At pilot selected decision height - may be AGL or barometric
Decision height	(alternate to minimums)

FIGURE 12.8   Examples of EGPWS Mode 6 callouts.

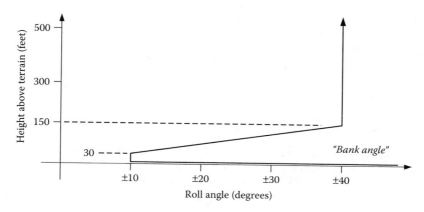

FIGURE 12.9 EGPWS Mode 6 overbank (excessive roll) alert.

12.3.7 Mode 7 — Flight into Windshear Conditions

Windshear is a sudden change in wind direction and/or windspeed over a relatively short distance in the atmosphere and can have a detrimental effect on the performance of an aircraft. The magnitude of a windshear is defined precisely in engineering terms by the sum of the rate of change of horizontal wind, and the vertical wind divided by the true airspeed of the aircraft:

$$F = -\left(\frac{w_{wind}}{V_A} + \frac{\dot{u}_{wind}}{g}\right)$$

where

F is expressed in units of g and is positive for increasing energy windshears

w_{wind} = vertical wind velocity (fps), positive for downdrafts

$\dot{u}_{wind} = \dfrac{du_{wind}}{dt}$ = rate of change of horizontal wind velocity

V_A = true airspeed (fps)

g = gravitational acceleration, 32.178 fps^2

There are various techniques for computing this windshear factor from onboard aircraft sensors (air data, inertial accelerations, etc.). The EGPWS performs this computation and alerts the crew with the aural "Windshear, Windshear, Windshear," when the factor exceeds predefined limits as required by TSO C117a.

12.3.8 Envelope Modulation

Early GPWS equipment was plagued by false and nuisance warnings, causing pilots to distrust the equipment when actual hazardous conditions existed. Many approach profiles and radar vectoring situations violated the best-selected warning curve designs. Even as GPWS algorithms were improved, there still existed some approaches that required close proximity to terrain prior to landing.

Modern GPWS equipment adapts to this problem by storing a table of known problem locations and providing specialized warning envelope changes when the aircraft is operating in these areas. This technique is known as GPWS envelope modulation.

An example exists in the southerly directed approaches to Glasgow Scotland, Runway 23. The standard approach procedures allow an aircraft flying level at 3000 ft barometric altitude to pass over mountain peaks with heights above 1700 ft when approaching this runway. At nominal airspeeds the difference in surrounding terrain height will generate closure rates well within the nominal curve of Figure 12.3. With the envelope modulation feature the GPWS, using latitude, longitude, and heading, notes that the aircraft is flying over this specific area and temporarily lowers the maximum warning altitude for Mode 2 from 2450 ft to the minimum 1250 ft AGL. This eliminates the nuisance warning while at the same time providing the minimum required DO-161A protection for inadvertent flight closer to the mountain peaks on the approach path.

12.3.9 "Enhanced Modes"

The enhanced modes provide terrain and obstacle awareness beyond the normal sensor-derived capabilities of the standard GPWS. Standard GPWS warning curves are deficient in two areas, even with the best designs. One area is immediately surrounding the airport; which is where a large majority of CFIT accidents occur. The other is flight directly into precipitous terrain, for which little or no Mode 2 warning time may occur.

The enhanced modes solve these problems by making use of a database of terrain and obstacle spot heights and airport runway locations arranged in a grid addressed by latitude and longitude. This combined terrain/airports/obstacle database — a virtual world within the computer — provides the ability to track the aircraft position in the real world given accurate x-y-z position combined with the aircraft velocity vector.

This database technique allows three improvements which overcome the standard GPWS modes shortcomings: terrain proximity display, terrain ahead alerting, and terrain clearance floor.

12.3.9.1 Terrain Proximity Display

The terrain proximity display is a particular case of a horizontal (plan view) moving map designed to enhance vertical and horizontal situational awareness. The basic display is based upon human factors studies recommending a minimum of contours and coloring. The display is purposely compatible with existing three-color weather radar displays, allowing economical upgrade of existing equipment.

Terrain well below the flight path of the aircraft is depicted in shades of green, brighter green being closer to the aircraft and sparse green-to-black for terrain far below the aircraft. Some displays additionally allow water areas to be shown in cyan (blue). Terrain in the proximity of the aircraft flight path, but posing no immediate danger (it can be easily flown over or around) is depicted in shades of yellow. Terrain well above the aircraft (nominally more than 2000 ft above flight level), toward which continued safe flight is not possible, is shown in shades of red.

12.3.9.2 Terrain Ahead Alerting

Terrain (and/or obstruction) alerting algorithms continually compare the state of the aircraft flight to the virtual world and provide visual and/or aural alerts if impact is possible or probable. Two levels of alerting are provided, a cautionary alert and a hard warning. The alerting algorithm design is such that, for a steady approach to hazardous terrain, the cautionary alert is given much in advance of the warning alert. Typical design criteria may try to issue caution up to 60 s in advance of a problem and a warning within 30 s.

Voice alerts for the cautionary alert are "Caution, Terrain" or "Terrain Ahead." For the warnings on turboprop and turbojet aircraft, the warning aural is "Terrain Terrain Pullup" or "Terrain Ahead Pullup," with the pullups being repeated continuously until the aircraft flight path is altered to avoid the terrain.

In conjunction with the aural alerts, yellow and red lamps may be illuminated, such as with the standard GPWS alerts. The more compelling visual alerts are given by means of the Terrain Awareness Display. Those areas that meet the criteria for the cautionary alert are illuminated in a bright yellow on the display. If the pullup alert occurs, those areas of terrain where an immediate impact hazard exists are illuminated in bright red. When the aircraft flight path is altered to avoid the terrain, the display returns to the normal terrain proximity depictions as the aural alerts cease.

12.3.9.3   Terrain Clearance Floor

The standard Modes 2 and 4 are desensitized when the aircraft is put in landing configuration (flaps down and/or gear lowered) and thus fail to alert for attempts at landing where there is no airport. Since the EGPWS database contains the exact position of all allowable airport runways, it is possible to define an additional alert, a terrain clearance floor, at all areas where there are no runways. When the aircraft descends below this floor value, the voice alert "Too Low Terrain" is given. This enhanced mode alert is also referred to as premature descent alert.

12.4   EGPWS Standards

ARINC 594 — Ground Proximity Warning System: This is the first ARINC characteristic for Ground Proximity Warning Systems and defines the original analog interfaced system. It applies to the original model (MkI and MkII) GPWS systems featuring Modes 1–5, manufactured by Sundstrand Data Control, Bendix, Collins, Litton and others. It also applies to the AlliedSignal (Honeywell) MkVII digital GPWS, which featured Modes 1–7 and a primarily analog interface for upgrading older models.

ARINC 743 — Ground Proximity Warning System: This characteristic applies to primarily digital (per ARINC 429 DITS) interfaced Ground Proximity Warning Systems, such as the AlliedSignal/Honeywell MkV series, which was standard on all newer Boeing aircraft from the 757/767 up through the introduction of the 777.

ARINC 762 — Terrain Avoidance and Warning System: This characteristic is an update of ARINC 743 applicable to the primarily digital interfaced (MkV) Enhanced GPWS.

ARINC 562 — Terrain Avoidance and Warning System: This proposed ARINC characteristic will be an update of ARINC 594, applicable to the primarily analog interfaced (MkVII) Enhanced GPWS.

RTCA DO-161A — Minimum Performance Standards, Airborne Ground Proximity Warning System: This 1976 document still provides the minimum standards for the classic GPWS Modes 1–5. It is required by both TSO C92c and the new TSO C151 for EGPWS (TAWS).

TSO C92c — Ground Proximity Warning, Glideslope Deviation Alerting Equipment: This TSO covers the classic Modes 1–6 minimum performance standards. It basically references DO-161A and customizes and adds features of the classic GPWS which were added subsequent to DO-161A, including voice callouts signifying the reason for the alert/warnings, Mode 6 callouts, and bank angle alerting.

CAA Specification 14 (U.K.) — Ground Proximity Warning Systems: This is the United Kingdom CAA standard for Modes 1–5 and also specifies some installation requirements. As with the U.S. TSOs, Spec 14 references DO-161A and customizes and augments features of the classic GPWS which are still required for U.K. approvals. Most notably, the U.K. version of Mode 5 is less stringent and requires a visual indication of Mode 5 cancellation. Spec 14 also requires that a stall warning inhibit the GPWS voice callouts, a feature which is found only on U.K.-certified installations.

TSO C117a — Airborne Windshear Warning and Escape Guidance Systems for Transport Airplanes: This TSO defines the requirements for EGPWS Mode 7, reactive low level windshear detection.

TSO C151a — Terrain Awareness and Warning System (TAWS): This TSO supersedes TSO C92c for certain classes of aircraft and requires the system to include the enhanced modes. It also extends coverage down to smaller aircraft, including a non-required functionality for small Part 91 piston powered aircraft. It describes three classes of TAWS equipment. Class A, the standard EGPWS, contains all the modes previously described here including the terrain display. Class B is for intermediate sized turbine powered aircraft. Class B requirements include the enhanced modes but do not require the old TSO C92c modes that depend upon a radio altimeter — altitude above ground level is instead computed from MSL altitude and the terrain display is optional. Class C is similar to Class B but the "must warn" and "must not warn" requirements are more appropriate for small general aviation operations. Class C is entirely voluntary but includes a requirement that a higher accuracy vertical source be used (e.g., GPS altitude) — as such the installation does not require an air data computer typically used on larger aircraft (although air data can be used if available).

AC 23-18 — Installation of Terrain Awareness and Warning System (TAWS) Approved for Part 23 Airplanes.

AC 25-23 — Airworthiness Criteria for the Installation Approval of a Terrain Awareness and Warning System (TAWS) for Part 25 Airplanes.

FAR 91.223 — **Terrain Awareness and Warning System:** Requires that at least Class B equipment is installed on all turbine-powered U.S.-registered aircraft with six passenger seats or more.

FAR 121.354 — **Terrain Awareness and Warning System:** Requires that Class A equipment is installed, including a terrain display, for all aircraft used in domestic, flag, and supplemental operations.

FAR 135.154 — **Terrain Awareness and Warning System:** For all aircraft used in commuter and on-demand operations, requires Class A equipment is installed, including a terrain display, for turbine-powered aircraft with ten passenger seats or more; and, requires at least Class B equipment is installed for all turbine-powered aircraft with six to nine passenger seats.

RTCA DO-200A — **Standards for Processing Aeronautical Data:** This standard is required by TSO C151c to be used for processing the TAWS terrain database.

Further Information

1. *Controlled Flight Into Terrain, Education and Training Aid* — This joint publication of ICAO, Flight Safety Foundation, and DOT/FAA consists of two loose-leaf volumes and an accompanying video tape. It is targeted toward the air transport industry, containing management, operations, and crew training information, including GPWS. Copies may be obtained by contacting the Flight Safety Foundation, Alexandria, Virginia.

2. *DOT Volpe NTSC Reports on CFIT and GPWS* — These may be obtained from the USDOT and contain accident analyses, statistics, and studies of the effectivity of both the classic and enhanced GPWS warning modes. There are a number of these reports which were developed in response to NTSB requests. Of the two most recent reports, the second one pertains to the Enhanced GPWS in particular:

   - Spiller, David — Investigation of Controlled Flight Into Terrain (CFIT) Accidents Involving Multi-engine Fixed-wing Aircraft Operating Under Part 135 and the Potential Application of a Ground Proximity Warning System (Cambridge, MA: U.S. Department of Transportation, Volpe National Transportation Systems Center) March 1989.

   - Phillips, Robert O. — Investigation of Controlled Flight Into Terrain Aircraft Accidents Involving Turbine Powered Aircraft with Six or More Passenger Seats Flying Under FAR Part 91 Flight Rules and the Potential for Their Prevention by Ground Proximity Warning Systems (Cambridge, MA: U.S. Department of Transportation, Volpe National Transportation Systems Center) March 1996.

13

Batteries

David G. Vutetakis
Concorde Battery Corporation

13.1 Introduction

The battery, an essential component of almost all aircraft electrical systems, is used to start engines and auxiliary power units (APUs), provide emergency backup power for essential avionics equipment, assure no-break power for navigation units and fly-by-wire computers, and provide ground power capability for maintenance and preflight checkouts. Many of these functions are mission-critical, so the performance and reliability of an aircraft battery is of considerable importance. Other important requirements include environmental ruggedness, a wide operating temperature range, ease of maintenance, rapid recharge capability, and tolerance to abuse.

Historically, only a few types of batteries have been suitable for aircraft applications. Up until the 1950s, vented lead-acid (VLA) batteries were used exclusively [Earwicker, 1956]. In the late 1950s, military aircraft began converting to vented nickel-cadmium (VNC) batteries, primarily because of their superior performance at low temperatures. The VNC battery subsequently found widespread use in both military and commercial aircraft [Fleischer, 1956; Falk and Salkind, 1969]. The only other type of battery used during this era was the vented silver-zinc battery, which provided an energy density about three times higher than VLA and VNC batteries [Miller and Schiffer, 1971]. This battery type was applied to several types of U.S. Air Force fighters (F-84, F-105, and F-106) and U.S. Navy helicopters (H-2, H-13 and H-43) in the 1950s and 1960s. Although silver-zinc aircraft batteries were attractive for reducing weight and size, their use has been discontinued due to poor reliability and high cost of ownership.

In the late 1960s and early 1970s, an extensive development program was conducted by the U.S. Air Force and Gulton Industries to qualify sealed nickel-cadmium (SNC) aircraft batteries for military and commercial applications [McWhorter and Bishop, 1972]. This battery technology was successfully demonstrated on a Boeing KC-135, a Boeing 727, and a UH-1F helicopter. Before the technology could be transitioned into production, however, Gulton Industries was taken over by SAFT, and a decision was made to terminate the program.

In the late 1970s and early 1980s, the U.S. Navy pioneered the development of sealed lead-acid (SLA) batteries for aircraft applications [Senderak and Goodman, 1981]. SLA batteries were initially applied to the AV-8B and F/A-18, resulting in a significant reliability and maintainability (R&M) improvement compared with VLA and VNC batteries. The Navy subsequently converted the C-130, H-46, and P-3 to

SLA batteries. The U.S. Air Force followed the Navy's lead, converting numerous aircraft to SLA batteries, including the A-7, B-1B, C-5, C-130, C-141, KC-135, F-4, F-16, and F-117 [Vutetakis, 1994]. The term "High Reliability, Maintenance-Free Battery" (HRMFB) was coined to emphasize the improved R&M capability of sealed-cell aircraft batteries. The use of HRMFBs soon spun off into the commercial sector, and numerous commercial and general aviation aircraft today have been retrofitted with SLA batteries.

In the mid-1980s, spurred by increasing demands for HRMFB technology, a renewed interest in SNC batteries took place. The U.S. Air Force initiated a program to develop advanced SNC batteries, and Eagle-Picher Industries was contracted for this effort [Flake, 1988; Johnson et al., 1994]. The B-52 bomber was the only aircraft to retrofit this technology, and Eagle-Picher later discontinued the SNC battery. SNC batteries were also developed by ACME Aerospace for several aircraft applications, including the F-16 fighter, Apache AH-64 helicopter, MD-90, and Boeing 777 [Anderman, 1994]. Only the Apache AH-64 and Boeing 777 continue to use ACME SNC batteries.

A recent development in aircraft batteries is the "low-maintenance" or "ultra-low-maintenance" nickel-cadmium battery [Scardaville and Newman, 1993], which is intended to be a direct replacement for conventional VNC batteries, avoiding the need to replace or modify the charging system. Although the battery still requires scheduled maintenance for electrolyte filling, the maintenance frequency can be decreased significantly. This type of battery was originally developed by SAFT and more recently by Marathon. Flight tests were successfully performed by the U.S. Navy on the H-1 helicopter, and most VNC batteries used by the Navy have been converted to the low-maintenance technology. This technology has also been applied to various commercial aircraft.

Determining the most suitable battery type and size for a given aircraft type demands detailed knowledge of the application requirements (load profile, duty cycle, environmental factors, and physical constraints) and the characteristics of available batteries (performance capabilities, charging requirements, life expectancy, and cost of ownership). With the various battery types available today, considerable expertise is needed to select the best type and size of battery for a given aircraft application. The information contained in this chapter will provide general guidance for original equipment design and for upgrading existing aircraft batteries. More detailed information can be found in the sources listed at the end of the chapter.

13.2   General Principles

13.2.1   Battery Fundamentals

Batteries operate by converting chemical energy into electrical energy through electrochemical discharge reactions. Batteries are composed of one or more cells, each containing a **positive electrode**, **negative electrode**, **separator**, and **electrolyte**. Cells can be divided into two major classes: primary and secondary. Primary cells are not rechargeable and must be replaced once the reactants are depleted. Secondary cells are rechargeable and require a DC charging source to restore reactants to their fully charged state. Examples of primary cells include carbon-zinc (Leclanche or dry cell), alkaline-manganese, mercury-zinc, silver-zinc, and lithium cells (e.g., lithium-manganese dioxide, lithium-sulfur dioxide, and lithium-thionyl chloride). Examples of secondary cells include lead-lead dioxide (lead-acid), nickel-cadmium, nickel-iron, nickel-hydrogen, nickel-metal hydride, silver-zinc, silver-cadmium, and lithium-ion. For aircraft applications, secondary cells are the most prominent, but primary cells are sometimes used for powering critical avionics equipment (e.g., flight data recorders).

Batteries are rated in terms of their **nominal voltage** and **ampere-hour capacity**. The voltage rating is based on the number of cells connected in series and the nominal voltage of each cell (2.0 volts for lead-acid and 1.2 volts for nickel-cadmium). The most common voltage rating for aircraft batteries is 24 volts. A 24-volt, lead-acid battery contains 12 cells, while a 24-volt, nickel-cadmium battery contains either 19 or 20 cells (the U.S. military rates 19-cell batteries at 24 volts). Voltage ratings of 22.8, 25.2, and 26.4 volts are also common with nickel-cadmium batteries, consisting of 19, 20, or 22 cells, respectively.

Twelve-volt lead-acid batteries, consisting of six cells in series, are also used in many general aviation aircraft.

The ampere-hour (Ah) capacity available from a fully charged battery depends on its temperature, rate of discharge, and age. Normally, aircraft batteries are rated at room temperature (25°C), the **C-rate** (1-hour rate), and beginning of life. Military batteries, however, often are rated based on the end-of-life capacity (i.e., the minimum capacity before the battery is considered unserviceable). Capacity ratings of aircraft batteries vary widely, generally ranging from as low as 3 Ah to about 65 Ah.

The maximum power available from a battery depends on its internal construction. High rate cells, for example, are designed specifically to have very low internal impedance as required for starting turbine engines and APUs. For lead-acid batteries, the peak power has traditionally been defined in terms of the cold-cranking amperes, or **CCA** rating. For nickel-cadmium batteries, the peak power rating has traditionally been defined in terms of the current at maximum power, or **Imp** rating. These ratings are based on different temperatures (−18°C for CCA, 23°C for Imp), making it difficult to compare different battery types. Furthermore, neither rating adequately characterizes the battery's initial peak current capability, which is especially important for engine start applications. A standard method of defining the peak power capability of an aircraft battery has recently been published in IEC 60952-1 (2004). This standard defines two peak power ratings, the **Ipp** and the **Ipr**, at three different temperatures (23, −18 and −30°C). The Ipp rating is the current at 0.3 seconds when discharged at half the nominal voltage, and the Ipr rating is the current at 15 seconds when discharged at half the nominal voltage. Other peak power specifications have been included in some military standards. For example, MIL-B-8565/15 specifies the initial peak current, the current after 15 seconds, and the capacity after 60 seconds during a 14-volt constant voltage discharge at two different temperatures (24 and −26°C) (Vutetakis and Viswanathan, 1996).

The **state-of-charge** of a battery is the percentage of its capacity available relative to the capacity when it is fully charged. By this definition, a fully charged battery has a state-of-charge of 100 percent, and a battery with 20 percent of its capacity removed has a state-of-charge of 80 percent. The **state-of-health** of a battery is the percentage of its capacity available when fully charged relative to its rated capacity. For example, a battery rated at 30 Ah but only capable of delivering 24 Ah when fully charged will have a state-of-health of $24/30 \times 100 = 80$ percent. Thus, the state-of-health takes into account the loss of capacity as the battery ages.

13.2.2 Lead-Acid Batteries

13.2.2.1 Theory of Operation

The chemical reactions that occur in a lead-acid battery are represented by the following equations:

$$\text{Positive electrode: } PbO_2 + H_2SO_4 + 2H^+ + 2e^- \underset{\text{charge}}{\overset{\text{discharge}}{\rightleftarrows}} PbSO_4 + 2H_2O \tag{13.1}$$

$$\text{Negative electrode: } Pb + H_2SO_4 \underset{\text{charge}}{\overset{\text{discharge}}{\rightleftarrows}} PbSO_4 + 2H^+ + 2e^- \tag{13.2}$$

$$\text{Overall reaction: } PbO_2 + Pb + 2H_2SO_4 \underset{\text{charge}}{\overset{\text{discharge}}{\rightleftarrows}} 2PbSO_4 + 2H_2O \tag{13.3}$$

As the cell is charged, the sulfuric acid (H_2SO_4) concentration increases and becomes highest when the cell is fully charged. Likewise, when the cell is discharged, the acid concentration decreases and becomes most dilute when the cell is fully discharged. The acid concentration generally is expressed in terms of specific gravity (SG), which is weight of the electrolyte compared to the weight of an equal volume of pure water. The cell's SG can be estimated from its open circuit voltage (OCV) using the following equation:

$$SG = OCV - 0.84 \tag{13.4}$$

There are two basic cell types: vented and recombinant. Vented cells have a flooded electrolyte, and the hydrogen and oxygen gases generated during charging are vented from the cell container. Recombinant cells have a starved or gelled electrolyte, and the oxygen generated from the positive electrode during charging diffuses to the negative electrode where it recombines to form water by the following reaction:

$$Pb + H_2SO_4 + 1/2O_2 \rightarrow PbSO_4 + H_2O \tag{13.5}$$

The recombination reaction suppresses hydrogen evolution at the negative electrode, thereby allowing the cell to be sealed. In practice, the recombination efficiency is not 100 percent, and a resealable valve regulates the internal pressure at a relatively low value, generally below 5 psig. For this reason, SLA cells are often called "valve-regulated lead-acid" (VRLA) cells.

13.2.2.2 Cell Construction

Lead-acid cells are composed of alternating positive and negative plates interleaved with single or multiple layers of separator material. Plates are made by pasting active material onto a grid structure made of lead or lead alloy. The electrolyte is a mixture of sulfuric acid and water. In flooded cells, the separator material is porous rubber, cellulose fiber, or microporous plastic. In recombinant cells with starved electrolyte technology, a glass fiber mat separator is used, sometimes with an added layer of microporous polypropylene. Gell cells, the other type of recombinant cell technology, are made by adding powered silica to the electrolyte to form a gelatinous structure that surrounds the electrodes and separators.

13.2.2.3 Battery Construction

Lead-acid aircraft batteries are constructed using injection-molded, plastic **monoblocs** that contain a group of cells connected in series. Monoblocs typically are made of polypropylene, but acrylonitrile-butadiene-styrene (ABS) is used by at least one manufacturer. Normally, the monobloc serves as the battery case, similar to a conventional automotive battery. For more robust designs, monoblocs are assembled into a separate outer container made of steel, aluminum, or fiberglass-reinforced epoxy. Cases usually incorporate an electrical receptacle for connecting to the external circuit with a quick connect/disconnect plug. Two generic styles of receptacles are common: the "Elcon-style" and the "Cannon-style." The Elcon-style is equivalent to military type MS3509. The Cannon-style has no military equivalent, but it is produced by Cannon and other connector manufacturers. Batteries sometimes incorporate thermostatically controlled heaters to improve low temperature performance. The heater is powered by the aircraft's AC or DC bus. Figure 13.1 shows an assembly drawing of a typical lead-acid aircraft battery; this particular example does not incorporate a heater.

13.2.2.4 Discharge Performance

Battery performance characteristics usually are described by plotting voltage, current or power versus discharge time, starting from a fully charged condition. Typical discharge performance data for SLA aircraft batteries are illustrated in Figures 13.2 and 13.3. Figure 13.4 shows the effect of temperature on the capacity when discharged at the C-rate. Manufacturers' data should be obtained for current information on specific batteries of interest.

13.2.2.5 Charge Methods

Constant voltage charging at 2.3 to 2.4 volts per cell is the preferred method of charging lead-acid aircraft batteries. For a 12-cell battery, this equates to 27.6 to 28.8 volts, which generally is compatible with the voltage available from the aircraft's 28-volt DC bus. Thus, lead-acid aircraft batteries normally can be charged by direct connection to the DC bus, avoiding the need for a dedicated battery charger. If the voltage regulation on the DC bus is not controlled sufficiently, however, the battery will be overcharged or undercharged, causing premature failure. In this case, a regulated voltage source may be necessary to achieve acceptable battery life. Some aircraft use voltage regulators that compensate, either manually or

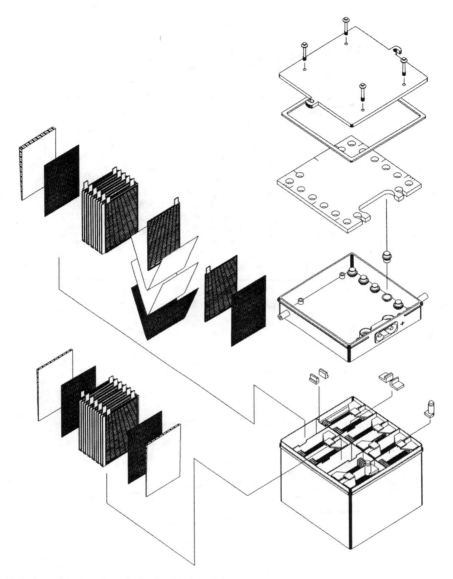

FIGURE 13.1 Assembly drawing of a lead-acid aircraft battery.

automatically, for the battery temperature by increasing the voltage when cold and decreasing the voltage when hot. Adjusting the charging voltage in this manner has the beneficial effect of prolonging the battery's service life at high temperature and achieving faster recharge at low temperatures.

13.2.2.6 Temperature Effects and Limitations

Lead-acid batteries generally are rated at 25°C (77°F) and operate best around this temperature. Exposure to low ambient temperatures results in performance decline, whereas exposure to high ambient temperatures results in shortened life. The lower temperature limit is dictated by the freezing point of the electrolyte. The electrolyte freezing point varies with acid concentration, as shown in Table 13.1. The minimum freezing point is a chilly −70°C (−95°F) at a SG of 1.300. Since fully charged batteries have SGs in the range of 1.28 to 1.33, they are not generally susceptible to freezing even under extreme cold conditions. However, when the battery is discharged, the SG drops and the freezing point rises. At low SG, the electrolyte first will turn to slush as the temperature drops; this is because the water content

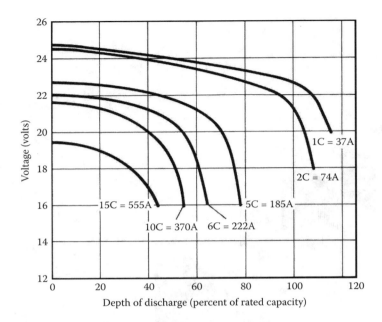

FIGURE 13.2 Discharge curves at 25°C for a 24V/37Ah SLA aircraft battery.

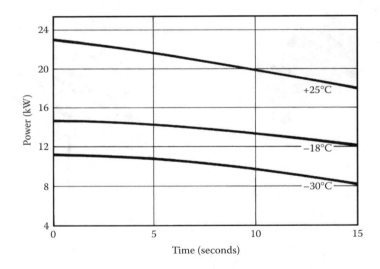

FIGURE 13.3 Maximum power curves (12V discharge) for a 24V/37Ah SLA Battery.

freezes first, gradually raising the SG of the remaining liquid so that it remains unfrozen. Solid freezing of the electrolyte in a discharged battery requires temperatures well below the slush point; a practical lower limit of −30°C is often specified. Solid freezing can damage the battery permanently (i.e., by cracking cell containers), so precautions should be taken to keep the battery charged or heated when exposed to temperatures below −30°C.

The upper temperature limit is generally in the range of 60 to 70°C. Capacity loss is accelerated greatly when charged above this temperature range due to vigorous gassing and rapid grid corrosion. The capacity loss generally is irreversible when the battery is cooled.

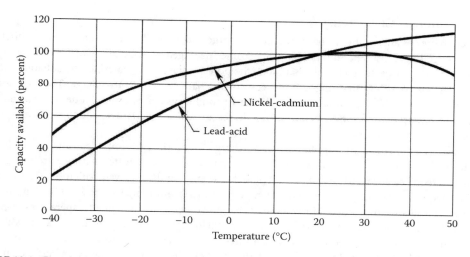

FIGURE 13.4 Capacity versus temperature for aircraft batteries at the C-rate.

TABLE 13.1 Freezing Points of Sulfuric Acid-Water Mixtures

Specific Gravity at 15°C	Cell OCV (Volts)	Battery OCV (Volts)	Freezing Point (°C)	Freezing Point (°F)
1.000	1.84	22.08	0	+32
1.050	1.89	22.68	−3	+26
1.100	1.94	23.28	−8	+18
1.150	1.99	23.88	−15	+5
1.200	2.04	24.48	−27	−17
1.250	2.09	25.08	−52	−62
1.300	2.14	25.68	−70	−95
1.350	2.19	26.28	−49	−56
1.400	2.24	26.88	−36	−33

13.2.2.7 Service Life

The service life of a lead-acid aircraft battery depends on the type of use it experiences (e.g., rate, frequency, and depth of discharge), environmental conditions (e.g., temperature and vibration), charging method, and the care with which it is maintained. Service lives can range from 1 to 5 years, depending on the application. Table 13.2 shows representative cycle life data as a function of the depth of discharge. Consult manufacturers' data for specific batteries.

TABLE 13.2 Cycle Life Data for SLA Aircraft Batteries

Depth of Discharge (% of Rated Capacity)	Number of Cycles to End of Life
10%	2000
30%	670
50%	400
80%	250
100%	200

Source: Hawker Energy Products, Technical Manual

13.2.2.8 Storage Characteristics

Lead-acid batteries always should be stored in the charged state. If allowed to remain in the discharged state for a prolonged time period, the battery becomes damaged by sulfation, which occurs when lead sulfate forms into large, hard crystals that block the pores in the active material. The sulfation creates a high impedance condition that makes it difficult for the battery to accept recharge. The sulfation may or may not be reversible, depending on the discharge conditions and specific cell design. The ability to recovery from deep discharge has been improved in recent years by electrolyte additives, such as sodium sulfate.

VLA batteries normally are supplied in a dry, charged state (i.e., without electrolyte), which allows them to be stored almost indefinitely (i.e., 5 years or more). Once activated with electrolyte, periodic charging is required to overcome the effect of self-discharge and to prevent sulfation. The necessary charging frequency depends on the storage temperature. At room temperature (25°C), charging every 30 days is typically recommended. More frequent charging is necessary at higher temperatures (e.g., every 15 days at 35°C), and less frequent charging is necessary at low temperatures (e.g., every 120 days at 10°C).

SLA batteries can be supplied only in the activated state (i.e., with electrolyte), so storage provisions are more demanding compared to dry charged batteries. As in the case of activated VLA batteries, periodic charging is necessary to overcome the effects of self-discharge and to prevent sulfation. The rate of self-discharge of SLA batteries varies widely from manufacturer to manufacturer, so the necessary charging frequency also varies widely. For example, recommended charging frequencies can range from 3 to 24 months.

13.2.2.9 Maintenance Requirements

Lead-acid aircraft batteries require routine maintenance to assure airworthiness and to maximize service life. For vented-cell batteries, electrolyte topping must be performed on a regular basis to replenish the water loss that occurs during charging. Maintenance intervals are typically 2 to 4 months. A capacity test or load test usually is included as part of the servicing procedure. For sealed-cell batteries, water replenishment obviously is unnecessary, but periodic capacity measurements generally are recommended. Capacity check intervals can be based either on calendar time (e.g., every 3 to 6 months after the first year) or operating hours (e.g., every 100 hours after the first 600 hours). Refer to the manufacturers' maintenance instructions for specific batteries of interest.

13.2.2.10 Failure Modes and Fault Detection

The predominant failure modes of lead-acid cells are summarized as follows:

- Shorts caused by growth of the positive grid, shedding or mossing of active material, or mechanical defects protruding from the grid that are manifested by inability of the battery to hold a charge (rapid decline in open circuit voltage)
- Loss of electrode capacity due to active material shedding, excessive grid corrosion, sulfation, or passivation that is manifested by low capacity and/or inability to hold voltage under load.
- Water loss and resulting cell dry-out due to leaking seal, repeated cell reversals, or excessive overcharge (this mode applies to sealed cells or to vented cells that are improperly maintained) that is manifested by low capacity and inability to hold voltage under load

Detection of these failure modes is straightforward if the battery can be removed from the aircraft because the battery capacity and load capability can be measured directly and the ability to hold a charge can be inferred by checking the open circuit voltage over time. However, detection of these failure modes while the battery is in service is more difficult. The more critical the battery is to the safety of the aircraft, the more important it becomes to detect battery faults accurately. A number of on-board detection schemes have been developed for critical applications, mainly for military aircraft [Vutetakis and Viswanathan, 1995].

13.2.2.11 Disposal

Lead, the major constituent of the lead-acid battery, is a toxic chemical. As long as the lead remains inside the battery container, no health hazard exists, however, improper disposal of spent batteries can result in exposure to lead. Environmental regulations in the United States and abroad prohibit the disposal of lead-acid batteries in landfills or incinerators. Fortunately, an infrastructure exists for recycling the lead from lead-acid batteries. The same processes used to recycle automotive batteries are used to recycle aircraft batteries. Federal, state, and local regulations should be followed for proper disposal procedures.

13.2.3 Nickel-Cadmium Batteries

13.2.3.1 Theory of Operation

The chemical reactions that occur in a nickel-cadmium battery are represented by the following equations:

$$\text{Positive electrode: } 2NiOOH + 2H_2O + 2e^- \underset{\text{charge}}{\overset{\text{discharge}}{\rightleftarrows}} 2Ni(OH)_2 + 2(OH)^- \qquad (13.6)$$

$$\text{Negative electrode: } Cd + 2(OH)^- \underset{\text{charge}}{\overset{\text{discharge}}{\rightleftarrows}} Cd(OH)_2 + 2e^- \qquad (13.7)$$

$$\text{Overall reaction: } 2NiOOH + Cd + 2H_2O \underset{\text{charge}}{\overset{\text{discharge}}{\rightleftarrows}} 2PbSO_4 + 2H_2O \qquad (13.8)$$

There are two basic cell types: vented and recombinant. Vented cells have a flooded electrolyte, and the hydrogen and oxygen gases generated during charging are vented from the cell container. Recombinant cells have a starved electrolyte, and the oxygen generated from the positive electrode during charging diffuses to the negative electrode where it recombines to form cadmium hydroxide by the following reaction:

$$Cd + H_2O + 1/2O_2 \rightarrow Cd(OH)_2 \qquad (13.9)$$

The recombination reaction suppresses hydrogen evolution at the negative electrode, thereby allowing the cell to be sealed. Unlike VRLA cells, recombinant nickel-cadmium cells are sealed with a high pressure vent that releases only during abusive conditions. Thus, these cells remain sealed under normal charging conditions. However, provisions for gas escape must still be provided when designing battery cases since abnormal conditions may be encountered periodically (e.g., in the event of a charger failure that causes an over-current condition).

13.2.3.2 Cell Construction

The construction of nickel-cadmium cells varies significantly, depending on the manufacturer. In general, cells feature alternating positive and negative plates with separator layers interleaved between them, a potassium hydroxide (KOH) electrolyte of approximately 31 percent concentration by weight (SG 1.300) and a prismatic cell container with the cell terminals extending through the cover. The positive plate is impregnated with nickel hydroxide and the negative plate is impregnated with cadmium hydroxide. The plates differ according to manufacturer with respect to the type of the substrate, type of plaque, impregnation process, formation process, and termination technique. The most common plate structure is made of nickel powder sintered onto a substrate of perforated nickel foil or woven screens. At least one manufacturer (ACME) uses nickel-coated polymeric fibers to form the plate structure. Cell containers typically are made of nylon, polyamide, or steel. One main difference between vented cells and sealed (recombinant) cells is the type of separator. Vented cells use a gas barrier layer to prevent gases from diffusing between adjacent plates. Recombinant cells feature a porous separator system that permits gas diffusion between plates.

13.2.3.3 Battery Construction

Nickel-cadmium aircraft batteries generally consist of a steel case containing identical, individual cells connected in a series. The number of cells depends on the particular application, but generally 19 or 20 cells are used. The end cells of the series are connected to the battery receptacle located on the outside of the case. The receptacle is usually a two-pin, quick-disconnect type; both Cannon and Elcon styles commonly are used. Cases are vented by means of vent tubes or louvers to allow escape of gases produced during overcharge. Some battery designs have provisions for forced-air cooling, particularly for engine start applications. Thermostatically controlled heating pads sometimes are employed on the inside or outside of the battery case to improve low temperature performance. Power for energizing the heaters normally is provided by the aircraft's AC or DC bus. Temperature sensors often are included inside the case to allow regulation of the charging voltage. In addition, many batteries are equipped with a thermal switch that protects the battery from overheating if a fault develops or if battery is exposed to excessively high temperatures. A typical aircraft battery assembly is shown in Figure 13.5.

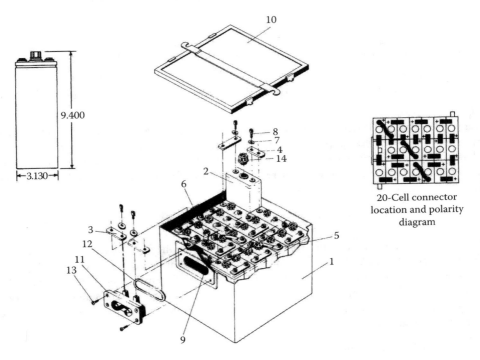

20-Cell connector
location and polarity
diagram

Item	Description	Quantity	Item	Description	Quantity
1	Can assembly	1	8	Socket head cap screw	42
2	Cell assembly	20	9	Spacer	1
3	Connector	12	10	Cover assembly	1
4	Connector	5	11	Receptacle assembly	1
5	Connector	3	12	Rectangular ring	1
6	Connector	1	13	Phillips head screw	4
7	Belleville spring	42	14	Filler cap and vent assembly	20

FIGURE 13.5 Assembly drawing of a nickel-cadmium aircraft battery.

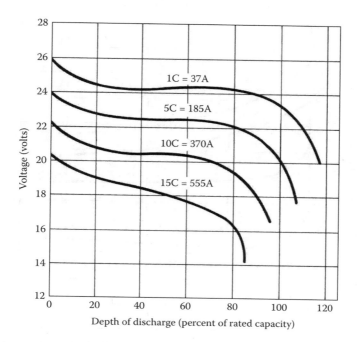

FIGURE 13.6 Discharge curves at 25°C for a 24V/37Ah VNC aircraft battery.

13.2.3.4 Discharge Performance

Typical discharge performance data for VNC aircraft batteries are illustrated in Figures 13.6 and 13.7. Discharge characteristics of SNC batteries are similar to VNC batteries. Figure 13.4 shows the effect of temperature on discharge capacity at the C-rate. Compared with lead-acid batteries, nickel-cadmium batteries tend to have more available capacity at low temperature, but less available capacity at high temperature. Manufacturers' data should be consulted for current information on specific batteries of interest.

13.2.3.5 Charge Methods

A variety of methods are employed to charge nickel-cadmium aircraft batteries. The key requirement is to strike an optimum balance between overcharging and undercharging while achieving full charge in the required time frame. Overcharging results in excessive water loss (vented cells) or heating (sealed cells). Undercharging results in capacity fading. Some overcharge is necessary, however, to overcome Coulombic inefficiencies associated with the electrochemical reactions. In practice, recharge percentages on the aircraft generally range between 105 and 120 percent.

For vented-cell batteries, common methods of charging include constant potential, constant current, or pulse current. Constant potential charging is the oldest method and normally is accomplished by floating a 19-cell battery on a 28-volt DC bus. The constant current method requires a dedicated charger and typically uses a 0.5 to 1.5 C-rate charging current. Charge termination is accomplished using a temperature-compensated voltage cutoff (VCO). The VCO temperature coefficient is typically (-) 4mV/ °C. In some cases, two constant current steps are used, the first step at a higher rate (e.g., C-rate), and the second step at a lower rate (e.g., 1/3 to 1/5 of the C-rate). This method is more complicated but results in less gassing and electrolyte spewage during overcharge. Pulse current methods are similar to the constant current methods, except the charging current is pulsed rather than constant.

For sealed-cell batteries, only constant current or pulse current methods should be used. Constant potential charging can cause excessive heating, resulting in thermal runaway. Special attention must be given to the charge termination technique in sealed-cell batteries, because the voltage profile is relatively

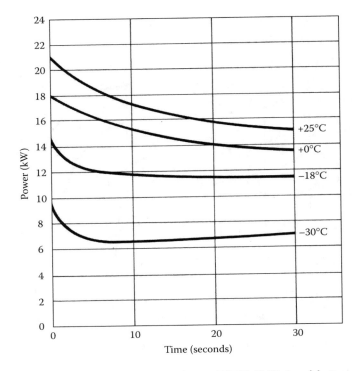

FIGURE 13.7 Maximum power curves (12V discharge) for a 24V/37Ah VNC aircraft battery.

flat as the battery becomes fully charged. For example, it may be necessary to rely on the battery's temperature rise rather than voltage rise as the signal for charge termination.

13.2.3.6 Temperature Effects and Limitations

Nickel-cadmium batteries, like lead-acid batteries, normally are rated at room temperature (25°C) and operate best around this temperature. Exposure to low ambient temperatures results in performance decline, and exposure to high ambient temperatures results in shortened life. The lower temperature limit is dictated by the freezing point of the electrolyte. Most cells are filled with an electrolyte concentration of 31 percent KOH, which freezes at –66°C. Lower concentrations will freeze at higher temperatures, as shown in Table 13.3. The KOH concentration may become diluted over time as a result of spillage or carbonization (reacting with atmospheric carbon dioxide), so the freezing point of a battery in service may not be as low as expected. As in the case of dilute acid electrolytes, slush ice will form

TABLE 13.3 Freezing Points of KOH-Water Mixtures

Concentration (Weight %)	Specific Gravity at 15°C	Freezing Point (°C)	(°F)
0	1.000	0	+32
5	1.045	–3	+27
10	1.092	–8	+18
15	1.140	–15	+5
20	1.188	–24	–11
25	1.239	–38	–36
30	1.290	–59	–74
31	1.300	–66	–87
35	1.344	–50	–58

TABLE 13.4 Cycle Life Data for SNC Aircraft Batteries

Depth of Discharge (% of Rated Capacity)	Number of Cycles to End of Life
30%	7500
50%	4500
60%	3000
80%	1500
100%	1000

Source: ACME Electric Corporation, Technical Manual

well before the electrolyte freezes solid. For practical purposes, a lower operating temperature limit of −40°C often is quoted. The upper temperature limit is generally in the range of 50 to 60°C; significant capacity loss occurs when batteries are operated (i.e., repeated charge and discharge cycles) above this temperature range. The battery capacity often is recoverable, however, when the battery is cooled to room temperature and subjected to several deep discharge cycles.

13.2.3.7 Service Life

The service life of a nickel-cadmium aircraft battery depends on many factors, including the type of use it experiences (e.g., rate, frequency, and depth of discharge), environmental conditions (e.g., temperature and vibration), charging method, and the care with which it is maintained and reconditioned. Thus, it is difficult to generalize the service life that can be expected. All things being equal, the service life of a nickel-cadmium battery is inherently longer than that of a lead-acid battery. Representative cycle life data for an SNC battery are listed in Table 13.4.

13.2.3.8 Storage Characteristics

Nickel-cadmium batteries can be stored in any state of charge and over a broad temperature range (i.e., −65 to 60°C). For maximum shelf life, however, it is best to store batteries between 0 and 30°C. Vented-cell batteries normally are stored with the terminals shorted together; however, shorting of sealed-cell batteries during storage is not recommended, since it may cause cell venting, cell reversal, or both.

When left on open circuit during periods of nonoperation, nickel-cadmium batteries will self-discharge at a relatively fast rate. As a rule of thumb, the self-discharge rate of sealed cells is approximately 1% per day at 20°C (when averaged over 30 days), and the rate increases by 1% per day for every 10°C rise in temperature (e.g., 2% per day at 30°C, 3 percent per day at 40°C, etc.). The self-discharge rate is somewhat less for vented cells. The capacity lost by self-discharge usually is recoverable when charged in the normal fashion.

13.2.3.9 Maintenance Requirements

Nickel-cadmium aircraft batteries require routine maintenance to assure airworthiness and to maximize service life. Maintenance intervals for vented-cell batteries in military aircraft are typically 60 to 120 days, and maintenance intervals for commercial aircraft can be as low as 100 and as high as 1000 flight hours, depending on the operating conditions. Maintenance procedures include capacity checks, cell equalization (deep discharge followed by shorting cell terminals for at least 8 hours), isolation and replacement of faulty cells (only if permitted; this practice generally is not recommended), cleaning to remove corrosion and carbonate build-up, and electrolyte adjustment. For sealed-cell batteries, maintenance requirements are much less demanding. Electrolyte adjustment is unnecessary, and the extent of corrosion is greatly reduced. However, some means of assuring airworthiness is still necessary, such as periodic capacity measurement. Manufacturers' recommendations should be followed for specific batteries of interest.

13.2.3.10 Failure Modes and Fault Detection

The predominant failure modes of nickel-cadmium cells are summarized as follows:

- Shorts caused by cadmium migration through the separator, swelling of the positive electrode, degradation of the separator, or mechanical defects protruding from the electrode, which are manifested by inability of the battery to hold a charge (soft shorts) or dead cells (hard shorts)
- Water loss and resulting cell dry-out due to leaking seal, repeated cell reversal or excessive over-charge (this mode applies to sealed cells or to vented cells that are improperly maintained), which are manifested by low capacity, inability to hold voltage under load, or both
- Loss of negative (cadmium) electrode capacity due to passivation or active material degradation, which is manifested by low capacity, inability to hold voltage under load, or both, and is usually reversible by deep discharge followed by shorting cell terminals or by "reflex" charging (pulse charging with momentary discharge between pulses)
- Loss of positive (nickel) electrode capacity due to swelling or active material degradation, which is manifested by low capacity that is nonrestorable

As discussed under section 13.2.2, detection of these failure modes is relatively straightforward if the battery can be removed from the aircraft. For example, the battery capacity and load capability can be directly measured and compared against pass/fail criteria. The occurrence of soft shorts (i.e., a high impedance short between adjacent plates) is more difficult to detect but often can be identified by monitoring the end-of-charge voltage of individual cells. Detection of these failure modes while the battery is in service is more difficult. As in the case of lead-acid batteries, a number of on-board detection schemes have been developed for critical applications [Vutetakis and Viswanathan, 1995]. The more critical the battery is to the safety of the aircraft, the more important it becomes to detect battery faults accurately.

13.2.3.11 Disposal

Proper disposal of nickel-cadmium batteries is essential because cadmium is a carcinogenic chemical. In the United States and abroad, spent nickel-cadmium batteries are considered to be hazardous waste, and their disposal is strictly regulated. Several metallurgical processes have been developed for reclaiming and recycling the nickel and cadmium from nickel-cadmium batteries, and these processes can be used for both vented and sealed cells. Federal, state, and local regulations should be followed for proper disposal procedures.

13.3 Applications

Designing a battery for a new aircraft application or for retrofit requires a careful systems engineering approach. To function well, the battery must be interfaced carefully with the aircraft's electrical system. The battery's reliability and maintainability depends heavily on the type of charging system to which it is connected; there is a fine line between undercharging and overcharging the battery. Many airframe manufacturers have realized that it is better to prepare specifications for a "battery system" rather than having separate specifications for the battery and the charger. This approach assures that the charging profile is tuned correctly to the specific characteristics of the battery and to the aircraft's operational requirements.

13.3.1 Commercial Aircraft

In general, aircraft batteries must be sized to provide sufficient emergency power to support flight essential loads in the event of failure of the primary power system. Federal Aviation Administration (FAA) regulations impose a minimum emergency power requirement of 30 minutes on all commercial airplanes, and some airlines impose a longer emergency requirement, such as 40 or 60 minutes, due to frequent bad weather on their routes or for other reasons. The emergency requirement for Extended Twin Operation (ETOPS) imposed on two-engine aircraft operating over water is a full 90 minutes, although 60 minutes is allowed with operating restrictions. The specified emergency power requirement may be satisfied by batteries or other backup power sources, such as a ram air turbine. If a ram air turbine is used, a battery still is required for transient fill-in. Specific requirements pertaining to aircraft batteries can be found in the Federal Aviation Regulations (FAR), Sections 25.1309, 25.1333, 25.1351, and 25.1353. FAA Advisory Circular No. 25.1333-1 describes specific methods to achieve compliance with applicable FAR sections. For international applications, Civil Aviation Authority (CAA) and Joint Airworthiness Authority (JAA) regulations should be consulted for additional requirements.

When used for APU or engine starting, the battery must be sized to deliver short bursts of high power, as opposed to the lower rates required for emergency loads. APU start requirements on large commercial aircraft can be particularly demanding; for instance, the APU used on the Boeing 757 and 767 airplanes has a peak current requirement of 1200 amperes [Gross, 1991]. The load on the battery starts out very high to deliver the in-rush current to the motor, then falls rapidly as the motor develops back electromotive force (EMF). Within 30 to 60 seconds, the load drops to zero as the APU ignites and the starter cutoff point is reached. The worst case condition is starting at altitude with a cold APU and a cold battery; normally, a lower temperature limit of –18°C is used as a design point. A rigorous design methodology for optimizing aircraft starter batteries was developed by Evjen and Miller [1971].

When nickel-cadmium batteries are used for APU or engine starting applications, FAA regulations require the battery to be protected against overheating. Suitable means must be provided to sense the battery temperature and to disconnect the battery from the charging source if the battery overheats. This requirement originated in response to numerous instances of battery thermal runaway, which usually occurred when 19-cell batteries were charged from the 28-volt DC bus. Most instances of thermal runaway were caused by degradation of the cellophane gas barrier, thus allowing gas recombination and resultant cell heating during charging. Modern separator materials (e.g., Celgard) have greatly reduced the occurrence of thermal runaway as a failure mode of nickel-cadmium batteries, but the possibility still exists if the electrolyte level is not properly maintained.

Sizes of commercially available aircraft batteries generally range from 12V/6.5Ah to 24V/65Ah. Detailed specifications for aircraft batteries are generally available on the Web site of those companies offering their product for sale (see listing at end of chapter).

13.3.2   Military Aircraft

Table 13.5 lists commonly used military aircraft batteries. However, this listing includes only those batteries that have been assigned a military part number based on an approved military specification; nonstandard batteries are not included. Detailed characteristics and performance capabilities can be found by referring to the applicable military specifications. A number of nonstandard battery designs have been proliferated in the military due to the unique form, fit, and functional requirements of certain aircraft. Specifications for these batteries normally are obtainable only from the aircraft manufacturer. Specific examples of battery systems used in present-day military aircraft were described by Vutetakis [1994].

Defining Terms

Ampere-Hour Capacity: The quantity of stored electrical energy, measured in ampere-hours, that the battery can deliver from its completely charged state to its discharged state. The dischargeable capacity depends on the rate at which the battery is discharged; at higher discharge rates, the available capacity is reduced.

C-Rate: The discharge rate, in amperes, at which a battery can deliver 1 hour of capacity to a fixed voltage endpoint (typically 18 or 20 volts for a 24-volt battery). Fractions or multiples of the C-rate also are used. C/2 refers to the rate at which a battery will discharge its capacity in 2 hours; 2C is twice the C-rate or the rate at which the battery will discharge its capacity in half an hour. This rating system helps to compare the performance of different sizes of cells.

TABLE 13.5 Military Aircraft Batteries

Military Part No.	Type	Rating[a] (Ah)	Max. Wt. (lb)	Applications	Notes
				MIL-B-8565 Series	
D8565/1-1	SNC	2.0 (26V)	8.6	AV-8A/C, CH-53D/E, MH-53E	Superceded by D8565/1-2
D8565/1-2	SLA	1.5	6.8	Same as D8565/1-1	Contains integral charger
D8565/2-1	VNC	30	88.0	OV-10D	Superceded by M81757/12-1
D8565/3-3	SLA	15	47.4	V-22(EMD)	MS3509 connector
D8565/4-1	SLA	7.5	26.0	F/A-18A/B/C/D, CH-46D/E, HH-46A, UH-46A/D, F-117A	MS27466T17B6S connector
D8565/5-1		30	80.2	C-1A, SP-2H, A-3B, KA-3B, RA-3B, ERA-3B, NRA-3B, UA-3B, P-3A/B/C, EP-3A/B/E, RP-3A, VP-3A, AC-130A/H/U, C-130A/B/E/F/H, DC-130A, EC-130E/H/G/Q, HC-130H/N/P, KC-130F/R/T, LC-130F/H/R, LC-130F/H/R, MC-130E/H, NC-130A/B/H, WC-130E/H, C-18A/B, EC-18B/D, C-137B/C, EC-137D, E-8A, TS-2A, US-2A/B, T-28B/C, QT-33A, MH-53J, MH-60G	Equivalent to D8565/5-2, except uses MS3509 connector
D8565/5-2	SLA	30	80.2	Same as D8565/5-1 (for aircraft equipped with Cannon-style mating connector)	Equivalent to D8565/5-1, except uses Cannon connector
D8565/6-1	SLA	1.5	6.4	V-22A, CV-22A, CH-47E, E-2C, S-3B	MS27466715B5S connector
D8565/7-1	SLA	24	63.9	AV-8B, TAV-8B, VH-60A, V-22A, CV-22A, MV-22B	MS3509 connector
D8565/7-2	SLA	24	63.9	Same as D8565/7-1	Replacement for D8565/7-1 with higher rate capability
D8565/8-1	SLA	15	43.0	T-45A	Cannon connector
D8565/9-1	SLA	24	63.0	T-34B/C, U-6A	MS3509 connector
D8565/9-2	SLA	24	63.0	None identified	Cannon connector
D8565/10-1	VNC	35	85.0	AH-1W	MS3509 connector; equipped with temperature sensor
D8565/11-1	SLA	10	34.8	F-4D/E/G, C-141B, MH-60E, NC-141A, YF-22A	Equivalent to D8565/11-2, except uses MS3509 connector
D8565/11-2	SLA	10	34.8	None identified	Equivalent to D8565/11-1, except uses Cannon connector
D8565/12-1	SLA	35	90.0	Developed for P-7, which got cancelled	MS3509 connector; includes heater circuit
D8565/13-1	SLA	10	31.0	Carousel IV, LTN-72 Inertial Navigation Systems (INS)	ARINC ∫ ATR case
D8565/14-1	SLA	15	45.2	F/A-18E/F	D38999/24YG11SN connector
D8565/15-1	SLA	35	90.0	C/KC-135 series	MS3509 connector
D8565/16-1	SLA	5	14.6	H-60	—
D8565/17-1	SLA	0.33	3.5	EA-6B	—
				MIL-B-8565 Specials	
MS3319-1	VNC	0.75	3.5	HH-2D, SH-2D/F	MS3106-12S-3P connector
MS3337-2	SNC	0.40	4.0	F-4S	Obsolete
MS3346-1	VNC	2.5	10.0	A-7D/E, TA-7C	Obsolete
MS3487-1	VNC	18	50.0	AH-1G	Equivalent to BB-649A/A
MS17334-2	SNC	0.33	3.5	E-1B, EA-6B, US-2D	MS3106R14S-7P connector
				MIL-B-83769 Series	
M83769/1-1	VLA	31	80.0	Same as D8565/5-1 (for aircraft equipped with Cannon style mating connector)	Supercedes AN3150; equivalent to BB-638/U; interchangeable with D8565/5-2 (Cannon connector)

TABLE 13.5 Military Aircraft Batteries (Continued)

Military Part No.	Type	Rating[a] (Ah)	Max. Wt. (lb)	Applications	Notes
M83769/2-1	VLA	18	56.0	AC-130H/U, NU-1B, U-6A	Supercedes AN3151; equivalent to BB-639/U; interchangeable with D8565/9-2 (Cannon connector)
M83769/3-1	VLA	8.4	34.0	C-141B, NC-141A	Supercedes AN3154; equivalent to BB-640/U; interchangeable with D8565/11-2 (Cannon connector)
M83769/4-1	VLA	18	55.0	T-34B/C	Supercedes MS18045-41; interchangeable with D8565/9-1 (MS3509 connector)
M83769/5-1	VLA	31	80.0	Same as D8565/5-1	Supercedes MS18045-42; interchangeable with D8565/5-1 (MS3509 connector)
M83769/6-1	VLA	31	80.0	Same as D8565/5-1 (for aircraft equipped with Cannon style mating connector)	For ground use only; equivalent to M83769/1-1 when filler caps are replaced with aerobatic vent plugs; equipped with Cannon connector
M83769/7-1	VLA	54(12V)	80.0	C-117D, C-118B, VC-118B, C-131F, NC-131H, T-33B	Supercedes MS90379-1; equipped with threaded terminals
				MIL-B-81757 Series (Tri-Service)	
M81757/7-2	VNC	10	34.0	CH-46A/D/E/F, HH-46A, UH-46A/D, U-8D/F, F-5E/F	Replaceable cells; supercedes MS24496-1 and MS24496-2
M81757/7-3	VNC	10	34.0	Same as M81757/7-2	Non-replaceable cells; supercedes MS18045-44, MS18045-48 and MS90221-66W
M81757/8-4	VNC	20	55.0	C-2A, T-2C, T-39A/B/D, OV-10A	Replaceable cells; supercedes MS24497-3, MS24497-5, and M81757/8-2
M81757/8-5	VNC	20	55.0	Same as M81757/8-4	Nonreplaceable cells; supercedes MS90365-1, MS90365-2, MS90321-68W, MS90321-77, MS90321-78W, MS18045-45, MS18048-49, and M81757/8-3
M81757/9-2	VNC	30	80.0	CT-39A/E/G, NT-39A, TC-4C, HH-1K, TH-1L, UH-1E/H/L/N, AH-1J/T, LC-130F/R, OV-1B/C/D	Replaceable cells; supercedes MS24498-1 and MS24498-2
M81757/9-3	VNC	30	80.0	Same as M81757/9-2	Nonreplaceable cells; supercedes MS18045-46, MS18045-50, MS90321-75W, MS90321-69W
M81757/10-1	VNC	6(23V)	24.0	A-6E, EA-6A, KA-6D	Non-replaceable cells; supercedes MS90447-2 and MS90321-84W
M81757/11-3	VNC	20	55.0	HH-2D, SH-2D/F/G, HH-3A/E, SH-3D/G/H, UH-3A/H, VH-3A/D	Nonreplaceable cells; supercedes MS90377-1, MS90321-79W, and M81757/11-1
M81757/11-4	VNC	20	55.0	None identified	Nonreplaceable cells with temperature sensor; supercedes MS90377-1, MS90321-79W and M81757/11-2
M81757/12-1	VNC	30	88.0	OV-10D	Nonreplaceable cells, air-cooled; supercedes D8565/2-1
M81757/12-2	VNC	30	88.0	C-2A (REPRO), OV-10D	Nonreplaceable cells, air-cooled, with temperature sensor

TABLE 13.5   Military Aircraft Batteries (Continued)

Military Part No.	Type	Rating[a] (Ah)	Max. Wt. (lb)	Applications	Notes
M81757/13-1	VNC	30	80.0	EA-3B, ERA-3B, UA-3B	Nonreplaceable cells; supercedes MS18045-75
M81757/14-1	VNC	5.5	17.5	SH-60	Low maintenance
M81757/15-1	VNC	25	53.5	H-2, H-3	Low maintenance
M81757/15-2	VNC	25	52.0	T-2	Low maintenance
M81757/16-1	VNC	35	78.0	A-10, UH-1N	Low maintenance
				MIL-B-26220 Series (U.S. Air Force)	
MS24496-1	VNC	11(C/2)	34.0	F-111A/D/E/F/G, EF-111A, FB-111A	Superceded by M81757/7-2
MS24496-2	VNC	11(C/2)	34.0	F-4D/E/G, NF-4C/D/E, NRF-4C, RF-4C, YF-4E	Superceded by M81757/7-2
MS24497-3	VNC	22(C/2)	55.0	None identified	Superceded by M81757/8-2
MS24497-4	VNC	22(C/2)	60.0	B-52H	Contains integral heater
MS24497-5	VNC	22(C/2)	55.0	B-52G, C-135, EC-135, KC-135, NC-135, NKC-135, RC-135, TC-135, WC-135, E-4B, CH-3E, NA-37B, OA-37B, OV-10A	Superceded by M81757/8-2
MS24498-1	VNC	34(C/2)	80.0	A-10A, C-20A, C-137A/B, EC-137D, OA-10A, T-37B, T-41A/B/C/D, HH-1H, UH-1N, CH-53A, MH-53J, NH-53A, TH-53A	Superceded by M81757/9-2
MS24498-2	VNC	34(C/2)	80.0	None identified	Superceded by M81757/9-2
MS27546	VNC	5	16.0	T-38A	Superceded by Marathon P/N 30030
				BB-Series (U.S. Army)	
BB-432A/A	VNC	10	34.0	CH-47A/B/C, U-8F	Equivalent to M81757/7-2
BB-432B/A	VNC	10	34.0	CH-47D	Equivalent to BB-432A/A, except includes temperature sensor
BB-433A/A	VNC	30	80.0	C-12C/D/F/L, OV-1D, EH-1H/X, UH-1H/V, RU-21A/B/C/H	Equivalent to M81757/9-2
BB-434/A	VNC	20	55.0	CH-54	Equivalent to M81757/8-4
BB-476/A	VNC	13	27.6	OH-58A/B/C	—
BB-558/A	VNC	17	38.5	OH-58D	—
BB-564/A	VNC	13	25.0	AH-64A	Superceded by BB-664/A
BB-638/U	VLA	31	80.0	None identified	Equvalent to M83769/1-1
BB-638A/U	VLA	31	80.0	None identified	Equivalent to M83769/6-1
BB-639/U	VLA	18	56.0	None identified	Equivalent to M83769/2-1
BB-640/U	VLA	8.4	34.0	None identified	Equivalent to M83769/3-1
BB-649A/A	VNC	18	50.0	AH-1E/F/P/S	Equivalent to MS3487-1
BB-664/A	VNC	13	27.0	AH-64A	
BB-678A/A	VNC	13	24.8	OH-6A	
BB-693A/U	VNC	30	83.0	Vulcan	
BB-708/U	VNC	5.5	15.0	OV-1D (Mission Gear Equipment)	
BB-716/A	VNC	5.5	17.5	EH-60A, HH-60H/J, MH-60S, SH-60B/F, UH-60A	

[a] Capacity rating is based on the one-hour rate unless otherwise noted. Voltage rating is 24 volts unless otherwise noted.

CCA: The numerical value of the current, in amperes, that a fully charged lead-acid battery can deliver at −18°C (0°F) for 30 seconds to a voltage of 1.2 volts per cell (i.e., 14.4 volts for a 24-volt battery). In some cases, 60 seconds is used instead of 30 seconds. CCA stands for cold cranking amperes.

Electrolyte: An ionically conductive, liquid medium that allows ions to flow between the positive and negative plates of a cell. In lead-acid cells, the electrolyte is a mixture of sulfuric acid (H_2SO_4)

and deionized water. In nickel-cadmium cells, the electrolyte is a mixture of potassium hydroxide (KOH) dissolved in deionized water.

Imp: The current, in amperes, delivered at 15 seconds during a constant voltage discharge at half of the nominal voltage of the battery (i.e., at 12 volts for a 24-volt battery). The Imp rating normally is based on a battery temperature of 23°C (75°F), but manufacturers generally can supply Imp data at lower temperatures as well.

Ipp: The current, in amperes, delivered at 0.3 seconds during a constant voltage discharge at half of the nominal voltage of the battery (i.e., at 12 volts for a 24-volt battery).

Ipr: Same definition as Imp.

Monobloc: A group of two or more cells connected in a series and housed in a one-piece enclosure with suitable dividing walls between cell compartments. Typical monoblocs come in 6-volt, 12-volt, or 24-volt configurations and are commonly used in lead-acid batteries but rarely in nickel-cadmium aircraft batteries.

Negative Electrode: The electrode from which electrons flow when the battery is discharging into an external circuit. Reactants are electrochemically oxidized at the negative electrode. In the lead-acid cell, the negative electrode contains spongy lead and lead sulfate ($PbSO_4$) as the active materials. In the nickel-cadmium cell, the negative electrode contains cadmium and cadmium hydroxide ($Cd(OH)_2$) as the active materials.

Nominal Voltage: The characteristic operating voltage of a cell or battery. The nominal voltage is 2.0 volts for lead-acid cells and 1.2 volts for nickel-cadmium cells. These voltage levels represent the approximate cell voltage during discharge at the C-rate under room temperature conditions. The actual discharge voltage depends on the state-of-charge, state-of-health, discharge time, rate, and temperature.

Positive Electrode: The electrode to which electrons flow when the battery is discharging into an external circuit. Reactants are electrochemically reduced at the positive electrode. In the lead-acid cell, the positive electrode contains lead dioxide (PbO_2) and lead sulfate ($PbSO_4$) as the active materials. In the nickel-cadmium cell, the positive electrode contains nickel oxyhydroxide (NiOOH) and nickel hydroxide ($Ni(OH)_2$) as the active materials.

Separator: An electrically insulating material that prevents metallic contact between the positive and negative plates in a cell but permits the flow of ions between the plates. In flooded cells, the separator includes a gas barrier to prevent gas diffusion and recombination of oxygen. In sealed cells, the separator is intended to allow gas diffusion to promote high recombination efficiency.

State-of-Charge: The available capacity of a battery divided by the capacity available when fully charged, normally expressed on a percentage basis. Sometimes referred to as "true state-of-charge."

State-of-Health: The available capacity of a fully charged battery divided by the rated capacity of the battery, normally expressed as a percentage. Sometimes referred to as "apparent state-of-charge," it can also be used in a more qualitative sense to indicate the general condition of the battery.

References

Anderman, M., Ni-Cd Battery for Aircraft: Battery Design and Charging Options, Proc. 9th Annual Battery Conf. on Applications and Advances, California State University, Long Beach, CA, 1994, pp. 12–19.

Earwicker, G.A., Aircraft batteries and their behavior on constant-potential charge, in *Aircraft Electrical Engineering*, G.G. Wakefield, Ed., Regel Aeronautical Society, United Kingdom, 1956, pp. 196–224.

Evjen, J.M. and Miller, L.D., Jr. Optimizing the Design of the Battery-Starter/Generator System, Society of Automotive Engineers (SAE) Pap. 710392, 1971.

Falk, S.U. and Salkind, A.J. Alkaline Storage Batteries, John Wiley & Sons, New York, NY, 1969, pp. 466–472.

Flake, R.A., Overview on the Evolution of Aircraft Battery Systems Used in Air Force Aircraft, Society of Automotive Engineers (SAE) Pap. 881411, 1988.

Fleischer, A., Nickel-Cadmium Batteries, Proc. 10th Annual Battery Research and Development Conf., 1956, pp. 37–41.

Gross, S., Requirements for Rechargeable Airplane Batteries, Proc. 6th Annual Battery Conference on Applications and Advances, California State University, Long Beach, CA, 1991.

Johnson, Z., Roberts, J., and Scoles, D., Electrical Characterization of the Negative Electrode of the USAF 20-Year-Life Maintenance-Free Sealed Nickel-Cadmium Aircraft Battery over the Temperature Range –40°C to +70°C, Proc. 36th Power Sources Conf., Cherry Hill, NJ, 1994, pp. 292–95.

McWhorter, T.A. and Bishop, W.S., Sealed Aircraft Battery with Integral Power Conditioner, Proc. 25th Power Sources Symposium, Cherry Hill, NJ, 1972, pp. 89–91.

Miller, G.H. and Schiffer, S.F. Aircraft Zinc-Silver Oxide Batteries, in *Zinc-Silver Oxide Batteries*, Fleischer, A., Ed., John Wiley & Sons, New York, NY, 1971, pp. 375–91.

Scardaville, P.A. and Newman, B.C., High Power Vented Nickel-Cadmium Cells Designed for Ultra Low Maintenance. Proc. 8th Annual Battery Conf. on Applications and Advances, California State University, Long Beach, CA, 1993.

Senderak, K.L. and Goodman, A.W. Sealed Lead-Acid Batteries for Aircraft Applications. Proc. 16th IECEC, 1981, pp. 117–122.

Vutetakis, D.G., Current Status of Aircraft Batteries in the U.S. Air Force, Proc. 9th Annual Battery Conference on Applications and Advances, California State University, Long Beach, CA, 1994, pp. 1–6.

Vutetakis, D.G. and Viswanathan, V.V., Determining the State-of-Health of Maintenance-Free Aircraft Batteries, Proc. 10th Annual Battery Conference on Applications and Advances, California State University, Long Beach, CA, 1995, pp. 13–18.

Vutetakis, D.G. and Viswanathan, V.V., Qualification of a 24-Volt, 35-Ah Sealed Lead-Acid Aircraft Battery, Proc. 11th Annual Battery Conference on Applications and Advances, California State University, Long Beach, CA, 1996, pp. 33–38.

Further Information

The following reference material contains further information on various as aspects of aircraft battery design, operation, testing, maintenance, and disposal.

IEC 60952-1, 2nd ed., Aircraft Batteries — Part 1: General Test Requirements and Performance Levels, 2004.

IEC 60952-2, 2nd ed., Aircraft Batteries — Part 2: Design and Construction Requirements, 2004.

IEC 60952-3, 2nd ed., Aircraft Batteries — Part 3: Product Specification and Declaration of Design and Performance (DDP), 2004.

Linden, D., Ed., *Handbook of Batteries*, 2nd ed., McGraw-Hill, New York, 1995.

NAVAIR 17-15BAD-1, Naval Aircraft and Naval Aircraft Support Equipment Storage Batteries. Request for this document should be referred to Commanding Officer, Naval Air Technical Services Facility, 700 Robbins Avenue, Philadelphia, PA 19111, 2004.

Rand, D.A.J., Moseley, P.T., Garche, J., and Parker, C.D., eds. Valve-Regulated Lead-Acid Batteries, Elsevier, Boston, 2004.

RTCA DO-293, Minimum Operational Performance Standards for Nickel-Cadmium and Lead-Acid Batteries, 2004.

Society of Automotive Engineers (SAE) Aerospace Standard AS8033, Nickel-Cadmium Vented Rechargeable Aircraft Batteries (Nonsealed, Maintainable Type), 1981.

The following companies manufacture aircraft batteries and can be contacted for technical assistance and pricing information:

Nickel-Cadmium Batteries

Acme Electric Corporation
Aerospace Division
528 W. 21st Street
Tempe, Arizona 85282
Phone (480) 894-6864
www.acmeelec.com

MarathonNorco Aerospace, Inc.
8301 Imperial Drive
Waco, Texas 76712
Phone (817) 776-0650
www.mptc.com

SAFT America Inc.
711 Industrial Boulevard
Valdosta, Georgia 31601
Phone (912) 247-2331
www.saftbatteries.com

Lead-Acid Batteries

Concorde Battery Corporation
2009 San Bernardino Road
West Covina, California 91790
Phone (800) 757-0303
www.concordebattery.com

Enersys Energy Products Inc.
(Hawker Batteries)
617 N. Ridgeview Drive
Warrensburg, MO 64093
Phone (800) 964-2837
www.enersysinc.com

Teledyne Battery Products
840 West Brockton Avenue
Redlands, California 92375
Phone (800) 456-0070
www.gillbatteries.com

14

ARINC Specification 653, Avionics Application Software Standard Interface

Paul J. Prisaznuk
ARINC

14.1 Introduction

This chapter acquaints the reader with the features of Integrated Modular Avionics (IMA) and, in particular, the avionics Real-Time Operating System (RTOS). It addresses the following questions: why use an avionics RTOS, what makes an avionics RTOS different from general computer RTOS, what are the basic concepts, and how do they work?

14.2 Why Use an Avionics Operating System?

The introduction of microelectronic technology into avionics has dramatically improved avionics equipment capability. Early Boeing 747 aircraft were produced with a three-man flight crew configuration. Twenty years later, advancements in technology have allowed the introduction of the Boeing 747-400 with a two-man crew with advanced flight deck automation. The direct benefit of weight, volume, and power savings afforded by microelectronics has been very attractive to the avionics development community.

The RTOS is defined as the interface between the microprocessor instructions and the software application that provides the avionics functionality. Microprocessor instructions are generally referred to as machine code, easy for a microprocessor to digest but difficult for even very experienced programmers to code. Even the earliest microprocessors had software development tools that allowed the programmer to compile the software code. These tool sets expanded to include linkers, loaders, code optimizers, and so forth.

Because each microprocessor is defined by its manufacture in a way that will provide greatest support for its customer base each has its own instructions, which required in-house engineering expertise for each microprocessor family. On the surface this may not seem very significant. However, when multiplied by the number of avionic computers on the airplane, the vast set of microprocessor instructions has been a problem for parties involved in aircraft design, manufacture, and operation. Airlines had avionics equipment on their airplanes each having their own unique RTOS and unique Application Program Interfaces (APIs).

A survey of modern digital avionics in the early 1990s revealed what we know today: all digital avionics have a common set of components that generally include the computer processing unit, memory, and input/output (I/O). In the same time period, it was recognized that the cost of avionics development was no longer driven by hardware; rather, cost in avionics is attributed largely to software development, which is estimated to be up to 85 percent of the overall development cost of complex avionics equipment.

14.3   Why Develop an Operating System Interface?

A standardized avionics RTOS interface is viewed to be necessary and desirable for two primary reasons. First, it establishes a known interface boundary for avionics software development, thus allowing independence of the avionics software applications and the underlying RTOS. This enables concurrent development of both RTOS and application software. Second, the standard RTOS interface allows the RTOS and the underlying hardware platform to evolve independent of the software applications. Together, this enables cost-effective upgrades over the life of the airplane.

The air transport industry has developed a standard RTOS interface definition, which is specified as the interface boundary between the avionics software applications and the RTOS proper. The standardization effort was sponsored by the airline user community, through Aeronautical Radio Inc. (ARINC), and involved many interested parties. Software specialists convened and identified the specific needs for avionics equipment:

- Safety-critical (as defined by Federal Aviation Regulations [FAR] Part 25.1309)
- Real-time (responses must occur within a prescribed time period)
- Deterministic (the results must be predictable and repeatable)

Determinism is the ability to produce a predictable outcome generally based on the preceding operations. The outcome occurs in a specified period of time with some degree of repeatability. (RTCA DO-297/ EUROCAE ED-124).

The IMA concept encourages the integration of many software functions. Mechanisms are necessary to manage the communication flows within the system, between applications and the data on which they operate, and provide control routines to be performed as a result of system level incidents that affect the aircraft operation. Clear interface specifications are necessary between both the elements within the software architecture and at the interface between the software and the other physical elements.

Software functionality within IMA equipment is provided by software applications that rely on the available resources provided by the avionics platform. The use of application software offers greater flexibility to meet user and customer needs, including functional enhancement. As the size and complexity of embedded software systems increases, modern software engineering techniques (e.g., object-oriented analysis and design, functional decomposition with structured design, top-down design, bottom-up design, etc.) are desired to aid the development process and to improve productivity. There are many advantages; for instance, software can be defined, developed, managed and maintained as modular components.

Avionics software is qualified to the appropriate level of RTCA DO-178/EUROCAE ED-12, a process that requires rigor and attention to detail. Software must be demonstrated to comply with the appropriate

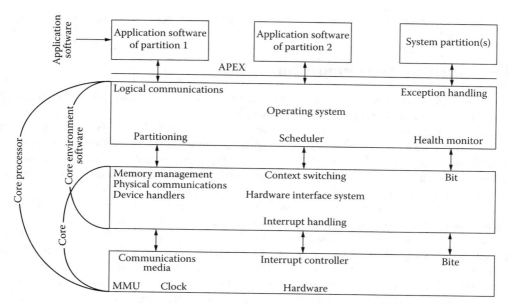

FIGURE 14.1 System architecture.

government standards set to assure safe operation. Therefore, the RTOS is designed to be simple and deterministic in its operational states.

14.4 Overall System Architecture

The ARINC 653 software interface specification has been developed for use with IMA systems primarily. However, the concepts also lend themselves to traditional federated equipment that may contain more than one partitioned function. An example of software architecture is shown in Figure 14.1. The primary components of the system are applications software that perform the avionics functions and core software that provides a standard and common environment for software applications, further divided into the following functions:

- The RTOS manages logical responses to applications and demands. It allocates processing time, communications channels, and a memory resource. This function maps application requests to systems level logical mechanisms and provides the uniform logical interface to the applications.
- The health monitor (HM) function initiates error recovery or reconfiguration strategies that perform a specific recovery action (See Chapter 22).
- The hardware interface system (HIS) manages the physical hardware resources on behalf of the RTOS. The HIS maps the logical requests made by the RTOS onto the particular physical configuration of the core hardware.

The partition is similar to that of a multitasking application within a general-purpose computer. Each partition consists of one or more concurrently executing processes sharing access to processor resources based upon the requirements of the application. All processes are uniquely identifiable, having attributes that affect scheduling, synchronization, and overall execution.

To enable software portability, communication between partitions is independent of the location of both the source and destination partition. An application sending a message to, or receiving a message from, another application will not contain explicit information regarding the location of its own host partition or that of its communications partner. The information required to enable a message to be correctly routed from source to a destination is contained in configuration tables that are developed and maintained by the system integrator, not the individual application developer. The system integrator configures the environment to ensure the correct routing of messages between partitions hosted on an IMA platform.

14.5   Software Decomposition

With the basic IMA architecture in mind, it is possible to develop software in well-defined modules. The process of structurally decomposing avionics software into well-partitioned units yields greater benefits than simply breaking the system down into manageable units; this includes grouping elements that are more likely to change, those that are dependent on specific temporal characteristics, and so on. Moreover, those elements susceptible to change can be further divided into those subject to functional enhancement and those that are implementation specific to the physical environment in which they run. This enables both functional enhancement of the software and its porting to other IMA platforms.

Multiple software applications may be hosted on an IMA platform. These applications, which can originate from different avionics sources, must be integrated into the selected implementation of the core hardware. It is necessary to create "brick walls" between applications to ensure reliable software partitioning, especially when these applications have different levels of software criticality. Detailed specifications are necessary to allow these applications to be integrated not only with each other but also with the IMA platform. Portability of applications software requires the standardization of the interface between the applications software and the IMA platform. This interface is defined as the RTOS interface.

14.6   The RTOS Interface

Among the many features of an IMA system is the RTOS interface, which defines the environment for application programs to communicate with the RTOS proper. The RTOS interface definition provides services and consists primarily of procedure calls. It places rigid bounds on the RTOS proper, the application software, and to some degree the entire IMA system.

The principal goal of the RTOS interface is to provide a general-purpose interface between the application software and the RTOS itself. Standardization of the RTOS interface allows the use of application code, hardware, and RTOS from a wide range of suppliers, encouraging competition and allowing reduction in the costs of system development and ownership:

1. The RTOS interface provides the minimum number of services consistent with fulfilling IMA requirements. The interface will consequently be easy for application software developers to efficiently use, assisting them in production of reliable products.
2. The RTOS interface is extendable to accommodate future system enhancements. It retains compatibility with previous versions of the software.
3. The RTOS interface satisfies the common real-time requirements for high-order programming languages such as Ada and C. Applications use the underlying services provided by the RTOS interface in accordance with their certification and validation criticality level.
4. The RTOS interface decouples application software from the actual processor architecture. Therefore, hardware changes will be transparent to the application software. The RTOS interface allows application software to access the executive services but insulates the application software from architectural dependency.
5. The RTOS interface specification is language-independent. This is a necessary condition for the support of application software written in different high-level languages or application software written in the same language but using different compilers. The fulfillment of this requirement allows flexibility in the selection of the following resources: compilers, development tools, and development platforms.

The RTOS interface places layering and partitioning demands on the RTOS to allow growth and functional enhancement. RTOS requests, particularly communication requests, are made in one of several modes. The application can request that the function be performed and be suspended awaiting completion of that request, or it can continue running and either poll the status of the request or merely be alerted that the transaction is complete. It is anticipated that the interface is split into groups or areas (e.g., memory management, data I/O, etc.) and that each area be further subdivided.

The processing of a communication request within the RTOS consists of setting up the appropriate I/O format and passing the message to the HIS. To ensure temporal determinism, each message definition includes a maximum and minimum time of response. Certain message types also include a timeout specification, which could be set up by the application. Hence, all communication (including requests, commands, responses, data I/O, etc.) between the applications and the RTOS is rigorously defined by this interface.

Partitioning between applications and between applications and the RTOS is controlled by standard hardware mechanisms. It can therefore be shown that if an application can be proved to correctly meet the RTOS interface then it will correctly interface with the RTOS. Given a rigorous definition of the RTOS, it is possible to integrate and test applications on a general-purpose computer that simulates the RTOS interface. It is also possible to host a variety of applications without need to modify the RTOS. Standardized RTOS message definitions allow a growth path for future enhancement of the interface. As long as any future RTOS definition is a superset of earlier definitions, no conflict will arise as a result of any enhancement.

14.7 Software Applications

Application software performs a specific avionics function. It is specified, developed, and verified to the level of criticality appropriate to its function DO-178B/ED-12B. Within an IMA platform, the method and the level of redundancy, fault detection, fault isolation, and reconfiguration is transparent to the application software.

The application software is responsible for the redundancy management for a specific function and for signal selection and failure monitoring of inputs from external sensors or other systems. Modular software design enables the implementation of software partitions to isolate avionics functions within a common hardware environment. Application software is independent of hardware, though some applications will require dedicated I/O sensors, such as a pitot-static probe.

Software modules may be developed by independent sources and integrated into the IMA platform; therefore, it is necessary for software within a partition to be completely isolated from other partitions so that one partition cannot cause adverse affects in another. To ensure partition integrity, partition load images are statically and separately built. These applications stand alone as independent program modules with no interdependencies with other program modules.

Process scheduling attributes are used by the partition code to change the execution or state of a process within that partition. In this way, the integrity of one application is not compromised by the behavior of another, whether the other application is deemed more or less critical. All communications between applications are performed through the RTOS, whose mechanisms ensure that there is no violation of the interface and that no application either monopolizes a resource or leaves another permanently suspended.

I/O handling is one significant area that is very specific to the aircraft configuration of sensors in traditional applications development. In the interest of portability and reuse, the partitioning of the applications software architecture should identify the aircraft-specific I/O software and partition it from the functional and algorithmic elements of the application. Such sensor I/O conditioning is logically defined as a separate function associated with the sensor.

Most applications require functional data at specific rates. Much of the design of applications that ties them specifically to a particular aircraft system is the sensor handling. The removal of this function from the application increases portability and, furthermore, concentrates the sensor handling into a single area, allowing sensor data with specific characteristics to be generally available to more than one application. Changes to sensor characteristics are confined to sensor data managers, thereby increasing portability and reuse of applications software. Applications software is invoked by the scheduling component of the RTOS in a deterministic manner.

14.8   The RTOS Proper

The main role of the RTOS is to ensure functional integrity in scheduling and dispatching application programs. It should be possible to prove that the level of temporal determinism (i.e., specific behavior at a specific time) is unaffected by the addition of other applications to the IMA platform. The RTOS ensures partition isolation, allocates processing time to the partitions, and dispatches processes for execution.

One method of achieving this is to implement a method of time slicing and to split applications into "strict tempo" and "scheduled" groups. Each strict tempo application is assured a specific amount of processing time in each time slice so that it can perform a certain number of defined algorithms in each time slice. If an application attempts to overrun its time slot, then it is timed out by the RTOS. The scheduling provides a sufficient amount of time to each time slice.

The RTOS is capable of recognizing both periodic and aperiodic processes and scheduling and dispatching all processes. It provides health status information and fault data for each partition. Since the RTOS will need to perform with a high degree of integrity for critical applications, it will have certification criteria that are commensurate with the collection of functions. It should, therefore, be as simple as possible.

The RTOS also manages the allocation of logical and physical resources on behalf of the applications. It is responsible for the management of memory and communications and receives interrupts associated with power failure and hardware error, relaying these incidents to the health monitor function, which in turn directs the necessary actions to enable recovery or otherwise. It should also channel application-specific software interrupts or events to the appropriate application to a defined time scale. As manager of all physical resources in this multiprogramming environment, the RTOS monitors requests for resources and controls access for those resources that cannot be used concurrently by more than one application. It has access to all memory, interrupts, and hardware resources.

The RTOS software should manage the redundancy within the IMA platform and be capable of reporting faults and executing subsequent actions. In its allocation and management of resource requests and communication interfaces from and to applications it has limited access to the applications memory. The RTOS monitors the hardware responsible for the integrity of the software partitions and communicates with the health monitor function about software and hardware integrity failures. The health monitoring function is specific to the IMA platform and to the specific aircraft installation; therefore, this software is partitioned into an RTOS health monitor and a recovery strategy table. The latter requires configuration definition and recovery strategies embedded within it. The RTOS contains an error-reporting capability that can be accessed by applications. If an application detects a fault in its operation, it is able to report this to the RTOS, which in turn invokes the health-monitoring function. It is the role of the application to inquire about health status and any reconfiguration that might have been performed. The RTOS interface enables interoperability to be achieved among application software products developed by different teams.

There are combinations of applications that have varying overall complexity. Simple applications use only the basic features of a full RTOS.

14.9   The Health Monitor Software Function

The health monitor function resides with the RTOS and interfaces with a recovery strategy table defined by the aircraft designer or system integrator. The health monitor is responsible for monitoring hardware faults within the IMA platform and functional faults within the RTOS. Hardware faults detected by the built-in test (BIT) include memory and processor faults and faults with the backplane data bus interface.

Any local hardware reconfiguration is specific to the hardware implementation, takes place within the HIS, and is transparent to the rest of the IMA system. Faults are reported to the central maintenance system, but some faults need to be reported and acted upon at a higher level; therefore, an interface between the built-in test equipment (BITE) functions and the health monitor is recommended. Faults detected within the RTOS include application violations, communication failures with remote devices, and faults detected by applications and reported back to the RTOS. The recovery table of faults is used to specify the action to be taken in response to the particular fault. This action would be initiated by the health monitor and could include the termination of an application and the initiation of an alternative application together with an appropriate level of reporting. The recovery action is largely dependent on the design of the IMA system. The health monitor should determine the need for action and initiate the recovery process.

14.10   Summary

Avionics software requirements place performance and integrity demands on the IMA platform. In addition to the underlying hardware, the IMA platform includes the core software and RTOS. The core software supports one or more avionics applications and allows independent execution of those applications. This is correctly achieved when partitioning is in place (i.e., a functional separation of the avionics applications, usually for fault containment). Partitioning is accomplished in memory space and time, which prevents any partitioned function from causing adverse affects on another. It will also ease software verification, validation, and certification.

Partitioning and functional isolation is the key to IMA. A partition can be thought of as a software program running in a single application environment. For systems that require large software applications the concept of multiple partitions providing a single application should be recognized. The RTOS interface provides a set of services to application software for control of scheduling, communication, and status information of its internal processing elements. It can be seen from the viewpoint of the application software as a High-Order Language (HOL) specification. Together, these principles enable millions of lines of avionics software code to be readily developed, approved, and updated as necessary to meet operational needs of an aircraft over its entire lifetime.

References

ARINC Report 651, *Design Guidance from Integrated Modular Avionics*, Published by ARINC, 2551 Riva Road, Annapolis, MD 21401 www.arinc.com.

ARINC Specification 653, *Avionics Application Software Standard Interface*, Published by ARINC, 2551 Riva Road, Annapolis, MD 21401 www.arinc.com.

RTCA DO-178/EUROCAE ED-12, *Software Considerations in Airborne Systems and Equipment Certification*, Published by RTCA, Washington, DC, www.rtca.com; Published by EUROCAE, Paris, France www.eurocae.org.

RTCA DO-248B/ED-94-B, *Final Report for Clarification of DO-178B, Software Considerations in Airborne Systems and Equipment Certification*, Published by RTCA, Washington, DC, www.rtca.org; Published by EUROCAE, Paris, France www.eurocae.org.

RTCA DO-255/EUROCAE ED-96, *Requirements Specification for an Avionics Computer Resource (ACR)*, Published by RTCA, Washington, DC, www.rtca.org; Published by EUROCAE, Paris, France www.eurocae.org.

RTCA DO-297/EUROCAE ED-124, *Integrated Modular Avionics (IMA) Guidance and Certification Considerations*, Published by RTCA, Washington, DC, www.rtca.org; Published by EUROCAE, Paris, France www.eurocae.org.

15
Ada

John G.P. Barnes

John Barnes Infomatics

15.1 Introduction

The world is becoming more and more concerned about both safety and security. There has been a long tradition of concern for safety going back to the development of railroads and more recently with aviation. Software now pervades all aspects of the workings of society; accordingly, it is important that software concerned with systems for which safety or security are a major concern should be correct. Vital software systems such as those that control aircraft navigation and landing have to meet well-established certification and validation criteria. More recently, there has been growing concern with security in systems such as banking and communications generally, which has been heightened by concerns about terrorist activities.

Safety and security are intertwined throughout communication. An interesting characterization of the difference is that in the case of safety, the software must not harm the world; in the case of security, the world must not harm the software. A safety-critical system is one in which the program must be correct, otherwise it might wrongly change some external device such as an aircraft flap or a railroad signal. A security-critical system is one in which it must not be possible for some incorrect or malicious input from the outside to violate the integrity of the system by, for example, corrupting a password-checking mechanism. To guard against both problems, the software must be correct in every detail. By "correct" we mean that it meets its specification. Of course, if the specification is incomplete or incorrect itself, then the system will be vulnerable.

One of the trends of the second half of the twentieth century was a concern with freedom. However, there are two aspects of freedom: the ability of individuals to do whatever they want conflicts with the rights of others to be protected from their actions. Maybe A would like the freedom to smoke in a pub, whereas B wants the freedom from smoke in a pub. Concern with health in this example is changing the balance between these freedoms. Maybe the twenty-first century will see further shifts from "freedom to" to "freedom from."

In terms of software, the languages Ada and C have very different attitudes toward freedom. Ada introduces restrictions and checks with the goal of providing freedom from errors. On the other hand, C gives the programmer more freedom to make errors. One of the historical guidelines in C was "trust the programmer"; this would be fine were it not for the fact that programmers, like all humans, are frail and fallible beings. Experience shows that whatever techniques are used, it is hard to write correct software. It is good advice, therefore, to use tools that can find and prevent bugs. Ada was specifically designed to help in this respect. There have been three versions of Ada—Ada 83, Ada 95, and now Ada 2005. Ada 83 in particular was seen by some as old-fashioned, rigid, or politically incorrect by obstructing freedom. Luckily, many key industries such as aviation have recognized the importance of the correctness of software and the contribution that these tools can make.

As the twenty-first century progresses, we will see software becoming even more pervasive. It would be nice to think that software in automobiles, for example, was developed with the same care as that in airplanes, but that is not so. My wife recently had an experience where her car displayed two warning icons. One said "stop at once," the other said "drive immediately to your dealer." Another motor story is of a driver attempting to select channel 5 on the radio; instead, the car changed into fifth gear! Luckily, he did not try "Replay."

At the time of writing, Ada is being updated, and the new version, known as Ada 2005, is expected to become a formal International Standards Organization (ISO) standard sometime during 2006. Ada offers many advantages over popular languages such as C and C++. This chapter provides a brief survey of some aspects of Ada with a strong emphasis on why Ada is good for writing software that matters and how Ada 2005 can help by illustrating some aspects of its features in comparison with other languages. It must be stressed that the discussion is not complete. Each section selects a particular topic under the banner of Safe X; however, Safe is just a brief token to designate both safety and security. For a fuller description of Ada or the related SPARK tools mentioned in the final chapter please consult the references.

15.2 Safe Syntax

Syntax is often considered to be a rather boring mechanical detail; the argument is that it is what you say that matters, not so much how you say it. That, of course, is not true. Being polite and clear are important aids to any communication in a civilized world, and a computer program is a communication between the writer and the reader, whether that reader is the compiler, another team member, a reviewer, or some other human. Indeed, most communication regarding a program is between two people. Clear and unambiguous syntax is a great aid to communication and, as we shall see, avoids a number of common errors.

An important aspect of good syntax design is to ensure that simple typographical errors cause the program to become illegal and fail to compile rather than to change its meaning. Of course, it is hard to prevent the accidental typing of X rather than Y or + rather than *, but many structural risks can be prevented. Note, incidentally, that it is best to avoid short identifiers for just this reason. If we have a financial program about rates and times, using identifiers R and T is risky since we could easily type the wrong identifier by mistake since the letters are next to each other on the keyboard. But if the identifiers are Rate and Time, then inadvertent typing of Tate or Rime will be caught by the compiler. This applies to any language, of course.

15.2.1 Equality and Assignment

It is obvious that assignment and equality are different things. If we do an assignment, we change the state of some variable. On the other hand, equality is simply an operation to test some state. Changing state and testing state are very different things, and understanding the distinction is important; however, poor programming language design has confused these fundamentally different logical operations.

In the earliest days of Fortran and autocodes one wrote:

X = X + 1

However, this is really rather peculiar. In mathematics, it is never the case that x equals $x + 1$. What the Fortran statement means, of course, is "replace the current value of X with the old value plus one." But why misuse the equals sign in this way when society has been using the equals sign to mean equals for hundreds of years? (The equals sign dates from 1543 when it was introduced by the English mathematician Robert Record.) The designers of Algol 60 recognized the problem and used the combination of a colon followed by an equals sign to mean assignment, thus:

X := X + 1;

This has the helpful consequence that the equals sign can unambiguously be used to mean just that as in

if X = 0 **then** ...

The C language (like Fortran) hijacked the equals sign for assignment and, as a consequence, C uses a double equals (==) to mean equality. This can cause much confusion.

Here is a fragment of C program controlling the crossing gates on a railroad:

```
if (the_signal == clear)
{
  open_gates( ... );
  start_train( ... );
}
```

The same program in Ada might be

```
if The_Signal = Clear then
  Open_Gates( ... );
  Start_Train( ... );
end if;
```

Now consider what happens if a programmer gets confused and accidentally forgets one of the equals signs in C:

```
if (the_signal = clear)
{
  open_gates( ... );
  start_train( ... );
}
```

This still compiles, but instead of just testing the_signal, it actually assigns the value clear to the_signal. Moreover, C confuses expressions, which have values, with assignments, which change state, so the assignment also acts as an expression, and the result of the assignment is then used in the test. If the encoding is such that clear is not zero, then the result will be true and the gates will always be opened, the_signal set to clear, and the train started on its perilous journey.

If the Ada programmer were to accidentally use an assignment in the test,

```
if The_Signal:= Clear then    -- illegal
```

then the program will simply fail to compile and all will be well.

It is important to note that C really has two problems here. It misuses = for assignment and so uses == for equality. That would not be so bad if it did not also confuse expressions with statements, which is very similar to the confusion between evaluating state (using an expression) and changing state (using a statement). But, as we have seen with the railroad example, using assignment with expressions and a simple syntax error regarding the equals sign leads to a lethal combination. Ada does not permit expressions and statements to be confused and so this very dangerous combination cannot arise.

15.2.2 Statement Groups

It is often necessary to group a sequence of statements together — for example, following a test using a keyword such as "if." There are two typical ways of doing this:

- Bracketing the group of statements so that they act as one (as in C)
- Closing the sequence with something matching the "if" (as in Ada)

These are also illustrated by the railroad example. The statements to open the gates and to start the train both need to be obeyed if the condition is true.

Now, suppose we inadvertently add a semicolon at the end of the first line of the C program (easily done). The program becomes

```
if (the_signal == clear) ;
{
  open_gates( ... );
  start_train( ... );
}
```

We now find that the condition is governing the null statement, which is between the test and the newly inserted semicolon. We cannot see it because a null statement is just nothing. Therefore, no matter what the state of the signal, the gates are always opened and the train set to going.

In Ada, the corresponding error would result in

```
if The_Signal = Clear then ;    -- illegal
  Open_Gates( ... );
  Start_Train( ... );
end if;
```

This is syntactically incorrect, so the error is safely caught by the compiler and the train wreck cannot occur.

15.2.3 Named Notation

Another feature of Ada, which is of a syntactic nature and can detect many unfortunate errors, is the possibility of using named associations in various situations. Dates provide a good illustration because the order of the components varies according to local culture. Thus, 12 January 2006 is written in Europe as 12/01/06, but in the United States it is usually written as 01/12/06, and the ISO standard gives the year first, so the date would be 06/01/12.

In C, we might declare a structure for manipulating dates as

```
struct {
  int day, month, year;
  } date;
```

which corresponds to the following type declaration in Ada

```
type Date is
  record
  Day, Month, Year: Integer;
  end record;
```

In C, we might write

```
date today = (01, 12, 06);
```

However, without looking at the type declaration, we do not know whether this means 1 December 2006, 12 January 2006, or even 6 December 2001.

In Ada we have the option of writing

```
Today: Date := (Day => 1, Month => 12, Year => 06);
```

which uses the so-called named association. Now it will be crystal clear if we write the values in the wrong order. We can also write the declaration as

```
Today: Date := (Month => 12, Day => 1, Year => 06);
```

which has the correct meaning and has the advantage that we do not need to remember the order in which the fields are declared.

Named associations can be used in other contexts in Ada as well. We might make similar errors with a function in which several parameters are of the same type. Suppose we have a function to compute a person's obesity index. The two parameters would be the height and the weight, which could be given as floating point values in pounds and inches (or kilograms and centimeters). So we might have

```
float index(float height, weight) {
  ...
  return ... ;
}
```

or in Ada

```
function Index(Height, Weight: Float) return Float is
  ...
  return ... ;
end;
```

In the case of the author, the appropriate call of the index function in C might be

```
my_index = index(68.0, 168.0);
```

But if by mistake the call were reversed

```
my_index = index(168.0, 68.0);
```

then we would have a very thin and very tall giant! Such an unhealthy disaster can be avoided in Ada by using named parameter calls:

```
My_Index := Index(Height => 68.0, Weight => 168.0);
```

Again, we can give the parameters in whatever order we wish, and no error will occur if we forget the order in the declaration of the function.

Named notation is a very valuable feature of Ada. Its use is optional, but it is well worth using freely since not only does it help to prevent errors but it also makes the program more readable.

15.3   Safe Typing

Safe typing (also known as strong typing) is about using the type structure of the language to prevent errors. Early languages such as Fortran and Algol treated everything as numeric types. Of course, at the end of the day, everything is indeed held in the computer as a numeric of some form, usually as an integer or floating point value encoded using a binary representation. Later languages, starting with Pascal, began to recognize that there was merit in taking a more abstract view of the objects being manipulated. Even if they were ultimately integers, there was much benefit in treating colors as colors and not as integers by using enumeration types. Modern languages such as Ada take this idea much further, as we shall see; sadly, however, C and C++ still treat scalar types as raw numeric types.

15.3.1   Using Distinct Types

Suppose we are monitoring some engineering production and checking for faulty items by counting the number of good ones and bad ones. We want to stop production if the number of bad ones reaches some limit and perhaps also stop when the number of good ones reaches some other limit. In C we might have variables

```
int badcount, goodcount;
int b_limit, g_limit;
```

and then perhaps

```
badcount = badcount + 1;
...
if (badcount == b_limit) { ... };
```

and similarly for the good items. Since everything is really an integer, there is nothing to prevent us from writing by mistake

```
if (goodcount == b_limit) { ... }
```

where we really should have written g_limit. This may have been a cut and paste error or a simple typo (g is next to b on a qwerty keyboard). We could do the same in any language. However, Ada gives us the opportunity to be more precise. We can write

```
type Goods is new Integer;
type Bads is new Integer;
```

These declarations introduce new types that have all the properties of the predefined type `Integer` (such as operations +, −) and indeed are implemented in the same way. We can now write

```
Good_Count, G_Limit: Goods;
Bad_Count, B_Limit: Bads;
```

Now we have distinct groups of entities for our manipulation, and any accidental mixing will be detected by the compiler and prevent the incorrect program from running. So we can happily write

```
Bad_Count := Bad_Count + 1;
if Bad_Count = B_Limit then
```

but we are prevented from writing

```
if Good_Count = B_Limit then    -- illegal
```

since this is a type mismatch. If we did indeed want to mix the types, perhaps to compare the bad items and good items, then we can do a type conversion (a cast) to change type. Thus, both of the following are permitted:

```
if Good_Count = Goods(B_Limit) then
if Integer(Good_Count) = Integer(Bad_Count) then
```

In the first case, we have converted the bad limit to type `Goods`; in the second case, we have converted both sides of the test to the parent type `Integer`. We could use the same technique to avoid accidental mixing of floating types. Thus, when dealing with weights and heights in Section 15.2, rather than

```
My_Height, My_Weight: Float;
```

it would better to write

```
type Inches is new Float;
type Pounds is new Float;
My_Height: Inches := 68.0;
My_Weight: Pounds := 168.0;
```

In this case, confusion between the two would be trapped by the compiler.

15.3.2 Enumerations and Integers

Section 15.2 provided an example of a railroad crossing, which included a test in C and Ada respectively:

```
if (the_signal == clear) { ... };
if The_Signal = Clear then ... end if;
```

In C, the variable `the_signal` and associated states such as `clear` will be of type `int` and could be declared thus:

```
int the_signal;
int danger = 0, caution = 1, clear = 2;
```

where we have given code values 0, 1, and 2 to the three possible states of the signal. But this is very unwise. Nothing prevents us from inadvertently changing the value of `caution` to 2. We can overcome this in C by using const or define. But nothing can prevent us from assigning a silly value such as 4 to the_signal, and so on. Moreover, other parts of the program might be concerned with chemistry and use the states anion and cation; nothing could prevent confusion between cation and caution. We might also be dealing with girls names such as Betty and Clare or weapons such as dagger and spear. Nothing prevents confusion between dagger and danger or clare and clear.

In Ada we would write

```
type Signal is (Danger, Caution, Clear);
The_Signal: Signal := Danger;
```

and no confusion could ever arise. If we also had

```
type Ions is (Anion, Cation);
type Names is (Anne, Betty, Clare, ... );
type Weapons is (Arrow, Bow, Dagger, Spear);
```

then the compiler would prevent the compilation of a program that mixed these things up. Moreover, the compiler would prevent us from assigning to Clear or Danger since these are literals, and this would be as stupid as, for example, trying to change the value of a literal integer such as 5 by writing

```
5 := 2 + 2;
```

At the machine level, the various enumeration types are indeed encoded as integers and we can access the encodings if we really need to by, for example,

```
Danger_Code: Integer := Signal'Pos(Danger);
```

We can even specify our own encodings. Incidentally, a very important built-in type in Ada is the type Boolean, which can be thought of as having the declaration

```
type Boolean is (False, True);
```

The result of tests such as The_Signal = Clear is of the type Boolean, and there are operations such as **and**, **or**, and **not**, which operate on Boolean values. It is never possible in Ada to treat an integer value as a Boolean or vice versa. In C, it will be recalled, tests just give integer values, and zero is treated as false and nonzero as true. Again, one can see the danger in

```
if (the_signal == clear)
{
...
};
```

Omitting one equals turns the test into an assignment; because C permits an assignment to act as an expression, the syntax is acceptable. The further sin is to treat the integer result as a Boolean for the test. So altogether, C has several flaws in this one example. It is the combination of all these flaws that makes this particular construction so unsafe in C:

- Equals should not have been hijacked for assignment.
- Assignments should not have been permitted as expressions.
- Integers and Booleans should not have been confused.

15.3.3 Constraints and Subtypes

It is often the case that we know that a certain variable is not ever going to be outside some range of sensible values. If so, we should say so, thereby making explicit in the program some assumption about the external world. Thus, My_Weight could never be negative and would hopefully never exceed 300 pounds. So we can declare

```
My_Weight: Float range 0.0 .. 300.0;
```

or, if we had been good and declared a type Pounds:

```
My_Weight: Pounds range 0.0 .. 300.0;
```

If, by mistake, the program generates a value outside this range and attempts to assign it to My_Weight:

```
My_Weight := Compute_Weight( ... );
```

then the exception Constraint_Error will be raised (or thrown). We might handle (or catch) this exception in some other part of the program and take remedial action. If we do not, the program will stop and the runtime system will produce an error message saying where the violation occurred. This all happens automatically — appropriate checks are inserted into the compiled code.

This idea of subranges was first introduced in Pascal and improved in Ada, but it is not available in C and most other languages. We would have to program our own checks all over the place; it is more likely, however, that we wouldn't bother and any error will be that much harder to detect.

If we knew that every weight to be dealt with by the program was in a restricted range then rather than putting a constraint on every variable declaration, we can impose it on the type Pounds in the first place.

```
type Pounds is new Float range 0.0 .. 300.0;
```

On the other hand, if some weights in the program are unrestricted and it is only the weight of people that are known to lie in a restricted range, we might write

```
type Pounds is new Float;
subtype People_Pounds is Pounds range 0.0 .. 300.0;
My_Weight: People_Pounds;
```

We can also apply constraints and declare subtypes of integer types and enumeration types. Thus, when counting good items, we would assume that the number was never negative and perhaps that it would never exceed 1000. So we might have:

```
type Goods is new Integer range 0 .. 1000;
```

If we just wanted to ensure that it was never negative but did not wish to impose an upper limit then we could write:

```
type Goods is new Integer range 0 .. Integer'Last;
```

where Integer'Last gives the upper value of the type Integer. The restriction to positive or non-negative values is so common that the Ada language has the following subtypes built in:

```
subtype Natural is Integer range 0 .. Integer'Last;
subtype Positive is Integer range 1 .. Integer'Last;
```

The type Goods could then be declared as

```
type Goods is new Natural;
```

and this would just impose the lower limit of zero as required.

Inserting constraints as in the above examples may seem to be tiresome, but it makes the program clearer. Moreover, it enables the compiler and runtime system to verify that the assumptions being expressed by inserting the constraints are indeed correct.

15.3.4 Arrays and Constraints

An array is an indexable set of things. We might have an array of 12 integers. Perhaps we are playing with a pair of dice and wish to record how many throws of each value have been obtained. (There are actually only 11 possibilities because the lowest throw is 2 — we will come back to this in a moment.) In C, we might write:

```
int counters[13];
int throw;
```

and this will in fact declare 13 variables referred to as counters[0] to counters[12] and a single integer variable throw. If we wish to record another throw, we might have something like

```
throw = ... ;
counters[throw] = counters[throw] + 1;
```

Now suppose the counting mechanism goes wrong, and a throw of 13 is generated. The C program does not trap the error but simply computes where the thirteenth element would be and adds one to that location. Almost certainly this will be the variable throw itself since it is declared after the array and it will become 14. The program just goes hopelessly wrong. This is an example of the infamous buffer overflow problem at the heart of the majority of serious programming problems.

Now consider the same program in Ada. We can write

```
Counters: array (1 .. 12) of Integer;
Throw: Integer;
```

and then

```
Throw := ... ;
Counters(Throw) := Counters(Throw) + 1;
```

Now if Throw has a rogue value such as 13, since Ada has runtime checks to ensure that we cannot read or write to a part of an array that does not exist, the exception Constraint_Error is raised, and the program does not run wild.

Note that Ada gives control over the lower bound of the array as well as the upper bound. Arrays in Ada do not all start at zero; lower bounds in real programs are more often one than zero. Only internal computer structures typically have a lower bound of zero. Indeed; in this particular example, the lower bound should really be 2, thus

```
Counters: array (2 .. 12) of Integer;
```

The reader might feel that the checks in Ada must surely slow down the program. In fact, various experiments have shown that the impact is very small. However, you can turn the checks off selectively if you are really, really sure that the program is correct by writing

```
pragma Suppress(Index_Check);
```

The problem with the dice program was not so much that the array was violated (that was the symptom) but rather that the value in Throw was incorrect, so we can catch the mistake earlier by declaring a constraint on Throw:

```
Throw: Integer range 2 .. 12;
```

Now Constraint_Error is raised when we try to assign 13 to Throw. The compiler is able to deduce that Throw always has a value appropriate to the range of the array, and no checks will actually be necessary for accessing the elements of the array. As a consequence, placing a constraint on variables used for indexing typically reduces the number of checks overall. Incidentally, we can reduce the double appearance of the range 2 .. 12 by writing

```
Throw: Integer range 2 .. 12;
Counters: array (Throw'Range) of Integer;
```

or even

```
subtype Dice_Range is Integer range 2 .. 12;
Throw: Dice_Range;
Counters: array (Dice_Range) of Integer;
```

The advantage of only writing the range once is that if we need to change the program (perhaps adding a third die so that the range becomes 3 ... 18), this only has to be done in one place.

Range checks in Ada are of enormous practical benefit during testing and can be turned off for a production program. Ada compilers are not unique in applying runtime checks in programs, but Ada is perhaps unique in specifying the checks in the language definition itself and not leaving it as an implementation nicety.

Perhaps it should also be mentioned that we can give names to array types as well. If we had several sets of counter values, it would be better to write

```
type Counter_Array is array (Dice_Range) of Integer;
Counters: Counter_Array;
Old_Counters: Counter_Array;
```

and if we wanted to copy all the elements of the array Counters into the corresponding elements of the array Old_Counters, then we simply write

```
Old_Counters := Counters;
```

Note that no checks are required because the ranges are bound to match.

Giving names to array types is not possible in many languages such as C and C++. The advantage of naming types is that it introduces more abstractions, as when counting the good and bad items. By telling the compiler more about what we are doing, we provide it with more opportunities to check our program.

15.4 Safe Pointers

Software made a big leap forward in its possibilities when the notion of pointers or references was introduced. But playing with pointers is like playing with fire; pointers can bring enormous benefits but, if misused, can bring immediate disaster with a blank screen or even a rampaging program destroying data or the loophole through which a virus can invade. High integrity software typically avoids pointers like the plague, but Ada provides pointers in the form of access types, and the language is constructed in such a way that they are safe.

In the basic C model, the machine works with addresses that are made directly available to the user in the form of pointers. Thus, given an address, we can do arithmetic upon it, thereby creating an address that does not point to a sensible location. A fundamental difficulty is that the addressing of arrays is treated simply as equivalent to the manipulation of pointers. C++ follows the C philosophy and is equally unsafe. Java works entirely by pointers, but this is hidden from the user to a large extent. It is fairly safe but inflexible for many purposes. Ada 95 and Ada 2005 have managed to combine safety and flexibility through a combination of strong typing and the notion of accessibility.

There are two dangers in manipulating pointers. One is that we might point to an object of the wrong type; the other is that we might point to something that no longer exists (we are then pointing to where it existed and that space might be now occupied by nothing in particular). Ada prevents the first problem by the typing rules and the second by accessibility rules. Perhaps because pointers and references have a bad history, Ada uses the term "access types." This presents the abstract view that values of access types give access to other objects and should not be thought of as raw pointers, addresses, or references.

15.4.1 Access Types and Strong Typing

We can declare a variable whose values give access to objects of type T by

```
X: access T;
```

If we do not give an initial value, then a special value **null** is assumed. X can refer to a normal declared object of type T (which must be marked **aliased**) by

```
Obj: aliased T;
...
X := Obj'Access;
```

Typically T will be a record type such as

```
type Date is
  record
    Day: Integer range 1 .. 31;
    Month: Integer range 1 .. 12;
    Year: Integer;
  end record;
```

So, we might have

```
Birthday: aliased Date := (Day => 10, Month => 12, Year => 1815);
AD: access Date := Birthday'Access;
```

Then, to retrieve the individual components of the date referred to indirectly by AD we can write, for example,

```
The_Day: Integer := AD.Day;
```

A variable such as AD can also refer to an object allocated on the heap (called a storage pool in Ada). We can write

```
AD := new Date'(Day => 27, Month => 11, Year => 1852);
```

A common application is to create linked lists — we might declare

```
type Cell is
  record
    Next: access Cell;
    Value: Integer;
  end record;
```

Then we can create chains of objects of the type Cell linked together. Sometimes it is convenient to give a name to an access type:

```
type Date_Ptr is access all Date;
```

Here, the **all** refers to the fact that this named type can refer to both objects on the heap and to those declared normally and marked as **aliased**. Having to mark objects as **aliased** is a useful safeguard that alerts the programmer to the fact that the object might be referred to indirectly (good for walkthrough reviews), and it also tells the compiler that the object should not be optimized into a register where it would be difficult to access indirectly.

Access types always give the type of the object they refer to and so the strong typing prevails. Moreover, an access object always has a legitimate value, although it could be **null**. So, whenever we attempt to access an object referred to by an object of the type Date_Ptr, there is a check to ensure that the value is not null, and the exception Constraint_Error is raised if this check fails.

We can explicitly state that an access object cannot have the null value by writing, for example,

```
WD: not null access Date := Wedding_Day'Access;
```

Of course, then it must be given an initial value that is not null. The advantage of a so-called null exclusion is that we are guaranteed that an exception cannot occur when accessing the indirect object.

Finally, note that we can refer to a compound structure provided the components are marked as aliased. For example,

```
A: array (1 .. 10) of aliased Integer := (1,2,3,4,5,6,7,8,9,10);
P: access Integer := A(4)'Access;
```

But we cannot perform any incremental operations on P such as P++ or P+1 to make it refer to A(5) as can be done in C. This sort of thing in C is prone to errors since nothing prevents us from pointing beyond the end of the array.

15.4.2   Access Types and Accessibility

Strong typing ensures that an access value can never refer to an object of the wrong type. The other problem is to ensure that the object referred to cannot cease to exist while access objects still refer to it. The notion of accessibility ensures this. Consider

```
package Data is
  type AI is access all Integer;
  Ref1: AI;
end Data;
procedure P is
  K: aliased Integer;
  Ref2: AI;
```

```
begin
   Ref2 := K'Access;      -- illegal
   Ref1 := Ref2;
   ...
end P;
```

This is clearly a very artificial example, but it illustrates the key points in a small space. The package Data has an access type AI and an object of that type called Ref1. The procedure P declares a local variable K and a local access variable Ref2, also of the type AI, and attempts to assign an access to K to the variable Ref2. This is forbidden. It is not so much that the reference to Ref2 is dangerous — because both Ref2 and K will cease to exist when we return from a call of the procedure P — the danger is that we might assign the value in Ref2 to the global variable Ref1, and that would then contain a reference to K after K had ceased to exist.

The basic rule is that the lifetime of the accessed object (such as K) must be at least as long as the lifetime of the access type (in this case, AI). If this is not true, the assignment to Ref2 (which is of type AI) is illegal. The rules are phrased in terms of accessibility levels and are mostly static; that is, they are checked by the compiler and incur no cost at run time. But the rules concerning the parameters of subprograms that are of anonymous access types are dynamic, and this gives more flexibility than would otherwise be possible. In this short introduction to Ada, it is not feasible to go into the details; however, suffice it to say that the accessibility rules of Ada prevent dangling references, which can be a source of many errors in other languages.

15.4.3 References to Subprograms

Ada permits references to procedures and functions to be manipulated in a similar way to references to objects. Both strong typing and accessibility rules apply. For example, we can write

```
A_Func: access function (X: Float) return Float;
```

and A_Func is then an object that can only refer to functions that take an argument of the type Float and return an argument of type Float (such as the predefined function Sqrt), or it can refer to **null** as it does by default until we assign a value to it. So we can write

```
A_Func := Sqrt'Access;
...
X: Float := A_Func(4.0);    -- indirect call
```

and this will call Sqrt with argument 4.0 and hopefully produce 2.0.

Ada thoroughly checks that the parameters and result always match properly, so we cannot call a function indirectly that has the wrong number or types of parameters. Thus, consider the predefined function Arctan (the inverse tangent). It takes two parameters

```
function Arctan(Y: Float; X: Float) return Float;
```

where X and Y are the two sides of the right-angled triangle concerned. If we attempt to write

```
A_Func := Arctan'Access;    -- illegal
Z := A_Func(A);             -- indirect call prevented
```

then we are stopped because the profile of Arctan does not match that of A_Func. This is just as well because otherwise the function Arctan would remove two items from the machine stack whereas the indirect call via A_Func placed only one parameter on the stack. This would result in a disaster. Such checks in Ada also occur across compilation unit boundaries. Equivalent mismatches are not prevented in C, and this is a common cause of serious errors.

More complex situations arise because a subprogram can have another subprogram as a parameter. Thus, we might have a function to solve an equation $F(x) = 0$, where the function to be solved is itself a parameter. Thus:

```
function Solve(Trial: Float; Acc: Float;
               Fn: access function (X: Float) return Float) return Float;
```

The parameter `Trial` is the initial guess, the parameter `Acc` is the accuracy required, and the third parameter `Fn` represents the equation to be solved.

As an example, suppose we invest 1000 dollars today and 500 dollars in a year's time; what would the interest rate have to be for the final value two years from now to be exactly 2000 dollars? If the interest rate is $X\%$ then the final value will be

$$1000 \times (1 + X/100)^2 + 500 \times (1 + X/100)$$

We can do this by declaring

```
function Npw(X: Float) return Float is
   Factor: constant Float := 1.0 + X/100.0;
begin
   return 1000.0 * Factor**2 + 500.0 * Factor - 2000.0;
end Npw;
```

and then

```
Answer: Float := Solve(Trial => 5.0, Acc => 0.01, Fn => Npw'Access);
```

We are guessing that the answer might be 5%, we need two decimal figures of accuracy, and, of course, `Npw'Access` identifies the problem. The reader is invited to estimate the interest rate — the answer is at the end of Section 15.4.4. (Note that npw is short for Net Present Worth, a term familiar to financial professionals.)

The point of this discussion is to emphasize that Ada checks the matching of the parameters of the slave function as well. Indeed, the nesting of profiles can continue to any degree and Ada matches all levels thoroughly. Many languages give up after one level.

Note that the parameter `Fn` was actually of an anonymous type. Access to subprogram types can be named or anonymous just like access to object types. They can also have a null exclusion. Thus, we should really have written

```
A_Func: not null access function (X: Float) return Float := Sqrt'Access;
```

The advantage of using a null exclusion is that there is no check that the value of A_Func is not null when the function is called indirectly. Similarly, we should really add **not null** to the profile in `Solve`:

```
function Solve(Trial: Float; Acc: Float;
               Fn: not null access function (X: Float) return Float) return Float;
```

The advantage is that `Solve` will be slightly faster because it will not have to check whether the actual function is null on every call of `Fn`.

15.4.4 Downward Closures

We mentioned that accessibility rules also apply to access to subprogram values. Suppose we had declared Solve so that the parameter Fn was of a named type and that Fn and Solve are in some library package

```
package Algorthms is
  type A_Function is not null access function (X: Float) return Float;
  function Solve(Trial: Float; Acc: Float; Fn: A_Function) return Float;
  ...
end Algorthms;
```

Suppose we now attempt the interest example and pass the target value as a parameter. We try

```
with Algorithms;  use Algorithms;
function Compute_Interest(Target: Float) return Float is
  function Npw(X: Float) return Float is
    Factor: constant Float := 1.0 + X/100.0;
  begin
  return 1000.0 * Factor**2 + 500.0 * Factor - Target;
  end Npw;
begin
  return Solve(Trial => 5.0, Acc => 0.01, Fn => Npw'Access); -- illegal
end Compute_Interest;
```

This is not allowed because it breaks the accessibility rules. The trouble is that the function Npw is at an inner level to the type A_Function. (It has to be in order to get hold of the parameter Target.) If this had been allowed, then we could have assigned the value Npw'Access to a global variable of the type A_Function so that when Compute_Interest had returned, we still would have had a reference to F even after it had ceased to exist, thus

```
Dodgy_Fn: A_Function := Default'Access;
function Compute_Interest(Target: Float) return Float is
  ...
begin
  Dodgy_Fn := Npw'Access;        -- illegal
end Compute_Interest;
Answer := Dodgy_Fn(99.9); -- would be nasty
```

The call of Dodgy_Fn (if we got that far) would attempt to call Npw, but that no longer exists since it is local to Compute_Interest. Anything could happen if Ada did not prevent it. However, if we use an anonymous type for the parameter as in the previous section, all is well. Assignment of the local access value is prohibited by accessibility rules, so the solution is just to change the package Algorithms:

```
package Algorthms is
  function Solve(Trial: Float; Acc: Float;
  Fn: not null access function (X: Float) return Float) return Float;
end Algorthms;
```

The function Compute_Interest is now exactly as before (except that the comment -- illegal needs to be removed).

Downward closures are used in several parts of the Ada predefined library for applications such as iterating over a data structure. The nesting of subprograms is a natural requirement for these applications because of the need to pass nonlocal information, which is not possible in flat languages such as C, C++, and Java. It should be observed that there are contorted workarounds in some flat languages using type extension, but these are very hard to understand and can be a problem for program maintenance.

Finally, the interest rate that turns the investment of 1000 dollars and 500 dollars into 2000 dollars in two years is about 18.6%.

15.5   Safe Architecture

Good architecture in a program, like that in a builidng, should provide unobtrusive safety for the detailed workings of the inner parts within a clean framework and permit interaction where appropriate and prevent unrelated activities from accidentally interfering with each other. And a good language should enable the writing of programs with good architecture.

The structure of an Ada program is based primarily around the concept of a package, which groups related entities together and provides a natural framework for hiding implementation details from the clients.

15.5.1   Package Specifications and Bodies

Early languages such as Fortran have a flat structure with everything essentially at the same level. As a consequence, all data (other than that local to a subprogram) is visible everywhere. This same flat structure applies to C. Other languages, such as Algol and Pascal, have a simple block structure, which is a bit better but still causes big communication problems.

Consider the simple problem of a stack of numbers. We want to have a protocol in which an item can be added to the stack by calling a procedure Push, and the top item can be removed from the stack by calling a function Pop, and perhaps we also want a procedure Clear to set the stack to an empty state. We do not want any other means of manipulating the stack since this might be confusing.

Now, consider the following implementation of a stack written in Pascal. The stack is represented by an array of reals and there are three operations: Push and Pop to add items and remove items respectively and Clear to set it to empty. We also declare a constant max and give it a suitable value such as 100, which avoids the need to write 100 in several places (this would be bad if we changed our minds about the size of the stack).

```
const max = 100;
var top : 0 .. max;
     a : array[1..max] of real;
procedure Clear;
begin
     top := 0
end;
procedure Push(x : real);
begin
     top := top + 1;
     a[top] := x
end;
function Pop : real;
begin
     top := top - 1;
     Pop:= a[top + 1]
end
```

The main trouble with this is that max, top, and a have to be declared outside Push, Pop, and Clear so that they can all be accessed. And from any part of the program from which we can call Push, Pop, and Clear, we can also change a and top directly and bypass the protocol. This is a source of danger. If we want to monitor how many times the stack is changed, then adding statements to Push, Pop, and Clear to do this is not adequate. Similarly, if we are reviewing a large program and are looking for all places where the stack is changed, then we have to track all references to top and a as well as the calls of Push, Pop, and Clear.

This problem applies to C as well as to Fortran and Pascal. These overcome the problem languages to some extent by adding some form of separate compilation facility. Those entities which are to be visible to other separately compiled units can then be marked by special statements such as EXTRN or by using a header file. However, by its very nature, separate compilation is itself flat and unstructured.

The technique in Ada is to use a package that encapsulates and hides the data shared by Push, Pop, and Clear so that only those procedures can see it. A package comes in two parts: its specification, which describes its interface to other units, and its body, which describes how it is implemented. We can paraphrase this by saying that the specification says what it does and the body says how it does it. The specification would simply be

```
package Stack is
  procedure Clear;
  procedure Push(X: Float);
  function Pop return Float;
end Stack;
```

This just describes the interface to the outside world, so on the outside, all that is available are the three subprograms. The specification gives just enough information for the external client to write calls of the subprograms and for the compiler to compile the calls. The body could then be written as

```
package body Stack is
  Max: constant := 100;
  Top: Integer range 0 .. Max := 0;
  A: array (1 .. Max) of Float;
  procedure Clear is
  begin
    Top := 0;
  end Clear;
  procedure Push(X: Float) is
  begin
    Top := Top + 1;
    A(Top) := X;
  end Push;
  function Pop return Float is
  begin
    Top := Top - 1;
    return A(Top + 1);
  end Pop;
end Stack;
```

The body gives the full details of the subprograms and also declares the hidden objects Max, Top, and A.

Now we know that the required protocol is enforced — the client cannot accidentally or malevolently interfere with the inner workings of the stack. Although this has nothing to do with separate compilation, nevertheless it is often convenient to compile the specification and body separately. So, we have three entities to consider: the specification, the body, and of course, the client.

There are important rules concerning the compilation of these entities. The client cannot be compiled without the specification being available, and the body also cannot be compiled without the specification being available. But there are no similar constraints relating to the client and the body. If we decide to change the details of the implementation and this does not require the specification to be changed, then the client does not have to be recompiled.

In order to use a package such as Stack, the client code must mention the Stack in a **with clause**. For example

```
with Stack;
procedure Client is
begin
  Stack.Clear;
  Stack.Push(37.4);
  ...
```

The package is mentioned each time an entity in it is used; this ensures that the client code is very clear as to what it is doing. Sometimes repeating the package name is tedious, so we can add a **use clause**:

```
with Stack;   use Stack;
procedure Client is
begin
  Clear;
  Push(37.4);
  ...
```

In conclusion, the specification defines a contract between the client and the body. The body promises to implement the specification, the client promises to use the package as described by the specification, and the compiler ensures that both sides stick to the contract. We will come back to these thoughts in Section 15.9, when we look into the ideas behind the SPARK toolset. A vital point about Ada is that the strong type matching is enforced across compilation unit boundaries.

15.5.2 Private Types

Another feature of a package is that part of the specification can be hidden from the client by using a so-called private part. The package Stack only implements a single stack, but it might be more useful to declare a package that enables us to declare many stacks. To do this, we need to introduce the concept of a stack as a type. We might write

```
package Stacks is            -- visible part
  type Stack is private;     -- private type
  procedure Clear(S: out Stack);
  procedure Push(S: in out Stack; X: in Float);
  procedure Pop(S: in out Stack; X: out Float);
private                      -- private part
  Max: constant := 100;
  type Vector is array (1 .. Max) of Float;
  type Stack is              -- full type
    record
      A: Vector;
      Top: Integer range 0 .. Max:=0;
```

```
      end record;
   end Stacks;
```

The body would then be

```
   package body Stacks is
     procedure Clear(S: out Stack) is
     begin
       S.Top := 0;
     end Clear;
     procedure Push(S: in out Stack; X: in Float) is
     begin
       S.Top := S.Top + 1;
       S.A(Top) := X;
     end Push;
       -- procedure Pop similarly
   end Stacks;
```

The user can now declare lots of stacks and act on them individually thus

```
   with Stacks; use Stacks;
   procedure Main is
     This_One: Stack;
     That_One: Stack;
   begin
     Clear(This_One);   Clear(That_One);
     Push(This_One, 37.4);
     ...
```

The detailed information about the type Stack is given in the private part of the package and, although visible to the human reader, is not directly accessible to the code written by the client. So, the specification is logically split into two parts, the visible part and the private part. If the private part alone is changed, then the text of the client will not need changing, but the client code will need recompiling because the object code might change even though the source code does not.

Any necessary recompilation is ensured by the compilation system and can be performed automatically if desired. Note that this is required by the Ada language and is not simply a property of a particular implementation. It is never left to the user to decide when recompilation is necessary, and so there is no risk of attempting to link together a set of inconsistent units — a big hazard in many languages.

Finally, note the modes **in**, **out**, and **in out** on the parameters. These refer to the flow of information and are explained in Section 15.7.1.

15.5.3 Generic Contract Model

Templates, an important feature of languages such as C++ and Java, correspond to generics in Ada, and in fact, C++ based its templates on Ada generics. Ada generics are safe because of the so-called contract model.

We can extend the stack example to enable us to declare stacks of any type and any size (we can do the latter other ways as well). Consider

```
   generic
     Max: Integer;    -- formal generic parameters
     type Item is private;
```

```
package Generic_Stacks is
   type Stack is private;
   procedure Clear(S: out Stack);
   procedure Push(S: in out Stack; X: in Item);
   procedure Pop(S: in out Stack; X: out Item);
private    -- private part
   type Vector is array (1 .. Max) of Item;
   type Stack is
     record
       A: Vector;
       Top: Integer range 0 .. Max:=0;
     end record;
end Generic_Stacks;
```

with an appropriate body obtained simply by replacing `Float` by `Item`. The generic is just a template and has to be instantiated with appropriate actual parameters corresponding to the two generic formal parameters `Max` and `Item`. For example, if we want stacks of integers with a maximum size of 50, we write

```
package Integer_Stacks is new Generic_Stacks(Max => 50, Item => Integer);
```

This creates a normal package called `Integer_Stacks`, which we can then use in the normal way. The essence of the contract model is that if we provide parameters that correctly match, then the package obtained from the instantiation will compile and execute correctly, and in particular, no strange errors will ever occur in the generic body.

Other languages such as C++ do not have this desirable property. In C++, some mismatches are caught by the linker rather than the compiler, and others are even left until execution and throw an exception.

There are extensive forms of generic parameters in Ada. Writing **type** `Item` **is private**; permits the actual type to be almost any type at all. Another example is that writing **type** `Item` **is** (<>); permits the actual type to be any integer type (such as `Integer` or `Long_Integer`) or an enumeration type (such as `Signal`). Within the generic, we can then use all the properties common to all integer and enumeration types with the certainty that the actual type will indeed provide these properties.

The generic contract model is very important. It enables the development of flexible but safe general-purpose libraries. An important goal is that the Ada user should never need to pore over the code of the generic body to puzzle out what went wrong. C++ users are not so lucky.

15.5.4 Child Units

The overall architecture in Ada permits a hierarchical structure of units giving extensive and flexible information-hiding properties. Child units can be public or private. Given a package called `Parent`, we can declare a public child as

```
package Parent.Child is ...
```

and a private child as

```
private package Parent.Slave ...
```

Both have bodies and can have private parts as usual. The key difference is that a public child essentially extends the specification of the parent (and is thus visible to clients), whereas a private child extends the body of the parent (and thus is not visible to clients). The structure permits grandchildren and so on to any depth.

There are various rules concerning visibility. Children do not need an explicit "with" clause for their parent; visibility is automatic. The parent body can have a "with" clause for the children, but since the specification of the parent must be available before the children are compiled (since the children share the name of the parent), the parent specification cannot have a normal "with" clause for its children. More on this later. Another rule is that the visible part of a private child has visibility of the private part of its parent, just as the body of the parent does. But in the case of a public child, only its private part and not its visible part has such visibility.

A special form of the "with" clause — the **private with** clause — is permitted on a package specification but only allows the private part to have visibility of the unit concerned. This might be useful when the private part of a public child needs information provided by a private child. Ada carefully blocks all attempts to bypass the strict visibility control.

15.5.5 Mutually Dependent Types

Many languages have the equivalent of private types, especially in connection with object-oriented programming (OOP). Basically, the intrinsic operations (methods) belonging to a type are those declared in a package along with the type. Thus, the intrinsic operations of the type Stack are Clear, Push, and Pop. The same structure in C++ would be written as

```
class Stack {
... /*  details of stack structure  */
public:
  void  Clear();
  void  Push(float);
  float Pop();
};
```

The C++ approach is neat in that it only has one level of naming Stack, whereas in Ada we have both package name and type name, thus Stacks.Stack. However, in practice, the Ada style is not a burden, especially if we apply a use clause.

On the other hand, if we have two types that wish to share private information, it is very easy in Ada. We write

```
package Twins is
  type Dum is private;
  type Dee is private;
  ...
private
  ...    -- shared private part
end Twins;
```

and the private part defines both Dum and Dee and so they have mutual access to anything in the private part. This is not so easy in other languages and involves constructs such as friends in C++. However, this is not foolproof since it relies upon the programmer naming the friends correctly. In Ada, there is no possibility of getting it wrong.

Other examples exhibit mutual recursion. Suppose we wish to study patterns of points and lines where each point has three lines through it and each line has three points on it. (Two of the most fundamental theorems of projective geometry, those of Pappus and Desargues concern such structures.) We use access types. A simple approach is a single package

```
package Points_and_Lines is
  type Point is private;
```

```
   type Line is private;
   ...
 private
   type Point is
     record
        L, M, N: access Line;
     end record;
   type Line is
     record
        P, Q, R: access Point;
     end record;
 end Points_and_Lines;
```

If we decided that each deserved its own package, we could still do that using a limited "with" clause. (Two packages cannot have a normal "with" clause for each other because that creates a forbidden circularity.) We write

```
 limited with Lines;
 package Points is
   type Point is private;
   ...
 private
   type Point is
   record
   L, N, N: access Lines.Line;
   end record;
 end Points;
```

and similarly for the package Lines. A limited "with" clause gives a so-called incomplete view of the types in the package concerned, which means roughly that they can only be used to form access types. We could actually write **limited private with** Lines; because it is only the private part that needs to mention the type Line.

15.6 Safe Object-Oriented Programming

OOP took programming by storm about twenty years ago. Its supreme merit is said to be its flexibility, but flexibility is somewhat like the freedom discussed in the introduction — the wrong kind of flexibility can be an opportunity for dangerous errors to intrude.

The key idea of OOP is that the objects dominate the programming and routines (methods, subprograms) that manipulate objects are properties of objects. The other, older, view — sometimes called SP (Structured Programming) — is that programming is primarily about functional decomposition, that it is the subprograms that dominate, and that objects are merely things being manipulated by the subprograms. Both views have their place, and fanatical devotion to just a strict object view is often inappropriate.

Ada strikes an excellent balance and enables either approach to be taken according to the needs of the application. Indeed, Ada has incorporated the idea of objects right from its inception in 1980 through the concept of packages, which encapsulate types and the operations upon them.

15.6.1 OOP versus SP

We will look at two examples that can be used to illustrate various points. They are chosen for their familiarity, which avoids the need to explain particular application areas. The examples concern geometrical objects (of which there are lots of kinds) and people (of which there are only two kinds, male and female). Consider the geometrical objects first. For simplicity, we will consider just flat objects in a plane. Every object has a position. In Ada, we can declare a root object, which has properties common to all objects:

```
type Object is tagged
  record
    X_Coord: Float;
    Y_Coord: Float;
  end record;
```

The word **tagged** distinguishes this type from a static structure (such as Date in Section 15.2.3) and indicates that it can be extended. Moreover, objects of this type carry a tag with them at execution time, and this tag identifies the type of the object. We are going to declare various specific object types such as Circle, Triangle, Square, and so on, and these will all have distinct values for the tag.

We can declare various properties of geometrical objects such as area and moment of inertia about the center. Every object has such properties, but they vary according to shape. These properties can be defined by functions, and they are declared in the same package as the corresponding type. We can start with

```
package Geometry is
  type Object is abstract tagged
    record
      X_Coord, Y_Coord: Float;
    end record;
  function Area(Obj: Object) return Float is abstract;
  function Moment(Obj: Object) return Float is abstract;
end Geometry;
```

We have declared the type and the operations as abstract. We don't actually want any objects of type Object, and making it abstract prevents us from inadvertently declaring any. We want real objects such as a Circle, which have properties such as Area. The functions Area and Moment have been declared as abstract also, which ensures that, when we declare a genuine type such as Circle, we are forced to declare concrete functions Area and Moment with appropriate code.

We can now declare the type Circle. It is best to use a child package for this:

```
package Geometry.Circles is
  type Circle is new Object with
    record
      Radius: Float;
    end record;
  function Area(C: Circle) return Float;
  function Moment(C: Circle) return Float;
end;
with Ada.Numerics;  use Ada.Numerics;  -- to give access to π
package body Geometry.Circles is
  function Area(C: Circle) return Float is
  begin
    return π * C.Radius**2;   -- uses Greek letter π
  end Area;
```

```
   function Moment(C: Circle) return Float is
   begin
      return 0.5 * C.Area * C.Radius**2;
   end Moment;
end Geometry.Circles;
```

Note that the code defining the `Area` and `Moment` is in the package body. As discussed in Section 15.5, this means that the code can be changed and recompiled as necessary without forcing recompilation of the description of the type itself and, consequently, all those programs that use it. We could then declare other types such as `Square` (which has an extra component giving the length of the side), `Triangle` (three components giving the three sides), and so on without disturbing the existing abstract type `Object` and the type `Circle` in any way.

The various types form a class rooted at `Object`, and this class of types is denoted by `Object'Class`. Ada very carefully distinguishes between a specific type such as `Circle` and a class of types such as `Object'Class`. This distinction avoids confusion that can occur in other languages. If we defined other types as extensions of the type `Circle` then we could usefully talk about the class `Circle'Class`.

The function `Moment` illustrates the use of the prefixed notation. We can write either:

```
C.Area      -- prefix notation
Area(C)     -- functional notation
```

The prefix notation emphasizes the object model and the fact that we are thinking of the object C as a predominant entity rather than the function `Area`.

Suppose now that we have declared various objects, perhaps

```
A_Circle: Circle := (1.0, 2.0, Radius => 4.5);
My_Square: Square := (0.0, 0.0, Side => 3.7);
The_Triangle: Triangle := (1.0, 0.5, A => 3.0, B => 4.0, C => 5.0);
```

By way of illustration, we have used named notation for components other than the x and y coordinates common to all the types. We might have a procedure to output the properties of a general object. We might write

```
procedure Print(Obj: Object'Class) is
begin
   Put("Area is ");   Put(Obj.Area);   -- dispatching call of Area
   ...   -- and so on
end Print;
```

and then

```
Print(A_Circle);
Print(My_Square);
```

The procedure `Print` can take any item in the class `Object'Class`. Within the procedure, the call of `Area` is dynamically bound and calls the function `Area` appropriate to the specific type of the parameter `Obj`. This always works safely since the language rules are such that every possible object in the class `Object'Class` is of a specific type derived ultimately from `Object` and will have a function `Area`. Note that the type `Object` itself was abstract and so no geometrical object of that type could be declared. Accordingly, it does not matter that the function `Area` for the type `Object` is abstract and has no code; it could never be called.

In a similar way, we might have types concerning persons. Consider

```
package People is
  type Person is abstract tagged
    record
      Birthday: Date;
      Height: Inches;
      Weight: Pounds;
    end record;
  type Man is new Person with
    record
      Bearded: Boolean;     -- whether he has a beard
    end record;
  type Woman is new Person with
    record
      Births: Integer;     -- how many children she has borne
    end record;
  ...    -- various operations
end People;
```

Since there is no possibility of any additional types of persons, we could use a variant record, which is more in the line of SP. Thus

```
type Gender is (Male, Female);
type Person (Sex: Gender) is
  record
    Birthday: Date;
    Height: Inches;
    Weight: Pounds;
    case Sex is
      when Male =>
        Bearded: Boolean;
      when Female =>
        Births: Integer;
    end case;
  end record;
```

and we might then declare various operations on this version of the type Person. Each operation would need a case statement to take account of the two sexes. This might be considered rather old fashioned and inelegant; however, it has its own considerable advantages.

If we need to add another **operation** in the OOP formulation, then the whole type structure will need to be recompiled — each type will need to be revisited to implement the new operation. If we need to add another **type** (such as a Pentagon), then the existing structure can be left unchanged.

In the case of the SP formulation, the situation is reversed. If we need to add another **type** in the SP formulation, then the whole type structure will need to be recompiled — each operation will need to be revisited to implement the new type by adding another branch to its case statement. If we need to add another **operation**, then the existing structure can be left unchanged.

The OOP approach has been lauded as much safer than SP because there are no case statements to maintain. This certainly is true, but sometimes the maintenance is harder if new operations are added because they have to be added individually for every type. Ada offers both approaches.

15.6.2 Overriding Indicators

One of the dangers of OOP occurs with overriding. When we add a new type to a class, we can add new versions of all the appropriate operations. If we do not add a new operation, then the parent's operation is inherited. The danger is that we may attempt to add a new version but spell it incorrectly

```
function Aera(C: Circle) return Float;
```

or get a parameter or result wrong

```
function Area(C: Circle) return Integer;
```

In both cases, the existing function Area is not overridden, but a totally new operation is added. Then, when a classwide operation dispatches to Area, it will call the inherited version rather than the overridden one. Such bugs can be very difficult to find — the program compiles and seems to run but just produces curious answers. Actually, Ada has already provided a safeguard here because we declared Area for Object as abstract, and this is a further line of defence. But if we had a second generation or had not had the wisdom to make Area abstract, then we would be in trouble.

In order to guard against such mistakes we can write, for example,

```
overriding
function Area(C: Circle) return Float;
```

If we make an error, we will not get a new operation; instead, the program will fail to compile. However, if we did truly want to add a new operation, we could assert that also by

```
not overriding
function Aera(C: Circle) return Float;
```

Such overriding indicators are optional largely for compatibility with earlier versions of Ada. Languages such as C++ and Java do not provide any assistance in this area, and consequently, subtle errors can remain undetected for some time.

15.6.3 Dispatchless Programming

In safety-critical programming, the dynamic selection of code is usually forbidden. Safety is enhanced if we can prove that the flow of control follows a strict pattern with, for example, no dead code. Traditionally, this means that we have to use a more SP-type approach with visible "if" statements and "case" statements to select the appropriate flow path.

Although dynamic dispatching is at the heart of much of the power of OOP, it has other valuable structural features. Thus, we might value the ability to extend types and thereby share much coding but declare specific named operations where dynamic behavior is required. We might also wish to use the prefix notation, which has a number of advantages.

Ada has a facility known as pragma Restrictions that enables a programmer to ensure that various aspects of Ada are not used in a particular program. In this case, we write

```
pragma Restrictions(No_Dispatch);
```

and this ensures that no use is made of the construction X'Class, which in turn means that no dispatching calls are possible. Note that this exactly matches the requirements of SPARK, which is often required for critical software. SPARK permits type extension but does not permit classwide types and operations. Moreover, this meets the requirements of DO-178B in this area and also minimizes the need for the justification of deactivated code for level A certification.

15.6.4   Interfaces and Multiple Inheritance

Some have looked upon multiple inheritance as a Holy Grail — an objective against which languages should be judged. This is not the place to digress on the history of various techniques; rather, we will summarize the key problems.

Suppose that we were able to inherit arbitrarily from two parent types. Recall Edwin Abbott's fabulous book *Flatland*, a satire on class structure in a world in which people are flat geometrical objects. The working classes are triangles, the middle classes are other polygons, and the aristocracy are circles (Abbott, 1884). Sadly, all females are two-sided, and thus simply a line segment.

We could conceive of representing the inhabitants of Flatland by a type such as

```
type Flatlander is new Geometry.Object and People.Person;
```

The question now arises: what are the properties inherited from the two parent types? We might expect a `Flatlander` to have components `X_Coord` and `Y_Coord` inherited from `Object` and also a `Birthday` inherited from `Person`. However, we cannot be sure about `Height` and `Weight` for a two-dimensional person. We would expect an operation such as `Area` to be inherited because clearly a `Flatlander` has an area and a moment of inertia.

But there are potential problems in the general case. Suppose both parent types have an operation with the same identifier. This situation would typically arise with operations of a rather general nature such as `Print`, `Make`, `Copy`, and so on. Which one is inherited? Suppose both parents have components with the same identifier. Which one do we get? These problems particularly arise if both parents themselves have a common ancestor.

Some languages have provided multiple inheritance and devised somewhat lengthy rules to overcome these difficulties (C++ and Eiffel, for example). Possibilities include using renaming, mentioning the parent name for ambiguous entities, and giving precedence to the first parent type in the list. Sometimes the solutions have the flavor of unification for its own sake — one person's unification is often another person's confusion.

The difficulties are basically twofold: inheriting components and inheriting the **implementation** of operations from more than one parent. But there is generally no problem with inheriting the **specification** of operations. This solution, adopted by Java, has proved successful and is also the approach used by Ada 2005. So, the Ada rule is that we can inherit from more than one type, thus:

```
type T is new A and B and C with
   record
   ...   -- additional components
   end record;
```

However, only the first type in the list (A) can have components and concrete operations; the other types must be what are known as interfaces, which are essentially abstract types without components, all of whose operations are abstract or null procedures. (The first type could be an interface as well.)

We can reformulate the type `Object` as an interface as follows

```
package Geometry is
   type Object is interface;
   procedure Move(Obj: in out Object; New_X, New_Y: in Float) is abstract;
   function X_Coord(Obj: Object) return Float is abstract;
   function Y_Coord(Obj: Object) return Float is abstract;
   function Area(Obj: Object) return Float is abstract;
   function Moment(Obj: Object) return Float is abstract;
end Geometry;
```

Observe that the components have been deleted and replaced by further operations. The procedure Move enables an object to be moved — that is, it sets both the *x* and *y* coordinates, and the functions X_Coord and Y_Coord return its current position.

Note that the prefix notation means that we can still access the coordinates by, for example, A_Circle.X_Coord and The_Triangle.Y_Coord just as when they were visible components. Now when we declare a concrete type Circle, we have to provide implementations of all these operations:

```
package Geometry.Circles is
   type Circle is new Object with private;    -- partial view
   procedure Move(C: in out Circle; New_X, New_Y: Float);
   function X_Coord(C: Circle) return Float;
   function Y_Coord(C: Circle) return Float;
   function Area(C: Circle) return Float;
   function Moment(C: Circle) return Float;
   function Radius(C: Circle) return Float;
   function Make_Circle(X, Y, R: Float) return Circle;
private
   type Circle is new Object with    -- full view
     record
       X_Coord, Y_Coord: Float;
       Radius: Float;
     end record;
end;
package body Geometry.Circles is
   procedure Move(C: in out Circle; New_X, New_Y: Float) is
   begin
      C.X_Coord := New_X;
      C.Y_Coord := New_Y;
   end Move;
   function X_Coord(C: Circle) return Float is
   begin
     return C.X_Coord;
   end X_Coord;
   -- and similarly Y_Coord and Area and Moment as before
   -- also functions Radius and Make_Circle
end Geometry.Circles;
```

We have made the type Circle private so that all the components are hidden. Nevertheless, the partial view reveals that it is derived from the type Object and so must have all the properties of the type Object. Note the functions to create a circle and to access the radius component.

So, the essence of programming with interfaces is that we have to implement the properties promised. It is not so much multiple inheritance of existing properties but multiple inheritance of contracts to be satisfied. Returning now to Flatland, we can declare

```
package Flatland is
   type Flatlander is abstract new Person and Object with private;
   procedure Move(F: in out Flatlander; New_X, New_Y: Float);
   function X_Coord(F: Flatlander) return Float;
   function Y_Coord(F: Flatlander) return Float;
private
   type Flatlander is abstract new Person and Object with
```

```
  record
    X_Coord, Y_Coord: Float := 0.0;   -- at origin by default
    ...   -- any new components we wish
  end record;
end;
```

and the type `Flatlander` will inherit the components `Birthday`, etc., of the type `Person`, any operations of the type `Person` (we did not show any above), and the abstract operations of the type `Object`. However, it is convenient to declare the coordinates as components since we need to do that eventually, and we can then override the inherited abstract operations Move, X_Coord and Y_Coord with concrete ones. Note also that we have given the coordinates the default value of zero so that any Flatlander is by default at the origin. The package body is

```
package body Flatland is
  procedure Move(F: in out Flatlander; New_X, New_Y: Float) is
  begin
      F.X_Coord := New_X;
      F.Y_Coord := New_Y;
  end Move;
  function X_Coord(F: Flatlander) return Float is
  begin
    return F.X_Coord;
  end X_Coord;
  -- and similarly Y_Coord
end Flatland;
```

Making `Flatlander` abstract means that we do not have to implement all the operations such as `Area` just yet. And finally, we could declare a type `Square` suitable for Flatland (originally, the book was published anonymously and the author was designated as "A Square") as follows

```
package Flatland.Squares is
  type Square is new Flatlander with
    record
      Side: Float;
    end record;
  function Area(S: Square) return Float;
  function Moment(S: Square) return Float;
end;
package body Flatland.Squares is
  function Area(S: Square) is
  begin
    return S.Side**2;
  end Area;
  function Moment(S: Square) is
  begin
    return S.Area * S.Side**2 / 6.0;
  end Moment;
end Flatland.Squares.
```

and all the operations are thereby implemented. By way of illustration, we have made the extra component Side of the type Square directly visible, but we could have used a private type. So we can now declare Dr. Abbott as

```
A_Square: Square := (Flatlander with Side => 3.00);
```

and he will have all the properties of a square and a person. Note the extension aggregate, which takes the default values for the private components and gives the additional visible component explicitly.

There are other important properties of interfaces that can only be touched upon in this overview. An interface can have a null procedure as an operation. A null procedure behaves as if it has a null body — that is, it can be called but does nothing. If two ancestors have the same operation, then a null procedure overrides an abstract operation with the same parameters and results. If two ancestors have the same abstract operation with equivalent parameters and results, then these merge into a single operation to be implemented. If the parameters and results are different, then this results in overloading, and both operations have to be implemented. In summary, the rules are designed to minimize surprises and maximize the benefits of multiple inheritance.

15.7   Safe Object Construction

This section covers a number of aspects of the control of objects. By objects here we mean both small objects in the sense of simple constants and variables of an elementary type such as Integer and big objects in the sense of OOP. Ada provides more control and flexibility in this area than most languages.

15.7.1   Variables and Constants

As we have seen, we can declare a variable or a constant by writing

```
Top: Integer;                    -- a variable
Max: constant Integer := 100;    -- a constant
```

respectively. Top is a variable and we can assign new values to it, whereas Max is a constant and its value cannot be changed. Note that when we declare a constant we have to give it a value since we cannot assign to it afterward. Optionally, a variable can be given an initial value as well.

The advantage of using a constant is that it cannot be changed accidentally; thus, it is not only a useful safeguard, but it helps any person later reading the program and informs them of its status. An important point is that the value of a constant does not have to be static — that is, it does not need to be computed at compile time. An example of this was shown in the program for interest rates where we declared a constant called Factor

```
function Npw(X: Float) return Float is
  Factor: constant Float := 1.0 + X/100.0;
begin
  return 1000.0 * Factor**2 + 500.0 * Factor - 2000.0;
end Npw;
```

Each call of the function Npw has a different value for X and so a different value for Factor, but Factor is constant throughout each individual call of Npw. Although this is a trivial example, and it is clear that Factor is not changed during an individual call, nevertheless we should get into the habit of writing **constant** when possible.

Parameters of subprograms are another example of variables and constants. Parameters may have three modes: **in**, **in out**, and **out**. If no mode is shown, then it is **in** by default. All parameters of functions have to be of mode **in**.

A parameter of mode **in** is a constant whose value is given by the actual parameter. Thus, the parameter X of Npw has mode **in** and so is a constant — this means that we cannot assign to it and are assured that its value will not change. The actual parameter can be any expression of the type concerned.

Parameters of modes **in out** and **out** are variables, and the actual parameter must also be a variable. The difference concerns their initial value — a parameter of mode **in out** is a variable whose initial value is given by that of the actual parameter, whereas a parameter of mode **out** has no initial value (unless the type has a default value such as **null** in the case of an access type). Examples of all three modes occur in the procedures Push and Pop in Section 15.5:

```
procedure Push(S: in out Stack; X: in Float);
procedure Pop(S: in out Stack; X: out Float);
```

The rules regarding actual parameters ensure that constancy is never violated, thus we could not pass a constant such as Factor to Pop since the relevant parameter of Pop has mode **out**, and this would enable Pop to change Factor.

The distinction between variables and constants also applies to access types and objects. Thus if we have

```
type Int_Ptr is access Integer;
K: aliased Integer;
KP: Int_Ptr := K'Access;
CKP: constant Int_Ptr := K'Access;
```

then the value of KP can be changed, but the value of CKP cannot. This means that CKP will always refer to K. However, although we cannot make CKP refer to any other object, we can use CKP to change the value in K by

```
CKP.all := 47;   -- change value of K to 47
```

On the other hand, we might have

```
type Const_Int_Ptr is access constant Integer;
J: aliased Integer;
JP: Const_Int_Ptr := J'Access;
CJP: constant Const_Int_Ptr := J'Access;
```

where the access type itself has **constant**. This means that we cannot change the value of the object referred to indirectly whether we use JP or CJP. Note that JP could, of course, refer to different objects from time to time, but CJP could not.

15.7.2 Constant and Variable Views

Sometimes it is convenient to enable a client to read a variable but not to write to it. In other words, sometimes it is better to give the client a constant view of a variable. This can be done with a so-called deferred constant and the access types just described.

A deferred constant is one declared in the visible part of a package and for which we do not give an initial value. The initial value can then be given in the private part. Consider the following

```
package P is
   type Const_Int_Ptr is access constant Integer;
   The_Ptr: constant Const_Int_Ptr;   -- deferred constant
private
   The_Variable: aliased Integer;
```

```
    The_Ptr: constant Const_Int_Ptr := The_Variable'Access;
    ...
  end P;
```

The client can read the value of The_Variable indirectly through the object The_Ptr of type Const_Int_Ptr by writing

```
  K := The_Ptr.all;    -- indirect read of The_Variable
```

But since the access type Cons_Int_Ptr is declared as **access constant**, the value of the object referred to by The_Ptr cannot be changed by writing

```
  The_Ptr.all := K;    -- illegal, cannot change The_Variable indirectly
```

However, any subprogram declared in the package P can access The_Variable directly and so write to it. This technique is particularly useful with tables in which the table is computed dynamically, but we do not want the client to be able to change it.

15.7.3 Limited Types

The types we have met so far (Integer, Float, Date, Circle and so on) have had various operations. Some were predefined, such as the ability to compare two values (with =), and some also had user-defined operations such as Area in the case of the type Circle. They also all had the ability to perform assignment.

Sometimes assignment is not a good idea. There are two main reasons for this:

- The type might represent some resource such as an access right, and copying could imply theft or a violation of security.
- The type might be implemented as a linked data structure, and copying would simply copy the head of the structure and not all of it.

We can prevent assignment by declaring the type as **limited**. A good illustration of the second problem occurs if we implement the stack using a linked list. We might have

```
  package Linked_Stacks is
    type Stack is limited private;
    procedure Clear(S: out Stack);
    procedure Push(S: in out Stack; X: in Float);
    procedure Pop(S: in out Stack; X: out Float);
  private
    type Cell is
      record
        Next: access Cell;
        Value: Float;
      end record;
    type Stack is access all Cell;
  end Stacks;
```

The body might be

```
  package body Stacks is
    procedure Clear(S: out Stack) is
    begin
```

```
      S := null;
   end Clear;
   procedure Push(S: in out Stack; X: in Float) is
   begin
      S := new Cell'(S, X);
   end Push;
   procedure Pop(S: in out Stack; X: out Float) is
   begin
        X := S.Value;
        S := Stack(S.Next);
   end Pop;
end Stacks;
```

This uses the normal linked list style of implementation of a stack. Note that the type Stack is declared as limited private so that assignment of a stack as in

```
This_One, That_One: Stack;
...
This_One := That_One;    -- illegal, type Stack is limited
```

is prohibited. If assignment had been permitted, then all that would have happened is that This_One would end up pointing to the start of the list defining the value of That_One. Calling Pop on This_One would simply move it down the chain representing That_One. This sort of problem is known as aliasing — we would have two ways of referring to the same entity, and that is often very unwise.

In this example there is no problem with declaring a stack; it is automatically initialized to be null, which represents an empty stack. However, sometimes we need to create an object with a specific initial value (necessary if it is a constant). We cannot do this by assigning in a general way as in

```
type T is limited ...
...
X: constant T := Y;    -- illegal, cannot copy value in variable Y
```

because this involves copying, which is forbidden since the type is limited. Two techniques are possible. One involves aggregates, and the other uses functions. We will consider aggregates first. Suppose the type represents some sort of key with components giving the data of issue and the internal code number such as

```
type Key is limited
  record
    Issued: Date;
    Code: Integer;
  end record;
```

The type is limited so that keys cannot be copied. But we can write

```
K: Key := (Today, 27);
```

since, in the case of a limited type, this does not copy the value defined by the aggregate as a whole but rather the individual components are given the values Today and 27. In other words, the value for K is built *in situ*.

It would be more realistic to make the type private, and then, of course, we could not use an aggregate because the components would not be individually visible. Instead, we can use a constructor function. Consider

```
package Key_Stuff is
  type Key is limited private;
  function Make_Key( ... ) return Key;
  ...
private
  type Key is limited
    record
      Issued: Date;
      Code: Integer;
    end record;
end Key_Stuff;
package body Key_Stuff is
  function Make_Key( ... ) return Key is
  begin
    return New_Key: Key do
      New_Key.Issued := Today;
      New_Key.Code := ... ;
    end return;
  end Make_Key;
  ...
end Key_Stuff;
```

The external client (for whom the type is private) can now write

```
My_Key: Key := Make_Key( ... )    -- no copying involved
```

where we assume that the parameters of Make_Key are used to compute the internal secret code.

It is interesting to look carefully at the function Make_Key. It has an extended return statement that starts by declaring the result Ne_Key. When the result type is limited (as it is here), the return object is actually built in the final destination of the result of the call (such as the object My_Key). This is similar to the way in which the components of the aggregate were actually built *in situ* in the earlier example, so again, no copying is involved.

The net outcome is that Ada provides a way of creating initial values for objects declared by clients, yet prevents the client from making copies. The limited type mechanism gives the provider of resources such as the keys considerable control over their use.

15.7.4 Controlled Types

Ada provides a further mechanism for the safe control of objects through the use of controlled types. This enables us to write special code to be obeyed when an object is created, when it ceases to exist, and when it is copied in the case of nonlimited types.

The mechanism is based on types called Controlled and Limited_Controlled declared in a predefined package:

```
package Ada.Finalization is
  type Controlled is abstract tagged private;
  procedure Initialize(Object: in out Controlled) is null;
  procedure Adjust(Object: in out Controlled) is null;
  procedure Finalize(Object: in out Controlled) is null;
  type Limited_Controlled is abstract tagged limited private;
  procedure Initialize(Object: in out Limited_Controlled) is null;
  procedure Finalize(Object: in out Limited_Controlled) is null;
```

```
private
  . . .
end Ada.Finalization;
```

The general idea (for a nonlimited type) is that the user declares a type, which is derived from Controlled, and then provides overridings of the three procedures Initialize, Adjust, and Finalize. These procedures are called when an object is created, when it is copied, and when it ceases to exist, respectively. Note carefully that these calls are inserted automatically by the system and the programmer does not have to write explicit calls. The same mechanism applies to a limited type, which has to be derived from Limited_Controlled, but there is no procedure Adjust since copying is not permitted.

As an example, suppose we reconsider the stack and decide that we want to use the linked mechanism (this has the advantage of effectively no upper limit) but wish to allow copying. We can write

```
package Linked_Stacks is
  type Stack is private;
  procedure Clear(S: out Stack);
  procedure Push(S: in out Stack; X: in Float);
  procedure Pop(S: in out Stack; X: out Float);
private
  type Cell is
    record
      Next: access Cell;
      Value: Float;
    end record;
  type Stack is new Controlled with
    record
      Header: access Cell;
    end record;
  overriding
  procedure Adjust(S: in out Stack);
end Linked_Stacks;
```

The type Stack is now just private, and the full type shows that it is actually a tagged type and derived from the type Controlled and has a component Header, which effectively is the stack in the previous formulation. In other words, we have introduced a wrapper. Note that the user cannot see that the type is controlled and tagged. Since we want to make assignment work properly, we have to override the procedure Adjust. Note also that we have supplied the overriding indicator so that the compiler can double-check that Adjust does indeed have the correct parameters.

The package body might be

```
package body Linked_Stacks is
  procedure Clear(S: out Stack) is
  begin
    S := (Controlled with Header => null);
  end Clear;
  procedure Push(S: in out Stack; X: in Float) is
  begin
    S.Header := new Cell'(S.Header, X);
  end Push;
  procedure Pop(S: in out Stack; X: out Float) is
  begin
```

```
      X := S.Header.Value;
      S.Header := S.Header.Next;
   end Pop;
   procedure Adjust(S: in out Stack) is
      L: access Cell := S.Header;
   begin
      while L /= null loop
         L := new Cell'(L.Next, L.Value);
      end loop;
   end Adjust;
end Linked_Stacks;
```

Assignment will now work properly. Suppose we write

```
This_One, That_One: Stack;
...
This_One := That_One;   -- calls Adjust automatically
```

The raw assignment of That_One to This_One copies just the record containing the component Header. The procedure Adjust is then called automatically with This_One as parameter. Adjust works down the list, making a copy of each element of the chain in turn, and so copies the whole list. This is often called a deep copy.

Another notable point is that the procedure Clear sets the parameter S to a record whose header component is null; the structure is known as an extension aggregate. The first part of the extension aggregate just gives the name of the parent type (or the value of an object of that type), and the part after **with** gives the values of the additional components, if any. The procedures Pop and Push are straightforward.

The reader might wonder about reclamation of unused storage when Pop removes an item and also when Clear sets a stack to empty. This can be dealt with safely, but this is beyond the scope of this overview.

Note that Initialize and Finalize are not overridden and thus inherit the null procedure of the type Controlled. Nothing special happens when a stack is declared — this is correct since we just get a record whose Header is null by default and nothing else is required. Also, nothing happens when an object of type Stack ceases to exist on exit from a procedure and so on — this again raises the issue of reclamation of storage and can be dealt with in a reliable manner.

15.8 Safe Concurrency

Computers typically only do one thing at a time, and the operating system makes it look as if they can do several things in parallel. This is not quite so true these days, since many computers do truly have multiple processors, but it still does apply to the vast majority of small computers, including those used in process control.

15.8.1 Operating Systems and Tasks

Operating systems vary enormously in the amount of parallel activity that they support. Simple systems like early versions of Windows actually support very little. Operating systems such as POSIX provide the programmer with multiple threads of control, which can flow through the program quite independently and support parallel activities.

On some hardware, there will only be one processor, and the processor will be allocated to the different threads according to some scheduling algorithm. One approach is simply to give the processor to each thread in turn for a small amount of time; a more sophisticated approach is to use priorities or deadlines to ensure that the processor is used effectively. Some hardware might have multiple processors, in which case several threads can truly be active in parallel. Again, hopefully a scheduler will allocate the processors in an effective way to the active threads of control.

Programming languages take very different approaches to parallel activities which are often known as tasks. Some languages have intrinsic facilities for tasking built into the language itself, and others provide simple access to the underlying primitives of the operating system. Yet others ignore the subject completely. Ada is a prime example of a language with intrinsic tasking facilities. C just provides access to the operating system. C++ remains silent on the topic.

It is a fact that programs with tasking are much harder to get right than ordinary sequential programs. There are at least three advantages of having tasking within the language itself:

- Built-in syntactic constructions make it much easier to write correct programs because the language can prevent a number of errors. It is essentially the old story about abstraction. By hiding irrelevant detail, certain errors cannot be made.
- Portability is difficult if operating system facilities are used directly because they vary widely from system to system.
- General operating systems do not provide the range of timing and related facilities needed by many applications.

The following are operations typically required in a tasking program:

- Tasks must be prevented from violating the integrity of data if several tasks need access to the data.
- Tasks need to communicate with each other by sending messages of some kind.
- Tasks need to be controlled to meet specific timing requirements.
- Tasks need to be scheduled to use resources efficiently and to meet their overall deadlines.

This section will briefly examine these topics and illustrate how Ada addresses them in a reliable manner. First, we introduce the simple idea of an Ada task and the overall program structure.

An Ada program can have many tasks running in parallel. A task is written in two parts rather like a package. It has a specification, which describes the interface it presents to other tasks, and a body, which contains the code saying what it actually does. In simple cases, the specification simply names the task so we might have

```
task A;   -- task specification
task body A is   -- task body
begin
...   -- statements saying what the task does
end A;
```

Sometimes it is convenient to have several similar tasks, in which case we can introduce a task type and then declare objects in the usual way

```
task type Worker;
task body Worker is ...
Tom, Dick, Harry: Worker;
```

This creates three tasks called Tom, Dick, and Harry. We can also declare arrays of tasks and have task components inside records and so on. Tasks can be declared wherever other objects can be declared — in a package, in a subprogram, or even within another task.

The main subprogram of a complete program is called by the environment task. An overall program with library packages A, B, and C and main subprogram M can therefore be thought of as

```
task Environment_Task;
task body Environment_Task is
   ...      -- declarations of library packages A, B, C
   ...      -- and main subprogram M
begin
   ...      -- call of main subprogram M
end;
```

Tasks become active simply by being declared, and they finish by running into the end of the task body. An important rule is that a local task in a subprogram or other tasks must finish before the embracing unit can itself be left; the embracing unit will be suspended until this happens. This rule prevents dangling references to data that no longer exist.

15.8.2   Protected Objects

Suppose that the three tasks Tom, Dick, and Harry are using a stack as some sort of temporary storage device. From time to time, one of them pushes an item onto the stack, and from time to time one of them (perhaps the same one, perhaps a different one) pops an item off the stack. The three tasks run in parallel, and the runtime system gives the processor to each in turn according to some algorithm. Perhaps they each get 10 ms in turn.

Suppose the stack they are using is like the one declared in Section 15.5. Suppose that Harry is calling Push when his time slot expires, and control then passes to Tom, who calls Pop. To be precise, suppose Harry loses the processor just after he has executed the statement to increment Top in

```
procedure Push(X: Float) is
begin
   Top := Top + 1;   -- Harry loses processor just after this
   A(Top) := X;
end Push;
```

Now we have the situation that Top has been incremented, but the new value X has not been assigned to the component of the array. When Tom calls Pop, he gets the old and possibly junk value in the array component that was about to be overwritten by the new value. When Harry gets the processor back (assuming no other stack activity occurs in the meantime) he will write the value X into a component of the array that is a part of the stack that is not in use. In other words, the value X is lost.

A worse situation can occur if the processor is switched partway through a statement; thus, Harry might lose the processor just after he has picked up Top into a register but before he replaces Top with the new value. Suppose Dick now comes along and also does a Push, thereby adding 1 to the old value of Top. When Harry resumes, he will replace the value that Dick computed by the same value. In other words, the two calls of Push add just 1 to Top rather than 2 as expected. This problem is overcome in Ada by using a protected object for the stack. We write

```
protected Stack is
   procedure Clear;
   procedure Push(X: in Float);
   procedure Pop(X: out Float);
private
   Max: constant := 100;
   Top: Integer range 0.. Max:=0;
   A: array (1 .. Max) of Float;
end Stack;
protected body Stack is
```

```
procedure Clear is
begin
  Top := 0;
end Clear;
procedure Push(X: in Float) is
begin
  Top := Top + 1;
  A(Top) := X;
end Push;
procedure Pop(X: out Float) is
begin
  X := A(Top);
  Top := Top - 1;
end Pop;
end Stack;
```

Package has been changed to **protected**, the data that was in the body now goes in the private part, and the function Pop has been changed into a corresponding procedure Pop.

The three procedures Clear, Push, and Pop are now called protected operations and are called in the same way as procedures. The behavior is that only one task can access the operations of the object at a time. If a task such as Tom attempts to call the procedure Pop while Harry is obeying Push, then Tom is forced to wait until Harry returns from Push. This is all done automatically with no effort on the part of the programmer, so all the problems are solved.

Behind the scenes, the protected object has a lock, and a task attempting to enter the object has to acquire the lock first. If another task already has the lock, then it has to wait until that other task has finished and relinquishes the lock.

We can elaborate on this example to consider how we might cope with an attempt to push an item on the stack when it is full. In the package formulation, this would raise Constraint_Error on the attempt to assign the value Max+1 to Top. As it is written, the same thing would happen and the lock would be automatically relinquished as the exception terminates the call of the procedure.

But we can do better. We can modify the protected object as follows

```
protected Stack is
  procedure Clear;
  entry Push(X: in Float);
  entry Pop(X: out Float);
private
  Max: constant := 100;
  Top: Integer range 0 .. Max:=0;
  A: array (1 .. Max) of Float;
end Stack;
protected body Stack is
  procedure Clear is
  begin
    Top := 0;
  end Clear;
  entry Push(X: in Float) when Top < Max is
  begin
    Top := Top + 1;
    A(Top) := X;
  end Push;
```

```
entry Pop(X: out Float) when Top > 0 is
begin
  X := A(Top);
  Top := Top - 1;
end Pop;
end Stack;
```

The procedures Push and Pop are now entries, and they have barrier expressions such as Top > Max. The effect of a barrier is to prevent the entry from being entered if the barrier is False. Note that this does not prevent the entry from being called; all that happens is that the calling task is suspended until the barrier becomes True. So, if Harry tries to call Push when the stack is full, then he has to wait until some other task (Tom or Dick) calls Pop and removes the top item. Harry will then automatically proceed. The user does not have to program anything special.

Note that entries, like protected procedures, are also called in the same way as normal procedures:

```
Stack.Push(Z);
```

It is instructive to consider how we might program this example in a language using lower-level primitives as we might in C. The historic basic primitives are the operations P and V acting on objects called semaphores. The idea is that we put pairs of calls of P and V around the operations we wish to protect; thus, using the same Ada syntax, Push would become

```
procedure Push(X: in Float) is
begin
  P(Stack_Lock);     -- secure the lock
  Top := Top + 1;
  A(Top) := X;
  V(Stack_Lock);     -- release the lock
end Push;
```

with similar pairs of calls around the body of Clear and Pop. This is essentially a DIY operation or assembly type coding for tasking. The opportunities for errors are many:

- We might omit one of a P and V pair, creating an imbalance.
- We might forget them altogether around one group of statements that should be protected.
- We might use the wrong semaphore name.
- We might inadvertently bypass a closing V.

The last problem would arise if, in the model without barriers, Push was called when the stack was full. This causes Constraint_Error to be raised. If we fail to provide a local exception handler to call V, then the system will be permanently locked. None of these difficulties can arise when using Ada-protected objects because all these low-level mechanisms are automatic.

Although, with care, semaphores can be used successfully in simple situations, it is very difficult to use them correctly in more complicated situations such as the example with barriers. Not only is it difficult to program correctly with semaphores, but it is extremely difficult to prove that a program is correct.

15.8.3   The Rendezvous

The other important communication requirement between tasks is for one task to send a message to another. This is done with a mechanism known as a rendezvous. The two tasks have a client-server relationship. The client needs to know the identity of the server task, but the server task will accept a message from any client. The general pattern of the server is

```
task Server is
  entry Some_Service(Formal: in out Data);
end;
task body Server is
begin
...
  accept Some_Sevice(Formal: in out Data) is
  ...   -- statements providing the service
  end Some_Service;
  ...
end Server;
```

The specification of the server indicates that it has an entry Some_Service. This is called by a client task in the same way an entry of a protected object is called. The difference is that the code to be obeyed is given by an accept statement, which is only executed when the server task reaches the accept statement. Until that happens, the calling task is suspended. When the server reaches the accept statement, it executes it using any parameters supplied by the client. The client continues to be suspended and is only permitted to proceed after the accept statement is finished and any **out** parameters are updated. So, the body of a client might look like

```
task body Client is
  Actual: Data;
begin
  ...
  Server.Some_Sevice(Actual);
  ...
end Client;
```

If several clients call a server, then they are queued automatically. An accept statement can be placed anywhere, such as in the branch of an if statement, so the mechanism is very flexible. The rendezvous is a high-level abstract mechanism (like the protected object) and as such is relatively easy to use correctly; the corresponding queuing mechanisms programmed at a low level as in C are hard to write correctly.

15.8.4 Ravenscar

Section 15.6 mentioned the pragma Restrictions, which can be used to ensure that we do not use certain features of the language. Many of the restrictions in Ada 2005 relate to tasking. The tasking features in Ada are very comprehensive and provide a whole range of facilities necessary to meet the programming needs of a variety of real-time applications. However, some applications are quite simple and do not need many of these facilities. Here are some samples of the sort of restrictions that can be applied:

```
No_Task_Hierachy
No_Task_Termination
Max_Entry_Queue_Length => n
```

The restriction No_Task_Hierarchy prevents tasks from being declared inside other tasks or inside subprograms — all tasks are therefore inside packages. No_Task_Termination means that all tasks run for ever; this is common in many control applications where each task essentially has an endless loop doing some repetitive action. The restriction on entry queues places a limit on the number of tasks that can be queued on a single entry at any time.

The advantage of giving appropriate restrictions are two-fold:

- It might enable a somewhat simpler runtime system to be used. This could be smaller and faster and thus more appropriate for some time and space critical embedded applications.
- It might enable various properties of the application to be proved concerning matters such as freedom from deadlock and ability to meet deadlines. This may be vital for certain safety-critical applications.

A particularly important group of restrictions is imposed by the Ravenscar profile. In order to ensure that a program conforms to this profile we write the following in the program:

```
pragma Profile(Ravenscar);
```

The key purpose of the Ravenscar profile is to restrict the use of tasking facilities so that the effect of the program is predictable. The profile is simply defined to be equivalent to a number of restrictions plus a few other related pragmas concerning matters such as scheduling. The restrictions include those mentioned earlier so there are no task hierarchies, all tasks run forever, and entry queues have a limit of one. The combined effect of the restrictions is that it is possible to make statements about the ability of a particular program to meet stringent requirements for the purposes of certification.

No other programming language can offer the reliability of Ada as constrained by the Ravenscar profile. A general description of the principles and use of the profile in high integrity systems can be found in an ISO/IEC Technical Report (ISO/IEC TR 24718:2004).

15.8.5 Timing and Scheduling

No survey of Ada tasking, however brief, would be complete without a few words about timing and scheduling. There are statements to enable a program to be synchronized with a clock. We can delay a program for an absolute amount of time or until a specific time, thus:

```
delay 2*Minutes;
delay until Next_Time;
```

Small absolute delays might be useful for interactive use, whereas a delay until a particular time can be used to program regular events. Time itself can be measured either by a real-time clock, which is guaranteed to have a certain accuracy, or by the local wall clock, which might be subject to changes because of Daylight Savings, for example. In Ada it is even possible to take time zones and leap seconds into account.

The language also provides a number of standard timers whose expiry can be used to trigger actions defined by a protected procedure. There are three kinds of timers: one enables the monitoring of the CPU time used by an individual task, one concerns the CPU budget for a group of tasks, and the third concerns time as measured by the real-time clock.

Here is an amusing example concerning the boiling of an egg. We declare a protected object Egg, thus:

```
protected Egg is
  procedure Boil(For_Time: in Time_Span);
private
  procedure Is_Done(Event: in out Timing_Event);
  Egg_Done: Timing_Event;
end Egg;
protected body Egg is
  procedure Boil(For_Time: in Time_Span) is
  begin
    Put_Egg_In_Water;
```

```
          Set_Handler(Egg_Done, For_Time, Is_Done'Access);
      end Boil;
      procedure Is_Done(Event: in out Timing_Event) is
      begin
        Ring_The_Pinger;
      end Is_Done;
    end Egg;
```

The consumer can then write

```
Egg.Boil(Minutes(4));
-- now read newspaper whilst waiting for egg
```

and the pinger will ring when the egg is ready.

A number of different scheduling policies provided in Ada 2005 can be applied to all tasks in a program or just to those in certain priority ranges by the use of pragmas. The policies are as follows:

- *FIFO_Within_Priorities* — Within each priority level to which it applies, tasks are dealt with on a first-in-first-out (FIFO) basis. Moreover, a task may preempt a task of a lower priority.
- *Non_Preemptive_FIFO_Within_Priorities* — Within each priority level to which it applies, tasks run to completion or until they are blocked or execute a delay statement. A task cannot be preempted by one of higher priority. This sort of policy is widely used in high integrity applications.
- *·Round_Robin_Within_Priorities* — Within each priority level to which it applies, tasks are timesliced with an interval that can be specified. This is a very traditional policy widely used since the earliest days of concurrent programming.
- *EDF_Across_Priorities* — This provides Earliest Deadline First (EDF) dispatching. The general idea is that within a range of priority levels, each task has a deadline and the task with the earliest deadline is processed. This is a new policy and has mathematically provable advantages with respect to efficiency.

Ada also has comprehensive facilities concerning the setting and changing of task priorities and the so-called ceiling priorities of protected objects.

15.9   Even Safer with SPARK

The preceding sections aimed to show that a program written in Ada is less likely to have errors than a program written in popular languages such as C, C++, and Java. For some applications, especially those that are safety-critical or security-critical, it is really important that the program be correct. For the most severe safety-critical applications, the consequence of an error can be loss of life or damage to the environment. Similarly, for the most severe security-critical applications, the consequence of an error may be equally catastrophic, such as loss of national security, commercial reputation, or just plain theft.

Applications are graded into different levels according to the risk. The following are possible levels for avionics applications:

- Level 1 — Minor level, some inconvenience; e.g., entertainment system fails (could be a benefit!)
- Level 2 — Major level, some injuries; e.g., bumpy landing with cuts and bruises
- Level 3 — Hazardous level, some dead; e.g., nasty landing with fire
- Level 4 — Catastrophic level, aircraft lost, all dead; e.g., structural failure, plane crashes

For the most demanding applications, which require certification by an appropriate authority, it is not enough for a program to be correct; it also has to be shown to be correct, and that is much more difficult.

This section gives a very brief introduction to SPARK, a language based on a subset of Ada and specifically designed for the writing of high-integrity systems. Although technically just a subset of Ada with additional information provided through Ada comments, it is helpful to consider SPARK as a language in its own right, which, for convenience, uses a standard Ada compiler. Analysis of a SPARK program is carried out by a suite of tools, of which the most important are the Examiner, Simplifier, and Proof Checker. We start by considering the important concept of correctness and contracts.

15.9.1 Contracts

What do we mean by correct software? Perhaps a general definition is software that does what the user had in mind. And "had in mind" might literally mean just that for a simple one-off program written to do an ad-hoc calculation, or, for a large avionics application, it might be interpreted as the text of some contract between the ultimate client and the software developer.

This idea of a contract is not new. If we look at the programming libraries developed in the early 1960s, particularly in mathematical areas and perhaps written in Algol 60 (a language favored for the publication of material in respected journals such as the *Communications of the ACM* and the *Computer Journal*), we find that the manual tells us what parameters are required, any constraints on their range, and so on. In essence, there is a contract between the writer of the subroutine and the user; the user promises to hand over suitable parameters and the subroutine promises to produce the correct answer.

The decomposition of a program into various component parts is very familiar, and the essence of the programming process is to define what these parts do and identify the interfaces between them — a process that enables the parts to be developed independently of each other. If we write each part correctly (so that it satisfies its side of the contract implied by its interface) and if we have defined the interfaces correctly, then we are assured that when we put the parts together to create the complete system, it will work correctly.

Bitter experience shows that life is not quite like that. Two things can go wrong: on the one hand, the interface definitions are not usually complete (there are holes in the contracts); on the other hand, the individual components may not be correct or are used incorrectly (the contracts are violated). And of course, the contracts might not say what we meant to say anyway.

15.9.2 Correctness by Construction

SPARK encourages the development of programs in an orderly manner with the aim of producing a program that is correct by virtue of its construction techniques. This "correctness by construction" approach is in marked contrast to other approaches, which aim to generate as much code as quickly as possible to have something to demonstrate.

A number of years of use of SPARK in application areas such as avionics and railway signalling has produced strong evidence that, indeed, not only is the program more likely to be correct, but the overall cost of development is actually less after all the testing and integration phases are taken into account.

We will now look in a little more detail at the two problem areas introduced in Section 15.9.1: incomplete interface definitions and incorrect components.

Ideally, the definition of the interfaces between the software components should hide all irrelevant detail but expose all relevant detail. Alternatively, we might say that an interface definition should be both complete and correct. As a simple example of an interface definition, consider the interface to a subprogram. As just mentioned, the interface should describe the full contract between the user and the implementer. The details of how the subprogram is implemented should not concern us. In order that these two concerns be clearly distinguished, it is helpful to use a programming language in which they are lexically distinct. Unfortunately, not many languages are like this. Popular languages such as Java and C usually present subprograms as one lump, with the interface physically bound to the implementation. This is a nuisance because not only does it make checking the interface less straightforward since the compiler wants the whole code, but it also encourages the developer to hack the code at the same time that he writes the interface, and this confuses the logic of the development process.

However, Ada has such a structure separating interface (the specification) from the implementation (the body). This applies both to individual subprograms and to groups of entities encapsulated into packages, and this is a key reason why Ada forms such a good base for SPARK. SPARK requires additional information to be provided through the mechanism of annotations, which conveniently take the form of Ada comments. A key purpose of these annotations is to increase the amount of information about the interface without providing unnecessary information about the implementation. In fact, SPARK allows the information to be added at various levels of detail as appropriate to the needs of the application.

Consider the information given by the following Ada specification

```
procedure Add(X: in Integer);
```

Frankly, it tells us very little. It just says that there is a procedure called Add and that it takes a single parameter of type Integer whose formal name is X. This is enough to enable the compiler to generate code to call the procedure, but it says nothing about what the procedure does. It might not do anything at all. It certainly doesn't have to add anything nor does it have to use the value of X. It could, for example, subtract two unrelated global variables and print the result to some file. But now consider what happens when we add the lowest level of annotation. The specification might become

```
procedure Add(X: in Integer);
--# global in out Total;
```

This states that the only global variable that the procedure can access is that called Total. Moreover, the mode information tells us that the initial value of Total must be used (**in**) and that a new value will be produced (**out**). The SPARK rules also say more about the parameter X. Although in Ada a parameter need not be used at all, nevertheless an **in** parameter must be used in SPARK.

So now we know rather a lot. We know that a call of Add will produce a new value of Total and that it will use the initial value of Total and the value of X. We also know that Add cannot affect anything else — it certainly cannot print anything or have any other malevolent side effect.

Of course, the information regarding the interface is not complete since nowhere does it require that addition be performed to obtain the new value of Total. In order to do this, we can add optional annotations that concern proof and obtain

```
procedure Add(X: in Integer);
--# global in out Total;
--# post Total = Total~ + X;
```

The annotation commencing **post** is called a postcondition and explicitly says that the final value of Total is the result of adding its initial value (distinguished by ~) to that of X. So now the specification is complete. It is also possible to provide preconditions; thus, we might require X to be positive and we could express this by

```
--# pre X > 0;
```

An important aspect of the annotations is that they are all checked statically by the SPARK Examiner and other tools and not when the program executes.

It is especially important to note that the pre- and postconditions are checked before the program executes. If they were only checked when the program executes, then it would be a bit like bolting the door after the horse has bolted (which reveals a nasty pun caused by overloading in English!). We don't really want to be told that the conditions are violated as the program runs. For example, we might have a precondition for landing an aircraft

```
procedure Touchdown( ... );
--# pre Undercarriage_Down;
```

It is pretty unhelpful to be told that the undercarriage is not down as the plane lands; we really want to be assured that the program has been analyzed to show that the situation will not arise.

This thought leads into the other problem with programming — ensuring that the implementation correctly implements the interface contract, often called debugging. Generally there are four ways in which bugs are found:

1. **By the compiler** — These are usually easy because the compiler tells us exactly what is wrong.
2. **By a language check at run time** — This applies in languages that carry out checks that, for example, ensure that we do not write outside an array. Typically, we obtain an error message saying what structure was violated and where in the program this happened.
3. **By testing** — This means running various examples and poring over the (un)expected results and wondering where it all went wrong.
4. **By the program crashing** — The effect might be that the application has written all over the operating system and usually destroyed the evidence. Reboot and try again!

Type 1 should really be extended to mean "before the program is executed." Thus, it includes program walkthroughs and similar review techniques, and it includes the use of analysis tools such as those provided for Spark.

These four ways provide a progression of difficulty. Errors are easier to locate and correct if they are detected early. Good programming tools are those that move bugs from one category to a lower-numbered category. Thus, good programming languages are those which provide facilities enabling one to protect oneself against errors that are hard to find. Ada is a particularly good programming language because of its strong typing and runtime checks. Therefore, the enumeration type is a simple feature, which, correctly used, makes hard bugs of type 3 into easy bugs of type 1, as we saw in Section 15.3.

A major goal of Spark is to strengthen interface definitions (the contracts) and move all errors to a low category (ideally, to type 1) so that they are all found before the program executes. The global annotations do this by preventing us from writing a program that accidentally changes the wrong global variables. Similarly, detecting the violation of pre- and postconditions results in a type 1 error. However, to check that such a violation cannot happen requires mathematical proof; this is not always straight-forward, but the Spark tools automate much of the proof process.

15.9.3 The Kernel Language

Ada is a very comprehensive language, and the use of some features makes total program analysis difficult. Accordingly, the subset of Ada supported by Spark omits certain features, mostly those concerning dynamic behavior. For example, there are no access types, no dynamic dispatching, generally no exceptions, all storage is static, and hence all arrays must have static bounds (but subprogram parameters can be dynamic), and there is no recursion. Tasking, of course, is very dynamic, and although Spark does not support full Ada tasking, it does support the Ravenscar profile mentioned in Section 15.8.

Another restriction that helps analysis is that every entity has to have a name and each name should uniquely identify one entity. Hence, all types and subtypes have to be named, and overloading is generally prohibited. However, the traditional block structure is supported, so local names are not restricted. Moreover, tagged types are permitted, although classwide types are not.

The idea of state is crucial to analysis, and there is a strong distinction between procedures whose purpose is to change state and functions whose purpose is simply to observe state. This echoes the confusion between statements and expressions mentioned in Section 15.2 when discussing the problem of C and the railroad example. Functions in Spark are not permitted to have any side effects at all.

The resulting kernel has proved to be sufficiently expressive for the needs of critical applications, which would not want to use features such as dynamic storage.

15.9.4   Tool Support

There are three main SPARK tools: the Examiner, the Simplifier, and the Proof Checker. The Examiner is vital. It has two basic functions:

- It checks conformance of the code to the rules of the kernel language.
- It checks consistency between the code and the embedded annotations by flow analysis.

The Examiner analyzes the interfaces between components and ensures that the details on either side do indeed conform to the specifications of the interfaces. The interfaces are, of course, the specifications of packages and subprograms, and the annotations say more about these interfaces and thereby improve the quality of the contract between the implementation of the component and its users.

The core annotations ensure that a program cannot have certain errors related to the flow of information; thus, the Examiner detects the use of uninitialized variables and the overwriting of values before they are used. This means that care should be taken not to give junk initial values to variables "just in case" because that would hinder the detection of flow errors.

However, the core annotations do not address the issue of dynamic behavior. In order to address the issue of dynamic behavior, a number of proof annotations can be inserted (such as the pre- and post-conditions we saw earlier) that enable dynamic behaviour to be analyzed prior to execution. The general idea is that these annotations enable the Examiner to generate conjectures (potential theorems), which then have to be proved to verify that the program is correct with respect to the annotations. These proof annotations address pre- and postconditions of subprograms and assertions such as loop invariants and type assertions. The generated conjectures are known as verification conditions. These can then be verified by human reasoning, which is usually tedious and unreliable, or by using other tools such as the Simplifier and the Proof Checker.

Even without proof annotations, the Examiner can generate conjectures corresponding to the run-time checks of Ada such as range checks. As discussed in Section 15.3, these checks are automatically inserted to ensure that a variable is not assigned a value outside the range permitted by its declaration or that no attempt is made to read or write outside the bounds of an array. The proof of these conjectures shows that the checks would not be violated and therefore that the program is free of run-time errors that would raise exceptions.

Note that the use of proof is not necessary. SPARK and its tools can be used at various levels. For some applications, it might be appropriate just to apply the core annotations because these alone enable flow analysis to be performed. But for other applications, it might be cost-effective to use the proof annotations as well. Indeed, different levels of analysis can be applied to different parts of a complete program.

There are a number of advantages to using a distinct tool such as the Examiner rather than simply a front-end processor, which then passes its output to a compiler. It is possible to write pieces of SPARK complete with annotations and to have them processed by the Examiner even before they can be compiled. For example, a package specification can be examined even though its private part might not yet be written; of course, such an incomplete package specification cannot be compiled.

There is a temptation to take an existing piece of Ada code and then to add the annotations (often referred to as "Sparking the Ada"). This is to be discouraged because it typically leads to extensive annotations indicative of an unnecessarily complex structure. Although in principle it might then be possible to rearrange the code to reduce the complexity, it is often the case that such good intentions are overridden by the desire to preserve as much as possible of the existing code. The proper approach is to treat the annotations as part of the design process and to use them to assist in arriving at a design that minimizes complexity before the effort of detailed coding takes one down an irreversible path.

15.9.5   Examples

As a simple example, here is a version of the stack with full core annotations (but not proof annotations):

```
package Stacks is
  type Stack is private;
  function Is_Empty(S: Stack) return Boolean;
  function Is_Full(S: Stack) return Boolean;
  procedure Clear(S: out Stack);
  --# derives S from ;
  procedure Push(S: in out Stack; X: in Float);
  --# derives S from S, X;
  procedure Pop(S: in out Stack; X: out Float);
  --# derives S, X from S;
private
  Max: constant := 100;
  type Top_Range is range 0 .. Max;
  subtype Index_Range is Top_Range range 1 .. Max;
  type Vector is array Index_Range of Float;
  type Stack is
  record
    A: Vector;
    Top: Top_Range;
  end record;
end Stacks;
```

We have added functions Is_Full and Is_Empty, which just read the state of the stack. They have no annotations at all.

"Derives" annotations have been added to the various procedure specifications; these are not mandatory, but they can improve flow analysis. Their purpose is to say which outputs depend upon which inputs — in this simple example, they can in fact be deduced from the parameter modes. However, redundancy is one key to reliability, and if they are inconsistent with the modes, then that will be detected by the Examiner and perhaps thereby reveal an error in the specification.

The declarations in the private part have been changed to give names to all the subtypes involved. At this level, there are no changes to the package body at all — no annotations are required. This emphasizes that SPARK is largely about improving the quality of the description of the interfaces.

A difference from the earlier examples is that we have not given an initial value of 0 for Top but require that Clear be called first. When the Examiner looks at the client code, it will perform flow analysis to ensure that Push and Pop are not called until Clear has been called; this analysis will be performed without executing the program. If the Examiner cannot deduce this, then it will report that the program has a potential flow error. On the other hand, if it can actually deduce that Push or Pop are called before Clear, then it will report that the program is definitely in error.

In this brief overview, it is not feasible to give serious examples of the proof process; however, the following trivial example will illustrate the ideas. Consider

```
procedure Exchange(X, Y: in out Float);
--# derives X from Y &
--#         Y from X;
--# post X = Y~ and Y = X~;
```

which shows the specification of a procedure whose purpose is to interchange the values of the two parameters. The body might be

```
procedure Exchange(X, Y: in out Float) is
  T: Float;
```

```
begin
    T := X;   X := Y;   Y := T;
end Exchange;
```

Analysis by the Examiner generates a verification condition that has to be shown to the true. In this particular example, this is trivial and it is done automatically by the Simplifier. In more elaborate situations, the Simplifier will not be able to complete a proof, in which case the Proof Checker is used. The Proof Checker is an interactive program which, under human guidance, hopefully will be able to find a valid proof.

15.9.6 Conclusion

As earlier sections have shown, Ada is an excellent language for writing reliable software. Ada prevents many of the errors that are easy to make when using C, C++, and similar languages. Using SPARK allows even more errors to be detected without having to use the unpredictable process known as testing. For the highest level of safety-critical and security-critical applications, it is not enough for a program to be correct. It also has to be shown to be correct. SPARK enables this to be done.

Acknowledgments

This paper is based on material from the booklet *Safer Programming with Ada* from AdaCore (see www.adacore.com). I am grateful to AdaCore for both their support in writing that booklet and for their permission to use this extract.

For further information on Ada and SPARK, consult *Programming in Ada 2005* and *High Integrity Software* by the author. The *Rationale for Ada 2005*, a source of further information, was originally published in installments in the *Ada User Journal* (see www.ada-europe.org). Both the *Rationale* for Ada 2005 and the *Ada Reference Manual* are available at www.adaic.org.

References

Abbott, E.A., *Flatland*, Basil Blackwell, Oxford, 1884.
Barnes, J., *High Integrity Software — The Spark Approach to Safety and Security*, Addison-Wesley, 2003.
Barnes, J., *Programming in Ada 2005*, Addison-Wesley, 2006.
ISO/IEC TR 24718:2004, *Guide to the Use of the Ravenscar Profile in High Integrity Systems*. This is based on University of York Technical Report YCS-2003-348.

16

RTCA DO-178B/ EUROCAE ED-12B

Thomas K. Ferrell
Ferrell and Associates Consulting

Uma D. Ferrell
Ferrell and Associates Consulting

16.1 Introduction

This chapter provides a summary of the document RTCA DO-178B, *Software Considerations in Airborne Systems and Equipment Certification*,[1] with commentary on the most common mistakes made in understanding and applying DO-178B. The joint committee of RTCA Special Committee 167 and EUROCAE* Working Group 12 prepared RTCA DO-178B** (also known as EUROCAE ED-12B), and it was subsequently published by RTCA and by EUROCAE in December 1992. DO-178B provides guidance for the production of software for airborne systems and equipment such that there is a level of confidence in the correct functioning of that software in compliance with airworthiness requirements. DO-178B represents industry consensus opinion on the best way to ensure safe software. It should also be noted that although DO-178B does not discuss specific development methodologies or management activities, there is clear evidence that by following rigorous processes, cost and schedule benefits may be realized. The verification activities specified in DO-178B are particularly effective in identifying software problems early in the development process.

---

*European Organization for Civil Aviation Equipment.

**DO-178B and ED-12B are copyrighted documents of RTCA and EUROCAE, respectively. In this chapter, DO-178B shall be used to refer to both the English version and the European equivalent. This convention was adopted solely as a means for brevity.

16.1.1 Comparison with Other Software Standards

DO-178B is a mature document, having evolved over the last 20 years through two previous revisions, DO-178 and DO-178A. It is a consensus document that represents the collective wisdom of both the industry practitioners and the certification authorities. DO-178B is self-contained, meaning that no other software standards are referenced except for those that the applicant produces to meet DO-178B objectives. Comparisons have been made between DO-178B and other software standards such as MIL-STD-498, MIL-STD-2167A, IEEE/EIA-12207, IEC 61508, and U.K. Defense Standard 0-55. All of these standards deal with certain aspects of software development covered by DO-178B. None of them has been found to provide complete coverage of DO-178B objectives. In addition, these other standards lack objective criteria and links to safety analyses at the system level. However, organizations with experience applying these other standards often have an easier path to adopting DO-178B.

Advisory Circular AC 20-115B specifies DO-178B as an acceptable means, but not the only means, for receiving regulatory approval for software in systems or equipment being certified under a Technical Standard Order (TSO) authorization, Type Certificate (TC), or Supplemental Type Certificate (STC). Most applicants use DO-178B to avoid the work involved in showing that other means are equivalent to DO-178B. Even though DO-178B was written as a guideline, it has become the standard practice within the industry. DO-178B is officially recognized as a *de facto* international standard by the International Organization for Standardization (ISO).

16.1.2 Document Overview

DO-178B consists of 12 sections, 2 annexes, and 3 appendices as shown in Figure 16.1.

Section 2 and Section 10 are designed to illustrate how the processes and products discussed in DO-178B relate to, take direction from, and provide feedback to the overall certification process. Integral Processes detailed in Sections 6, 7, 8, and 9, support the software life cycle processes noted in Sections 3, 4, and 5.

Section 11 provides details on the life cycle data and Section 12 gives guidance to any additional considerations. Annex A, discussed in more detail below, provides a leveling of objectives. Annex B provides the document's glossary. The glossary deserves careful consideration since much effort and care was given to precise definition of the terms. Appendices A, B, C, and D provide additional material including a brief history of the document, the index, a list of contributors, and a process improvement form, respectively. It is important to note that with the exception of the appendices and some examples embedded within the text, the main sections and the annexes are considered normative, i.e., required to apply DO-178B.

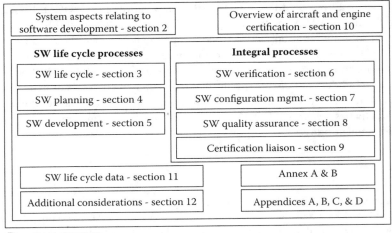

FIGURE 16.1   Document structure.

	Objective		Applicability by SW Level				Output		Control category by SW level			
	Description	Ref.	A	B	C	D	Description	Ref.	A	B	C	D
1	Low-level requirements Comply with high-level requirements.	6.3.2 a	●	●	○		Software Verification Results	11.14	②	②	②	②

FIGURE 16.2   Example objective from Annex A.

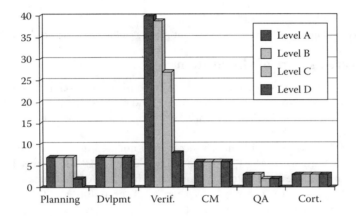

FIGURE 16.3   Objectives over the software development life cycle.

The 12 sections of DO-178B describe processes and activities for the most stringent level of software.* Annex A provides a level by level tabulation of the objectives for lower levels of software.** This leveling is illustrated in Figure 16.2 extracted from Annex A Table A-4, Verification of Outputs of Software Design Process.

In addition to the leveling of objectives, Annex A tables serve as an index into the supporting text by way of reference, illustrate where independence is required in achieving the objective, which data items should include the objective evidence, and how that evidence must be controlled. More will be said about the contents of the various Annex A tables in the corresponding process section of this text. If an applicant adopts DO-178B for certification purposes, Annex A may be used as a checklist to achieve these objectives. The FAA's position is that if an applicant provides evidence to satisfy the objectives, then the software is DO-178B compliant. Accordingly, the FAA's checklists for performing audits of DO-178B developments are based on Annex A tables.

Before discussing the individual sections, it is useful to look at a breakout of objectives as contained in Annex A. While DO-178B contains objectives for the entire software development life cycle, there is a clear focus on verification as illustrated by Figure 16.3. Although at first glance it appears that there is only one objective difference between levels A and B, additional separation between the two is accomplished through the relaxation of independence requirements. Independence is achieved by having the verification or quality assurance of an activity performed by a person other than the one who initially conducted the activity. Tools may also be used to achieve independence.

*Levels are described in Section 16.1.3, "Software as part of system."

**Software that is determined to be at level E is outside the scope of DO-178B.

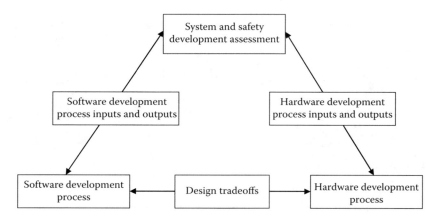

FIGURE 16.4   Relationship between system development process and the software development process.

16.1.3   Software as Part of the System

Application of DO-178B fits into a larger system of established or developing industry practices for systems development and hardware. The system level standard is SAE ARP4754, *Certification Considerations for Highly-Integrated or Complex Aircraft Systems.*[2] The relationship between system, software, and hardware processes is illustrated in Figure 16.4.

The interfaces to the system development process were not well defined at the time DO-178B was written. This gap was filled when ARP4754 was published in 1996. DO-178B specifies the information flow between system processes and software processes. The focus of the information flow from the system process to the software process is to keep track of requirements allocated to software, particularly those requirements that contribute to the system safety. The focus of information flow from the software process to the system process is to ensure that changes in the software requirements, including the introduction of derived requirements (those not directly traceable to a parent requirement), do not adversely affect system safety.

The idea of system safety, although outside the scope of DO-178B, is crucial to understanding how to apply DO-178B. The regulatory materials governing the certification of airborne systems and equipment define five levels of failure conditions. The most severe of these is *catastrophic*, meaning failures that result in the loss of ability to continue safe flight and landing. The least severe is *no effect*, where the failure results in no loss of operational capabilities and no increase in crew workload. The intervening levels define various levels of loss of functionality resulting in corresponding levels of workload and potential for loss of life. These five levels map directly to the five levels of software defined in DO-178B. This mapping is shown in Figure 16.5.

It is important to note that software is never certified as a standalone entity. A parallel exists for the hardware development process and flow of information between hardware processes and system

Failure condition	DO-178B software level
Catastrophic	A
Hazardous	B
Major	C
Minor	D
No effect	E

FIGURE 16.5   Software levels.

process. Design trade-offs between software processes and hardware processes are also taken into consideration at the system level. Software levels may be lowered by using protective software or hardware mechanisms elsewhere in the system. Such architectural methods include partitioning, use of hardware or software monitors, and architectures with built-in redundancy.

16.2   Software Life-Cycle Processes

The definition of how data are exchanged between the software and systems development processes is part of the software life-cycle processes discussed in DO-178B. Life-cycle processes include the planning process, the software development processes (requirements, design, code, and integration), and the integral processes (verification, configuration management, software quality assurance, and certification liaison). DO-178B defines objectives for each of these processes as well as outlining a set of activities for meeting the objectives.

DO-178B discusses the software life-cycle processes and transition criteria between life-cycle processes in a generic sense without specifying any particular life-cycle model. Transition criteria are defined as "the minimum conditions, as defined by the software planning process, to be satisfied to enter a process." Transition criteria may be thought of as the interface points between all of the processes discussed in DO-178B. Transition criteria are used to determine if a process may be entered or reentered. They are a mechanism for knowing when all of the tasks within a process are complete and may be used to allow processes to execute in parallel. Since different development models require different criteria to be satisfied for moving from one step to the next, specific transition criteria are not defined in DO-178B. However, it is possible to describe a set of characteristics that all well-defined transition criteria should meet. For transition criteria to successfully assist in entry from one life-cycle process to another, they should be quantifiable, flexible, well documented, and present for every process. It is also crucial that the process owners agree upon the transition criteria between their various processes.

16.2.1   Software Planning Process

DO-178B defines five types of planning data* for a software development. They are

- Plan for Software Aspects of Certification (PSAC)
- Software Development Plan
- Software Verification Plan
- Software Configuration Management Plan
- Software Quality Assurance Plan

These plans should include consideration of methods, languages, standards, and tools to be used during the development. A review of the planning process should have enough details to assure that the plans, proposed development environment, and development standards (requirements, design, and code) comply with DO-178B.

Although DO-178B does not discuss the certification liaison process until Section 9, the intent is that the certification liaison process should begin during the projects' planning phase. The applicant outlines the development process and identifies the data to be used for substantiating the means of compliance for the certification basis. It is especially important that the applicant outline specific features of software or architecture that may affect the certification process.

---

*The authors of DO-178B took great pains to avoid the use of the term "document" when referring to objective evidence that needed to be produced to satisfy DO-178B objectives. This was done to allow for alternative data representations and packaging as agreed upon between the applicant and the regulatory authority. For example, the four software plans-outlining development, verification, QA, and CM may all be packaged in a single plan, just as the PSAC may be combined with the System Certification Plan.

16.2.2 Software Development Process

Software development processes include requirements, design, coding, and integration. DO-178B allows for requirements to be developed that detail the system's functionality at various levels. DO-178B refers to these levels as high- and low-level requirements. System complexity and the design methodology applied to the system's development drive the requirements' decomposition process. The key to understanding DO-178B's approach to requirement's definition can be summed up as, "one person's requirements are another person's design." Exactly where and to what degree the requirements are defined is less important than ensuring that all requirements are accounted for in the resulting design and code, and that traceability is maintained to facilitate verification.

Some requirements may be derived from the design, architecture, or the implementation nuances of the software and hardware. It is recognized that such requirements will not have a traceability to the high-level requirements. However, these requirements must be verified and must also be considered for safety effects in the system safety assessment process.

DO-178B provides only a brief description of the design, coding, and integration processes since these tend to vary substantially between various development methodologies. The one exception to this is in the description to the outputs of each of the processes. The design process yields low-level requirements and software architecture. The coding process produces the source code, typically either in a high-order language or assembly code. The result of the integration effort is executable code resident on the target computer along with the various build files used to compile and link the executable. Each of these outputs is verified, assured, and configured as part of the integral processes.

16.3 Integral Processes

DO-178B defines four processes as integral, meaning that they overlay and extend throughout the software life cycle. These are the software verification process, software configuration management, software quality assurance, and certification liaison process.

16.3.1 Software Verification*

As noted earlier, verification objectives outnumber all others in DO-178B, accounting for over two thirds of the total. DO-178B defines verification as a combination of reviews, analyses, and testing. Verification is a technical assessment of the results of both the software development processes and the software verification process. There are specific verification objectives that address the requirements, design, coding, integration, as well as the verification process itself. Emphasis is placed at all stages to assure that there is traceability from high-level requirements to the final "as-built" configuration.

Reviews provide qualitative assessment of a process or product. The most common types of reviews are requirements reviews, design reviews, and test procedure reviews. DO-178B does not prescribe how these reviews are to be conducted, or what means are to be employed for effective reviews. Best practices in software engineering process states that for reviews to be effective and consistent, checklists should be developed and used for each type of review. Checklists provide:

- Objective evidence of the review activity
- A focused review of those areas most prone to error
- A mechanism for applying "lessons learned"
- A practical traceable means for ensuring that corrective action is taken for unsatisfactory items

Review checklists can be common across projects, but they should themselves be reviewed for appropriateness and content for a particular project.

---

*Software Verification is a complex topic, which deserves in-depth treatment. The reader is directed to References 4, 5, and 6 for detailed discussion on verification approaches and explanation of terms.

Analyses provide repeatable evidence of correctness and are often algorithmic or procedural in nature. Common types of analyses used include timing, stack, data flow, and control flow analyses. Race conditions and memory leakage should be checked as part of the timing and stack analysis. Data and control coupling analysis should include, a minimum, basic checks for set/use and may extend to a full model of the system's behavior. Many types of analyses may be performed using third-party tools. If tools are used for this purpose, DO-178B rules for tool qualification must be followed.

The third means of verification, testing, is performed to demonstrate that

- The software product performs its intended function
- The software does not demonstrate any unintended actions

The key to accomplishing testing correctly to meet DO-178B objectives in a cost-effective manner is to maintain a constant focus on requirements. This requirements-based test approach represents one of the most fundamental shifts from earlier versions of the document. As test cases are designed and conducted, requirements coverage analysis is performed to assess that all requirements are tested. A structural coverage analysis is performed to determine the extent to which the requirements-based test exercised the code. In this manner, structural coverage is used as a means of assessing overall test completion. The possible reasons for lack of structural coverage are shortcomings in requirements-based test cases or procedures, inadequacies in software requirements, compiler generated code, unreachable, or inactive code.

As part of the test generation process, tests should be written for both normal range and abnormal inputs (robustness). Tests should also be conducted using the target environment whenever possible.

Structural coverage and how much testing is required for compliance at the various levels are misunderstood topics. Level D software verification requires test coverage of high-level requirements only. No structural coverage is required.

Low-level requirements testing is required at level C. In addition, testing of the software structure to show proper data and control coupling is introduced. This coverage involves coverage of dependencies of one software component on another software component via data and control. Decision coverage is required for level B, while level A code requires Modified Condition Decision Coverage (MCDC).

For level A, structural coverage analysis may be performed on source code only to the extent that the source code can be shown to map directly to object code. The reason for this rule is that some compilers may introduce code or structure that is different from source code.

MCDC coverage criteria were introduced to retain the benefits of multiple-condition coverage while containing the exponential growth in the required number of test cases required. MCDC requires that each condition must be shown to independently affect the outcome of the decision and that the outcome of a decision changes when one condition is changed at a time. Many tools are available to determine the minimum test case set needed for DO-178B compliance. There is usually more than one set of test cases that satisfy MCDC coverage.[3] There is no firm policy on which set should be used for compliance. It is best to get an agreement with the certification authorities concerning the algorithms and tools used to determine compliance criteria.

16.3.2 Software Configuration Management

Verification of the various outputs discussed in DO-178B are only credible when there is clear definition of what has been verified. This definition or configuration is the intent of the DO-178B objectives for configuration management. The six objectives in this area are unique, in that they must be met for all software levels. This includes identification of what is to be configured, how baselines and traceability are established, how problem reports are dealt with, how the software is archived and loaded, and how the development environment is controlled.

While configuration management is a fairly well-understood concept within the software engineering community (as well as the aviation industry as a whole), DO-178B does introduce some unique terminology that has proven to be problematic. The concept of control categories is often misunderstood in a way that overall development costs are increased, sometimes dramatically. DO-178B defines two control categories (CC1 and CC2) for data items produced throughout the development.

The authors of DO-178B intended the two levels as a way of controlling the overhead costs of creating and maintaining the various data items. Items controlled as CC2 have less requirements to meet in the areas of problem reporting, baselining, change control, and storage. The easiest way to understand this is to provide an example. Problem reports are treated as a CC2 item. If problem reports were a CC1 item and a problem was found with one of the entries on the problem report itself, a second problem report would need to be written to correct the first one.

A second nuance of control categories is that the user of DO-178B may define what CC1 and CC2 are within their own CM system as long as the DO-178B objectives are met. One example of how this might be beneficial is in defining different retention periods for the two levels of data. Given the long life of airborne systems, these costs can be quite sizeable. Another consideration for archival systems selected for data retention is technology obsolescence of the archival medium as well as means of retrieval.

16.3.3   Software Quality Assurance

Software quality assurance (SQA) objectives provide oversight of the entire DO-178B process and require independence at all levels. It is recognized that it is prudent to have an independent assessment of quality. SQA is active from the beginning of the development process. SQA assures that any deviations during the development process from plans and standards are detected, recorded, evaluated, tracked, and resolved. For levels A and B, SQA is required to assure transition criteria are adhered to throughout the development process.

SQA works with the CM process to assure that proper controls are in place and applied to life cycle data. This last task culminates in the conduct of a software conformity review. SQA is responsible for assuring that the as-delivered products matches the as-built and as-verified product. The common term used for this conformity review in commercial aviation industry is "First Article Inspection."

16.3.4   Certification Liaison Process

As stated earlier, the certification liaison process is designed to streamline the certification process by ensuring that issues are identified early in the process. While DO-178B outlines twenty distinct data items to be produced in a compliant process, three of these are specific to this process and must be provided to the certifying authority. They are

- Plan for Software Aspects of Certification (PSAC)
- Software Configuration Index
- Software Accomplishment Summary

Other data items may be requested by the certification authority, if deemed necessary. As mentioned earlier, applicants are encouraged to start a dialogue with certification authorities as early in the process as possible to reach a common understanding of a means of achieving compliance with DO-178B. This is especially important as new technology is applied to avionics and as new personnel enter the field. Good planning up front, captured in the PSAC, should minimize surprises later in the development process, thus minimizing cost. Just as the PSAC states *what you intend to do*, the accomplishment summary captures *what you did*. It is used to gauge the overall completeness of the development process and to ensure that all objectives of DO-178B have been satisfied.

Finally, the configuration index serves as an overall accounting of the content of the final product as well as the environment needed to recreate it.

16.4 Additional Considerations

During the creation of DO-178B, it was recognized that new development methods and approaches existed for developing avionics. These included incorporation of previously developed software, use of tools to accomplish one or more of the objectives required by DO-178B, and application of alternate means to meet an objective such as formal methods. In addition, there are a small class of unique issues such as field-loadable and user-modifiable software. Section 12 collects these items together under the umbrella title of Additional Considerations. Two areas, Previously Developed Software (PDS) and Tool Qualification, are common sources of misunderstanding in applying DO-178B.

16.4.1 Previously Developed Software

PDS is software that falls in one of the following categories:

- Commercial off-the-shelf software (e.g., shrink-wrap)
- Airborne software developed to other standards (e.g., MIL-STD-498)
- Airborne software that predates DO-178B (e.g., developed to the original DO-178 or DO-178A)
- Airborne software previously developed at a lower software level

The use of one or more of these types of software should be planned for and discussed in the PSAC. In every case, some form of gap analysis must be performed to determine where specific objectives of DO-178B have not been met. It is the applicant's responsibility to perform this gap analysis and propose to the regulatory authority a means for closing any gaps. Alternate sources of development data, service history, additional testing, reverse engineering, and wrappers* are all ways of ensuring the use of PDS is safe in the new application.

In all cases, usage of PDS must be considered in the safety assessment process and may require that the process be repeated if the decision to use a PDS component occurs after the approval of PSAC. A special instance of PDS usage occurs when software is used in a system to be installed on an aircraft other than the one for which it was originally designed. Although the function may be the same, interfaces with other aircraft systems may behave differently. As before, the system safety assessment process must be repeated to assure that the new installation operates and behaves as intended.

If service history is employed in making the argument that a PDS component is safe for use, the relevance and sufficiency of the service history must be assessed. Two tests must be satisfied for the service history approach to work. First, the application for which history exists must be shown to be similar to the intended new use of the PDS. Second, there should be data, typically problem reports, showing how the software has performed over the period for which credit is sought. The authors of DO-178B intended that any use of PDS be shown to meet the same objectives required of newly developed code.

Prior to identifying PDS as part of a new system, it is prudent to investigate and truly understand the costs of proving that the PDS satisfies the DO-178B objectives. Sometimes, it is easier and cheaper to develop the code again!

16.4.2 Tool Qualification

DO-178B requires qualification of tools when the processes noted by DO-178B are eliminated, reduced, or automated by a tool without its output being verified according to DO-178B. If the output of a tool is demonstrated to be restricted to a particular part of the life cycle, the qualification can also be limited to that part of the life cycle. Only deterministic tools can be qualified.

---

*Wrappers is a generic term used to refer to hardward or software components that isolate and filter inputs to and from the PDS to protect the system from erroneous PDS behavior.

Tools are classified as development tools and verification tools. Development tools produce output that becomes a part of the airborne system and thus can introduce errors. Rules for qualifying development tools are fashioned after the rules of assurance for generating code. Once the need for development tool qualification is established, a tool qualification plan must be written. The rigor of the plan is determined by the nature of the tool and the level of code upon which it is being used. A tool accomplishment summary is used to show compliance with the tool qualification plan. The tool is required to satisfy the objectives at the same level as the software it produces, unless the applicant can justify a reduction in the level to the certification authority.

Verification tools cannot introduce errors but may fail to detect them or mask their presence. Qualification criterion for verification tools is the demonstration of its requirements under normal operational conditions. Compliance is established by noting tool qualification within PSAC and Software Accomplishment Summary. A tool qualification plan and a tool accomplishment summary are not required for verification tools by DO-178B although an applicant may find them useful for documenting the qualification effort.

16.5 Additional Guidance

RTCA SC-190/EUROCAE WG-52 was formed in 1997 to address issues that were raised by the industry and certification authorities in the course of applying DO-178B since its release in 1992. This committee produced two outputs. The first, DO-248B/ED-94B provides clarifications and corrections to DO-178B in the form of errata, frequently asked questions, and discussion papers. DO-248B/ED-94B is titled Final Report for Clarification of DO-178B "Software Considerations in Airborne Systems and Equipment." The second document, DO-278/ED-109 provides assurance objectives for ground and space-based systems. The document is titled Guidelines for Communication, Navigation Surveillance and Air Traffic Management (CNS/ATM) Systems Software Integrity Assurance. DO-278/ED-109 is based on and must be used in conjunction with DO-178B. DO-278/ED-109 introduces an additional assurance level for compatibility with existing ground-based safety regulations, as well as guidelines for the use of Commercial Off The Shelf (COTS) software and adaptation data.

RTCA SC-205/EUROCAE WG-71 was formed in 2005 for the purposes of updating DO-178B. The primary reason for this update cycle is to address industry concerns related to emerging technology including formal methods and model-based development. Specific areas of DO-178B are also targeted for improvement including the use of tools, object oriented technology, and coupling. While the format and content of DO-178C/ED-12C are not yet finalized, minimal change to the overall objective-based format of the current document is expected. Both DO-248B/ED-94B and DO-278/ED-109 are also expected to be updated given their close relationship to DO-178B. The final outputs of this update cycle are expected to be available in the 2009 timeframe.

16.6 Synopsis

DO-178B provides objectives for software life-cycle processes, activities to achieve these objectives, and outlines objective evidence for demonstrating that these objectives were accomplished. The purpose of software compliance to DO-178B is to provide considerable confidence that the software is suitable for use in airborne systems. DO-178B should not be viewed as a documentation guide.

Compliance data are intended to be a consequence of the process. Complexity and extent of the required compliance data depend upon the characteristics of the system/software, associated development practices, and the interpretation of DO-178B, especially when it is applied to new technology and no precedent is available.

Finally, it has to be emphasized that DO-178B objectives do not directly deal with safety. Safety is dealt with at the system level via the system safety assessment. DO-178B objectives help verify the correct implementation of safety-related requirements that flow from the system safety assessment. Like any standard, DO-178B has good points and bad points (and even a few errors). However, careful consideration of its contents, taken together with solid engineering judgment, should result in better and safer airborne software.

References

1. RTCA DO-178B, *Software Considerations in Airborne Systems and Equipment Certification,* RTCA Inc.,Washington, D.C, 1992. Copies of DO-178B may be obtained from RTCA, Inc., 1828 L St., NW, Suite 805, Washington, D.C. 20036-4001 U.S. (202) 833-9339. This document is also known as ED 12B, *Software Considerations in Airborne Systems and Equipment Certification,* EUROCAE, Paris, 1992. Copies of ED-12B may be obtained from EUROCAE, 17, rue Hamelin, 75783 PARIS CEDEX France, (331) 4505-7188.
2. SAE ARP4754, *Certification Considerations for Highly-Integrated or Complex Aircraft Systems,* SAE, Warrendale, PA, 1996.
3. Chilenski, J.J. and Miller, P.S., Applicability of modified condition/decision coverage to software testing, *Software Eng. J.,* 193, September 1994.
4. Myers, G. J., *The Art of Software Testing,* John Wiley & Sons, New York, 1979.
5. Beizer, B., *Software Testing Techniques,* 2nd ed., Coriolis Group, Scottsdale, AZ, 1990.
6. McCracken, D. and Passafiume, M., *Software Testing and Evaluation,* Benjamin/Cummings, Menlo Park, CA, 1987.
7. RTCA DO-248B, Final Report for Clarification of DO-178B "Software Considerations in Airborne Systems and Equipment Certification; RTCA, Inc., Washington, D.C., 2001, EUROCAE ED-9B, Final Report for Clarification of ED-12B "Software Considerations in Airborne Systems and Equipment Certification, EUROCAE, Paris, France."

Further Information

1. The Federal Aviation Administration Web Page: www.faa.gov.
2. The RTCA Web Page: www.rtca.org.
3. Spitzer, C.R., *Digital Avionics Systems Principles and Practice,* 2nd ed., McGraw-Hill, New York, 1993.
4. Wichmann, B.A., A Review of a Safety-Critical Software Standard, National Physical Laboratory, Teddington, Middlesex, U.K. (report is not dated).
5. Herrman, D.S., Software Safety and Reliability, IEEE Computer Society Press, Washington, D.C., 1999.

Section II

Functions

17

Communications

Roy T. Oishi
ARINC

17.1   Air-Ground Communications

17.1.1   History

Airplanes became a tool of war in World War I, and they became a tool of commerce in the following decade, the Roaring Twenties. In Europe and America, aircraft were used for entertainment and then as a means to carry mail from city to city. As soon as a single company acquired multiple aircrafts, the need to communicate to the pilots while in flight became a requirement.

Early attempts at air-ground communications used visual means: lights, flags, and even bonfires. But more was needed. Early radios used Morse code to communicate, but that was not practical in an open, bouncing cockpit. With practical voice radios, air-ground communications became a necessary element of the fledgling air transport industry, and it remains so to this day.

17.1.2   Introduction of Radios

Early on, the use of radio waves was recognized as an important resource. By its third decade, national and international bodies were allocating the radio spectrum. In the United States, the Federal Radio Commission (FRC), predecessor to the Federal Communications Commission (FCC), was licensing radio frequencies to operators. In 1929, the FRC directed the aircraft operating companies to band together to make consolidated frequency allocation requests. Aeronautical Radio Inc. (ARINC) was formed in that year specifically for that purpose, and it continues to serve that purpose today.

As would be expected, communications needs were greatest around airports. Since each aircraft operating company had its own frequency allocation and its own radio operators, as the industry grew, the need for more frequencies grew. This recurring problem, known as spectrum depletion, has been solved in various ways. In the early days, it was solved by banding together to use common frequencies and radio operators. ARINC was a natural choice to implement these common radio stations. At one point, ARINC had 55 such communication centers across the United States.

Another method of solving spectrum depletion was the introduction of new technology. The steady improvements in radio technology have opened higher and higher frequency bands to practical communications use, subject to physical limitations. The increased sophistication of radio circuitry allowed different ways of modulating (i.e., impressing information onto) the radio signals. Initially, these improvements allowed better, clearer, and more efficient and effective voice communications. Later technology permitted the evolution of data communications.

The combination of higher usable frequencies and improved modulation techniques has served to extend the useful life of air-ground voice communications for nearly a century. In the 1930s and 1940s, air-ground voice moved from the high frequency (HF) to the very high frequency (VHF) bands. Amplitude modulation (AM) was used and continues to be used as the basic means of air traffic control (ATC) communications in domestic airspace. Long-range voice communications relies on the properties of the HF band, as will be discussed later.

17.1.3 Introduction of Data Communications

As the airlines came to rely increasingly on information provided to and received from the aircraft in flight, voice gave way to data. In 1978, the airlines initiated air-ground VHF data link, which served two purposes: (1) to allow the messages to originate automatically on the aircraft, reducing crew workload, and (2) to allow the messages to be relayed to airline computer systems without any ground radio operators. The data link was initially called the ARINC Communications Addressing and Reporting System (ACARS), but "ARINC" was soon changed to "Aircraft" in recognition of the nonproprietary nature of the new medium.

Air-ground data link has become the mainstay of airline operations. In the early days, ACARS messages consisted primarily of four automated downlinks per flight segment: the so-called OOOI (pronounced oo-ee) or Out, Off, On, and In messages. These messages allowed the airlines to better track their flights and provided automated timekeeping on the crew. The original 50 aircraft participating in ACARS has grown to almost 10 thousand. The number of messages now tops 20 million in a month, and the types of messages cover every imaginable facet of airline operations: flight operations, administrative information such as crew scheduling, passenger information such as connecting gates, maintenance information such as engine performance and failure reports, airport and airline coordination such as de-icing and refueling, and the list goes on and on. Also, now many data link transactions are two-way, including uplinks as well as downlinks. In fact, some applications are interactive with requests and responses initiated by the ground or by the flight crew. Although these interactive transactions via data link lack the immediacy of voice conversations, they are asynchronous in that both requestor and responder need not be "on the line" at the same time. For non-time-critical applications, this is a significant advantage.

17.1.4 Introduction of ATC Data Link

While VHF voice remains the primary means of ATC communications in domestic airspace, ACARS was first approved for ATC in the South Pacific flight information regions (FIRs) in 1995. Initially, Boeing 747–400 aircraft flying between the U.S. west coast and Australia and New Zealand pioneered ATC data link by using controller-pilot data link communications (CPDLC) and, later, automatic dependent surveillance (ADS). Boeing offered this combination of features in the FANS 1 avionics package. FANS — Future Air Navigation System, originally an acronym coined by the International Civil Aviation Organization (ICAO) — was a term covering communication and navigation using satellites; however, the term has taken on a life of its own. FANS 1/A, as it is now known to acknowledge the Airbus offering of the same ADS and CPDLC applications, is being supported by air traffic service providers all over the world. The original air traffic service providers of the South Pacific have been joined by those of the North Atlantic, North Pacific, Indian Ocean, far east Russia, and other regions.

Prior to the use of ACARS for airborne ATC communications, two applications were implemented on the ground between the aircraft and ATC: Pre-Departure Clearance (PDC) and Digital Automatic Terminal Information Service (D-ATIS). The receipt and acknowledgement of these messages by the flight crew is mandatory prior to takeoff and landing. Receiving these messages via ACARS has several significant advantages. For the flight crew, the message need not be transcribed for later reference, and it can be requested and received without the effort of finding the proper voice channel and requesting the PDC or listening for the beginning of the recorded ATIS. The tower controller need not line up all of the PDC slips, call each aircraft, read the clearance, and verify the readback. Plus, the reduction in congestion on the clearance delivery channel is a significant advantage for situations such as peak departure times at a busy airport.

All of these ATC applications use the ACARS air-ground link, which was neither designed for nor initially approved as an ATC communication medium. For that purpose, ICAO developed standards and recommended practices (SARPs) for the Aeronautical Telecommunication Network (ATN), which was designed for both air-ground and ground-ground communication. In the latter role, it is intended to replace the Aeronautical Fixed Telecommunication Network (AFTN), which has served the industry well as a teletype-based, message-switching network for many years. However, technology has overtaken AFTN, and the more modern packet-switched technology of the ATN is seen as more appropriate. The ATN SARPs development began in the early 1990s and, other than trial implementations, has yet to be fully implemented. In the past 15 years, the Internet, which is based on a different packet-switched technology, i.e., Transmission Control Protocol/Internet Protocol (TCP/IP), has had unprecedented success. It may be necessary for ICAO to make some accommodation to TCP/IP lest the cost of implementing the ATN be prohibitive.

As ACARS became essential to airline operations, the limitations of the initial VHF link became intolerable, first because of coverage limitations and then because of speed. The former was solved in two different ways. Long-range data link was implemented first using Inmarsat satellites; this was the basis for initial FANS implementations in the South Pacific. The oceanic coverage provided by satellites and data link was an improvement over HF voice services. All of the advantages of data link over voice communications were highlighted in the initial FANS trials and operational use. The advantages included: (a) consistent and rapid delivery of messages; (b) standardized message texts, which were understood by all no matter what their native language; (c) automated delivery of position reports; and (d) integration of message content with flight management systems (FMSs). High frequency data link (HFDL) provided another long-range ACARS subnetwork that covered the north polar regions, which are not reached by Inmarsat signals. VHF Digital Link (VDL) Mode 2 provided a higher-speed subnetwork in continental airspace. These points will be elaborated in subsequent sections.

17.2 Voice Communications

17.2.1 VHF Voice

The modern VHF transceiver provides air-ground communications for all aircraft in controlled airspace. For transport aircraft (e.g., commercial airliners), the VHF transceiver is a minimum equipment list (MEL) item, meaning the aircraft cannot take off without the requisite number of operational units, in this case two. The reason for the dual-redundancy requirement is that the VHF transceiver is the primary means of communication with ATC.

The aeronautical VHF communication band covers the frequency range 118–136.975 MHz. VHF signals are limited to line-of-sight between the ground station and the aircraft, usually taken as a radius of approximately 120 nmi around the ground station. Aeronautical VHF voice operations are primarily limited by the radio horizon — the lowest unobstructed path angle between the aircraft and the ground station. Other factors include the altitude of the aircraft and the power of the transmission. Practically, aeronautical communications on a given voice frequency are limited to the extent of the ATC sector, as each new sector controller will be assigned a different channel. Reuse of a given channel is appropriate at about twice the usable radius.

The aeronautical VHF band is protected spectrum, which means that any transmission not related to the safety and regularity of flight is prohibited. The wavelength of these signals is about 2 meters or 90 inches, which drives antenna size. The VHF band is divided into 760 channels spaced 25 kHz apart from 118.000 to 137.000 MHz. There is a 12.5 kHz guard band on each end of the allocated band. For 8.33 kHz operation, the VHF transceiver must be capable of tuning to one of 2280 channels spaced 8.33 kHz apart in the same frequency band. This capability was developed for European airspace when the number of ATC sectors (and, therefore, the number of radio channels assigned) grew beyond the ability to assign usable 25 kHz channels. The universally recognized emergency frequency is 121.500 MHz, which is monitored by all ATC facilities.

ICAO Annex 10 to the Convention on International Civil Aviation "International Standards and Recommended Practices, Aeronautical Telecommunications" promulgates the SARPs for voice and data communications in support of international air traffic services. In the case of voice ATC, the ICAO SARPs are generally followed for domestic ATC services as well.

The VHF voice audio is impressed upon the radio frequency (RF) signal at the carrier frequency by using double-sideband (DSB) AM. This modulation method impresses the audio signal, typically 1–2 kHz, on the RF by varying the amplitude of the RF in proportion to the amplitude of the audio signal. In the frequency domain, the signal can be seen as a peak at the carrier frequency flanked by equal peaks above and below — the sidebands. Upon reception, this signal is reconverted to audio and distributed to the headsets and the cockpit voice recorder. Figure 17.1 shows an AM signal in both the time domain, where the audio signal can be seen as riding on the RF carrier, and in the frequency domain, where the peaks representing the carrier and the sidebands can be easily seen.

Older VHF radios were tuned to a frequency (channel) from a remote radio control panel that housed a set of dials used to select each digit. The remote control panel was connected to the radio by 19 wires. Five lines, two of which were grounded for any selection, represented each decade digit of the frequency. This method originated from a scheme whereby the digit selection grounded a power connection to a motor at the radio. When the motor had driven a similar dial switch to its corresponding position, the ground was removed and the motor stopped. Presumably, the motor turned a tuning device (typically a variable capacitor). Later, non-motor-driven tuning methods kept the two-out-of-five grounded wire scheme until a digital data bus replaced it.

A modern VHF radio (e.g., one specified by ARINC Characteristic 750), is connected to the radio control panel by the two wires of an ARINC 429 bus, which carry command words that perform all of the frequency select functions of the former 20 wires and others as well.

Frequency tuning is not the only element of the modern airborne VHF radio that has evolved. Whereas the motor-driven radio performed the DSB AM function using vacuum tubes or later transistors, the modern radio develops the same output signal in a radically different manner. Today's radio replaces the analog RF and modulator circuits with a high-speed microprocessor called a digital signal processor

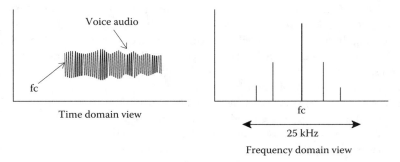

FIGURE 17.1   Double-sideband AM voice signals.

(DSP). The DSP works in conjunction with a high-speed analog-to-digital (abbreviated A to D or A/D) converter (ADC) that takes the voice audio input and converts it into a series of binary words, each representing the amplitude of the signal. A series of such samples, taken at a high enough rates, can faithfully represent the original analog waveform. This is the same method that records music onto a CD-ROM or MP3 file. A digital-to-analog (D/A) converter (DAC) performs the reverse function.

The DSP takes the digital representation of the audio input and algorithmically combines that information with the RF carrier signal and produces the DSB AM signal, which is sent to the power amplifiers. That was an easy sentence to write, but one that takes many lines of code to implement within the DSP. This method of combining information content, in this case voice audio, with the RF carrier at the selected frequency has tremendous flexibility. Within the constraints imposed by the power and speed of the DSP, the sample rates and sample size of the DAC/ADC, and other considerations, this architecture allows a great deal of flexibility. As will be seen, this type of radio is capable of not only producing DSB AM signals, but other voice and data signals as well.

The terms "digital" and "analog" must be used with care. It is true to say that the modern avionics VHF radio is a "digital radio." It is also true to say that it handles voice signals digitally. However, to imply that we have "digital voice" is misleading. The signals propagated from the ground radio to the aircraft are the same DSB AM signals as the motor-tuned, vacuum-tube radio sent and received. Later, we will briefly discuss methods of sending voice over the RF in a digital manner.

17.2.2 HF Voice

HF voice is used for ATC communications in oceanic and remote airspace at various frequencies between 2.850 MHz and 23.350 MHz with wavelengths between 100 m and 10 m, respectively. The nature of the propagation of HF signals is such that it provides reliable communications at ranges of thousands of miles. This is possible because HF signals can be reflected by the bottom of the ionosphere at heights of about 70 miles. This effectively permits over-the-horizon or sky wave reception, not unlike the service performed by a satellite. At these frequencies, RF signals propagate with both a ground wave and a sky wave. The ground wave can give useful communication over the horizon, and the sky wave is usable well beyond that. Multiple hops are possible but are not reliable enough for aeronautical voice communication. Other characteristics of HF signals detract from its effectiveness for voice communications. For example, HF signals fade in a diurnal cycle and are susceptible to interference from solar activity. The sunspot cycle of 11 years can have significant effects on the ionosphere and on HF propagation.

The long wavelengths, comparable in length or longer than the aircraft, provide challenges as far as antenna placement is concerned. In the propeller-driven age, long wires from the tail forward were used. Later, long probes were mounted on the wingtips or tail. Now the HF antenna is typically installed in the forward edge of the vertical stabilizer, which is commonly made of composite materials. Ground station antennas can cover a football field.

HF voice is modulated onto the RF carrier using single-sideband (SSB) suppressed carrier modulation. Figure 17.2 shows the frequency domain view of an SSB signal. Note that only the upper sideband is used for aeronautical voice communications. Reliable HF communications requires the aircraft to transmit at 200 watts peak envelope power (PEP) and the ground stations to transmit at as much as 5 kW PEP. SSB has the advantage over DSB AM of increased power in the intelligence-carrying signal as opposed to the carrier.

Flying in remote or oceanic airspace would require long hours of listening to the static-filled HF channel if it were not for selective calling or SELCAL, a technique that allows the flight crew to turn down the volume on the HF radio until the ground signals them using preselected tones. Special receive circuits are needed as these tones are not sent as SSB signals. When the preselected tones are recognized, the flight crew is alerted to come up on the HF channel.

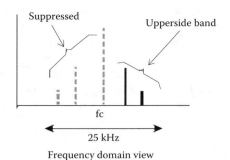

FIGURE 17.2 SSB voice signal.

17.2.3 Voice Developments

The proliferation of satellite telephone service on long-range aircraft has raised the question, "Why not just dial up ATC and talk to the controller?" Flight crews would like this method, but the reluctance is with the air navigation service providers. The management of frequencies in protected spectrum is, by now, a well-established procedure; the management of telephone numbers at each control position is not. There are other considerations, such as the use of nonprotected spectrum and the consequence of probable loss of protected spectrum, that inhibit a wholesale acceptance of telephone calls for ATC.

Europe has demonstrated that the aeronautical VHF voice band can be expanded by the use of 8.33 kHz voice. The ICAO SARPs for VDL Mode 3 also expand the number of voice channels possible; however, the costs of replacing all VF radios, both airborne and ground, has reduced support for this technique. VDL Mode 3 defines a true digital voice signal-in-space. The audio signal is converted to a digital representation, which is transmitted digitally across the air-ground VDL Mode 3 subnetwork and is not reconverted to analog audio signals until it reaches its destination.

The long-term possibility that broadband network connectivity to the aircraft may provide acceptable quality voice communication deserves some consideration for the far-term. Meanwhile, DSB AM voice will remain the primary method of ATC voice communications for the foreseeable future.

17.3 Data Communications

17.3.1 ACARS Overview

Today, ACARS provides worldwide data link coverage. Four distinct air-ground subnetworks are available for suitably equipped aircraft: original VHF, satcom, HFDL, and VDL Mode 2. It is impossible to understand the function of the avionics for ACARS without seeing the larger picture. Figure 17.3 shows an overview of the ACARS network showing the aircraft, the four air-ground subnetworks, the central message processor, and the ground message delivery network.

The ACARS message-passing network is an implementation of a star topology with the central message processor as the hub. The ground message network carries messages to and from the hub, and the air-ground subnetworks all radiate from the hub. There are a number of ACARS network service providers, and their implementations differ in some details, but all have the same star topology. Two data link service providers provide worldwide ACARS coverage, with several others providing regional coverage. Any given ACARS message can be carried over any of the air-ground subnetworks, a choice configured by the aircraft operator. It should be noted that ACARS is a character-oriented network, which means that only valid ASCII characters are recognized and that certain control characters are used to frame a valid message.

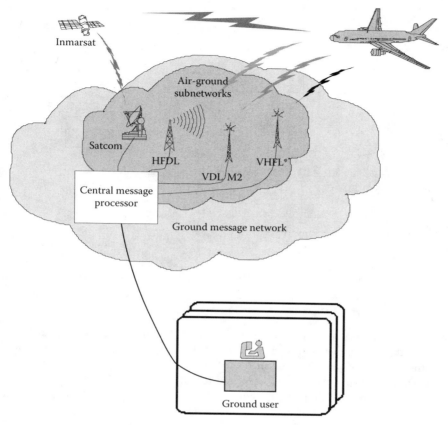

FIGURE 17.3 ACARS network overview. *VHFL:VHF data link; either ACARS or VDLM3 or VDLM4.

17.3.2 ACARS Avionics

The ACARS avionics architecture is centered on the management unit (MU) or communications management unit (CMU), which acts as an onboard router. All air-ground radios connect to the MU or CMU to send and receive messages. The CMU is connected to all of the various radios that communicate to the ground. Figure 17.4 illustrates the avionics architecture.

17.3.3 ACARS Management Unit

The MU or CMU acts as the ACARS router onboard the aircraft. All message blocks to or from the aircraft, over any of the air-ground subnetworks, pass through the MU. Although the MU handles all ACARS message blocks, it does not perform a message-switching function because it does not recombine multiple message blocks into a "message" prior to passing it along. It passes each message block in accordance with the "label" identifier, and it is up to the receiving end system to recombine message blocks into a complete message. The original OOOI messages were formatted and sent to the MU from an avionics unit that sensed various sensors placed around the aircraft and determined the associated changes of state. In the modern transport aircraft, many other avionics units send and receive routine ACARS messages.

The multifunction control and data unit (MCDU), along with the printer, is the primary ACARS interface to the flight crew. Other units, such as the FMS or the air traffic services unit (ATSU), will also interact with the crew for FANS messages. The vast majority of data link messages today are downlinks automatically generated by various systems on the airplane. The MU identifies each uplink message block

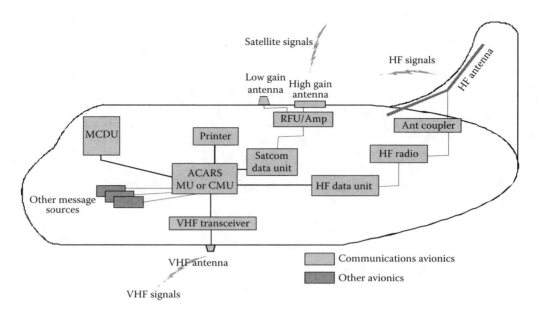

FIGURE 17.4 ACARS avionics architecture.

and routes it to the appropriate device. Similarly, it takes each downlink, adds associated aircraft information such as the tail number, and sends it to one of the air-ground subnetworks. The latest avionics for each of the four subnetworks accepts an ACARS block as a data message over a data bus, typically ARINC 429. The subnetwork avionics will then transform the message block into the signals needed to communicate with the ground radio. Each subnetwork has its own protocols for link layer and physical layer exchange of a data block.

17.3.4 VHF Subnetwork

The original VHF subnetwork that was pioneered in 1978 uses the same 25 kHz VHF channels used by ATC and aeronautical operational communication (AOC) voice; the signal-in-space is sometimes called plain-old-ACARS (POA) for reasons that will become clearer when we discuss VDL Mode 2. The VHF subnetwork uses a form of frequency shift keying (FSK) called minimum shift keying (MSK) wherein the carrier is modulated with either a 1200 Hz or 2400Hz tone. Each signaling interval represents one bit of information, so the 2400 baud rate (i.e., rate of change of the signal) equals the bit rate of 2400 BPS. After initial synchronization, the receiver then can determine whether a given bit is a one or a zero.

VHF ACARS uses the carrier-sensed multiple access (CSMA) protocol to reduce the effects of two transmitters sending a data block at the same or overlapping times. CSMA is nothing more than the automated version of voice radio protocols wherein the speaker first listens to the channel before initiating a call. Once a transmitter has begun sending a block, no other transmitter will "step on" that transmission. The VHF ACARS subnetwork is an example of a connectionless link layer protocol in that the aircraft does not "log in" to each ground station along its route of flight. The aircraft does initiate a contact with the central message processor, and it does transmit administrative message as it changes subnetworks. A more complete description of the POA signal and an ACARS message block as it is transmitted over a VHF channel can be found in ARINC 618, Appendix B.

In congested airspace, such as the northeastern United States or Europe, multiple VHF ACARS channels are needed to carry the message traffic load. For example, in the Chicago area, ten channels are needed and a sophisticated frequency management scheme has been put in place that automatically changes the frequency used by individual aircraft to balance the loads.

Initial ACARS MUs worked with VHF radios that were little modified from their voice-only cousins. The ACARS modulation signal was created as two-tone audio by the MU (e.g., ARINC 724 MU) and sent to the radio (e.g., ARINC 716 VHF radio), where it modulated the RF, just as voice did from a microphone. Later evolutions of the ACARS interface between the CMU (e.g., ARINC 758 CMU) and the latest radio (e.g., ARINC 750 VDR [VHF data radio]) sent ACARS message blocks between the CMU and the radio over a serial data bus (i.e., ARINC 429 Digital Information Transfer System (DITS)), and the radio modulated the RF directly from the data.

17.3.5 Satcom

The satellite ACARS subnetwork uses the Inmarsat constellation. Four satellites in geosynchronous orbit provide global beam coverage of the majority of the globe (up to about 82° latitude) with spot beam coverage over the continents. The Inmarsat constellation provides telephone circuits as well as data link, so it uses a complex set of protocols over several different types of channels using different signals-in-space. A circuit is established between the satellite data unit (SDU) and the ground earth station (GES). Any data link message block generated by the MU for transfer over the satcom subnetwork is sent to the SDU for transfer over this circuit to the GES, where it is forwarded to the central message processor. This is an example of a connection-oriented link layer protocol, which is common for modern subnetworks.

The Inmarsat constellation operates in the L-band, around 1 GHz on frequencies reserved for aeronautical mobile satellite (route) services, or AMS(R)S, which are protected for safety and regularity of flight. Satcom avionics have been purpose-built, meaning that they did not evolve from the previous use of L-band radios for voice as VHF ACARS and (as we shall see) HFDL radios evolved from voice radios. The RF unit (RFU), along with high-gain and low-noise amplifiers and the diplexer, sends and receives signals over the various L-band channels defined for Inmarsat services.

17.3.6 HFDL

The HFDL ACARS subnetwork uses channels in the HF voice band. The HFDL radio can be a slightly modified HF voice radio connected to the HF data unit (HFDU). Alternatively, an HF data radio (HFDR) can contain both voice radio and data link functions. In either case, the HF communication system must be capable of independent voice or data operation.

HFDL uses phase shift modulation (PSK) and time-division multiple access (TDMA). A 32-second frame is divided into 13 slots, each of which can communicate with a different aircraft at a different data rate. Four data rates (1800 BPS, 1200 BPS, 600 BPS, and 300 BPS) use three different PSK methods (8PSK, 4PSK, and 2PSK). The slowest data rate is affected by doubling the power of the forward error-correcting code. All of these techniques (i.e., multiple data rates, forward error correction, TDMA), are used to maximize the long-range properties of HF signals while mitigating the fade and noise inherent in the medium. Twelve HFDL ground stations provide worldwide coverage, including good coverage over the North Pole but excluding the south polar region. More details on HFDL may be found in *ARINC 753: HF Data Link System*.

The need for a large antenna, plus the fact that even a quarter-wavelength antenna is problematic, necessitates an antenna coupler that matches the impedance of the feed line to the antenna. The RFU, whether it is a separate unit or incorporated in the HFDR, combines the audio signal representing the data modulation with the carrier frequency, suppresses the carrier and lower sidebands with appropriate filtering, and amplifies the resultant signal.

17.3.7 VDL Mode 2

VDL Mode 2 operates in the same VHF band as POA. Four channels have been reserved worldwide for VDL Mode 2 services. Currently, the only operating frequency is 136.975 MHz. VDL Mode 2 and differential 8-phase shift keying (D8PSK) at a signaling rate of 10.5 Kbaud is used to modulate the carrier.

Since each phase change represents one of eight discernable phase shifts, three bits of information are conveyed by each baud or signal change; therefore, the data rate is 31.5 KBPS. With about ten times the capacity of a POA channel, VDL Mode 2 has the potential to significantly reduce channel congestion for ACARS. CSMA is used for media access, but a connection-oriented link layer protocol called the aviation VHF link control (AVLC) is established between the VDR and the ground station. ACARS over AVLC (AOA) is the term used to distinguish ACARS message blocks from other data packets that can also be passed over AVLC. It should be noted that VDL Mode 2 has been implemented in accordance with the ICAO SARPs as a subnetwork of the ATN. Therefore, VDL Mode 2 is a bit-oriented data link layer protocol, which, in the case of AOA, happens to be carrying ACARS message blocks. The ARINC 750 radio is capable of supporting 25Khz and 8.33 kHz voice, POA, and AOA. It may only be used for one of these functions at any given time.

17.3.8   Data Link Developments

ICAO has defined the ATN as the worldwide data communication network (air-ground and ground-ground) for air traffic services. The ATN is defined using the Open Systems Interconnection (OSI) seven-layer stack. ICAO SARPs have been developed in great detail for satcom, VDL, and HFDL air-ground subnetworks. The ACARS satcom, VDL Mode 2, and HFDL subnetworks take advantage of or, in the case of satcom, were used to define the ICAO SARPs. To this date, the Federal Aviation Administration (FAA) has implemented a limited subset of ATN CPDLC in one domestic region using VDL Mode 2; however, the trial was cancelled and does not look to be reinstated before 2011 at the earliest. Europe has implemented a larger subset of ATN CPDLC in the Maastricht upper air sector on an operational basis and has well-established plans for expansion of data link.

The implementation of broadband Internet connections in the aircraft while in flight has the potential to provide versatile, fast, and cheap connectivity between the aircraft and the ground. Since the earliest voice radio links, through all of the ACARS air-ground subnetworks, air-ground communications has been so specialized that the equipment has been specially designed and built at great cost. If broadband Internet (meaning TCP/IP) connectivity can be made reliable and secure, there is no reason this medium could not be used for air-ground data link communication.

The trend in the telecommunications industry is toward high-speed, high-capacity, general-purpose connectivity. For example, fiber optic links installed to carry cable TV are being used, without significant change, as Internet connections or telephone lines. Sophisticated high-capacity RF modulation techniques are permitting the broadcast of digital signals for high-definition TV and radio. Mobile telephone technology carries digital voice and data messages over the same network. The Internet itself carries far more than the text and graphics information it was originally designed to carry.

17.4   Summary

The airlines will continue to increase their dependence upon air-ground data link to send and receive information necessary to efficiently operate their fleets. ATC will increase its dependence upon air-ground communications, even as the number of voice transactions is reduced. Looking ten to twenty years ahead, data link will increasingly be used for ATC communications. If the concept of air traffic management (ATM) is to become the rule instead of the exception, the ground automation systems and the FMSs will no doubt be in regular contact, exchanging projected trajectory, weather, traffic, and other information. Voice intervention will be minimal, and likely still be over DSB AM in the VHF band.

The modern transport aircraft is becoming a flying network node that will inevitably be connected to the ground for seamless data communication. It's only a matter of time and ingenuity. When that happens, presuming there is sufficient bandwidth, availability, and reliability for each use, many applications will migrate to that link.

References

American Radio Relay League, *The Radio Amateur's Handbook*, 36th ed., The Rumford Press, Concord, NH, 1959.

ARINC Specification 410-1, Mark 2 Standard Frequency Selection System, Aeronautical Radio, Inc., Annapolis, MD, , October 1, 1965.

ARINC Characteristic 566A-9, Mark 3 VHF Communications Transceiver, Aeronautical Radio, Inc., Annapolis, MD, January 30, 1998.

ARINC Specification 618-5, Mark 2 Standard Frequency Selection System Air/Ground Character-Oriented Protocol Specification, Aeronautical Radio, Inc., Annapolis, MD, August 31, 2000.

ARINC Specification 619-2, ACARS Protocols for Avionic End Systems, Aeronautical Radio, Inc., Annapolis, MD, March 11, 2005.

ARINC Specification 620-4, Data Link Ground System Standard and Interface Specification, Aeronautical Radio, Inc., Annapolis, MD, November 24, 1999.

ARINC Characteristic 719-5, Airborne HF/SSB System, Aeronautical Radio, Inc., Annapolis, MD, July 6, 1984.

ARINC Characteristic 750-4, VHF Data Radio, Aeronautical Radio, Inc., Annapolis, MD, August 11, 2004.

ARINC Characteristic 753-3, HF Data Link System, Aeronautical Radio, Inc., Annapolis, MD, February 16, 2001.

ARINC Specification 720-1, Digital Frequency/Function Selection for Airborne Electronic Equipment, Aeronautical Radio, Inc., Annapolis, MD, July 1, 1980.

ARINC Characteristic 724-9, Aircraft Communications Addressing and Reporting System, Aeronautical Radio, Inc., Annapolis, MD, October 9, 1998.

ARINC Characteristic 724B-5, Aircraft Communications Addressing and Reporting System, Aeronautical Radio, Inc., Annapolis, MD, February 21, 2003.

ARINC Characteristic 741P2-7, Aviation Satellite Communication System Part 2 System Design and Equipment Functional Description, Aeronautical Radio, Inc., Annapolis, MD, December 24, 2003.

ARINC Characteristic 758-2, Communications Management Unit Mark 2, Aeronautical Radio, Inc., Annapolis, MD, July 8, 2005.

The ARINC Story, The ARINC Companies, Annapolis, MD, 1987.

Institute of Electrical and Electronics Engineers and Electronic Industries Association (IEEE and IEA), *Report on Radio Spectrum Utilization*, Joint Technical Advisory Committee, Institute of Electrical and Electronics Engineers, New York, 1964.

18
Navigation Systems

Myron Kayton
Kayton Engineering Company

18.1 Introduction

Navigation is the determination of the position and velocity of a moving vehicle on land, at sea, in the air, or in space. The three components of position and the three components of velocity make up a six-component state vector whose time variation fully describes the translational motion of the vehicle. With the advent of the Global Positioning System (GPS), surveyors use the same sensors as navigators but achieve higher accuracy as a result of longer periods of observation and more complex postprocessing.

In the usual navigation system, the state vector is derived on-board, displayed to the crew, recorded on-board, or transmitted to the ground. Navigation information is usually sent to other on-board subsystems, such as the waypoint steering, communication control, display, weapon-control, and electronic warfare (emission detection and jamming) computers. Some navigation systems, called position-location systems, measure a vehicle's state vector using sensors on the ground or in another vehicle (Section 18.7). The external sensors usually track passive radar returns or a transponder. Position-location systems usually supply information to a dispatch or control center.

The term guidance has two meanings, both of which are different than navigation. In the first, steering toward a destination of known position from the vehicle's present position, the steering equations are derived from a plane triangle for nearby destinations or from a spherical triangle for distant destinations. In the second definition, steering toward a destination without calculating the state vector explicitly, a guided vehicle hones in on radio, infrared, or visual emissions. Guidance toward a moving target is usually of interest to military tactical missiles in which a steering algorithm assures impact within the maneuver and fuel constraints of the interceptor. Guidance toward a fixed target involves beam-riding, as in the Instrument Landing System (ILS), Section 18.5. The term flight control refers to the deliberate rotation of an aircraft in three-dimensions around its mass center.

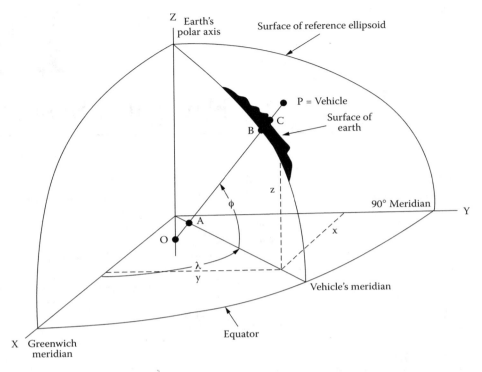

FIGURE 18.1 Latitude-Longitude-Altitude and X-Y-Z Coordinate Frames ϕ = geodetic latitude; OP is normal to the ellipsoid at B; λ = geodetic longitude; h = BP = altitude above reference ellipsoid = altitude above mean sea level.

18.2 Coordinate Frames

Navigation is with respect to a coordinate frame of the designer's choice. For navigation over hundreds of kilometers (e.g., helicopters), various map grids exist whose coordinates can be calculated from latitude-longitude. NATO helicopters and land vehicles use a Universal Transverse Mercator grid. Long-range aircraft navigate relative to an Earth-bound coordinate frame; the following are the most common (Figure 18.1):

1. **Latitude-longitude-altitude** — The most useful reference ellipsoid is described in WGS-84 [(U.S. Government 1991)]. Longitude becomes indeterminate in polar regions rendering these coordinates unsuitable there.
2. **Earth-centered rectangular (XYZ)** — These coordinates are valid worldwide, hence GPS calculates in them and often converts to latitude-longitude-altitude for readout.

Spacecraft in orbit around the Earth navigate with respect to an Earth-centered, inertially nonrotating coordinate frame whose Z-axis coincides with the polar axis of the Earth and whose X-axis lies along the equator.

18.3 Categories of Navigation

Navigation systems can be categorized as absolute navigation systems, dead-reckoning navigation systems, or mapping navigation systems. Absolute navigation systems measure the state vector without regard to the path traveled by the vehicle in the past. These are of two kinds: radio systems (Section 18.5) and celestial systems (Section 18.6). Radio systems consist of a network of transmitters (sometimes transponders) on the ground or in satellites. A vehicle detects the transmissions and computes its position relative to the known positions of the stations in the navigation coordinate frame. The vehicle's velocity is measured from the Doppler shift of the transmissions or from a sequence of position measurements. The second of the absolute navigation systems, celestial systems, measures the elevation and azimuth of celestial bodies relative to local-level and true north. Electronic star sensors are used in special-purpose high-altitude aircraft and in spacecraft. Manual celestial navigation was practiced at sea for millennia [Bowditch 1995].

Dead-reckoning navigation systems derive their state vector from a continuous series of measurements beginning at a known initial position. There are two kinds: those that measure vehicle heading and either speed or acceleration (Section 18.4) and those that measure emissions from continuous-wave radio stations whose signals create ambiguous "lanes" (Section 18.5). Dead reckoning systems must be updated as errors accumulate and if electric power is lost. The only dead reckoning radio system, Omega, was decommissioned in 1997.

Lastly, mapping navigation systems observe and recognize images of the ground, profiles of altitude, sequences of turns, or external features (Section 18.7). They compare their observations to a stored data base, often on compact disc.

18.4   Dead Reckoning

The simplest dead-reckoning systems measure aircraft heading and speed, resolve speed into the navigation coordinates, then integrate to obtain position (Figure 18.2). The oldest heading sensor is the magnetic compass: a magnetized needle or an electrically excited toroidal coil (called a flux gate), or an electronic magnetometer, as shown in Figure 18.3. It measures the direction of the Earth's magnetic field to an accuracy of 2 degrees at a steady speed below 60 degrees magnetic latitude. The horizontal component of the magnetic field points toward magnetic north. The angle from true to magnetic north is called magnetic variation and is stored in the computers of modern vehicles as a function of position over the region of anticipated travel [Quinn (1996)]. Magnetic deviations caused by iron and motors in the vehicle can exceed 30 degrees and must be compensated in the navigation computer.

A more complex heading sensor is the gyrocompass, consisting of a spinning wheel whose axle is constrained to the horizontal plane by a pendulous weight. The aircraft version (more properly called a directional gyroscope) holds any preset heading relative to Earth and drifts at more than 50 deg/hr. Inexpensive gyroscopes (some built on silicon chips as vibrating beams with on-chip signal conditioning) are often coupled to magnetic compasses to reduce maneuver-induced errors and long-term drift.

The usual speed-sensor on an aircraft or helicopter is a pitot tube that measures the dynamic pressure of the air stream from which airspeed is derived in an air-data computer. To compute ground speed, the velocity of the wind must be vectorially added to that of the aircraft (Figure 18.2). Hence, unpredicted wind will introduce an error into the dead-reckoning computation. Most pitot tubes are insensitive to the component of airspeed normal to their axis, called drift. Another speed sensor is Doppler radar that measures the frequency shift in radar returns from the ground or water below the aircraft, from which ground-speed is inferred directly. Multibeam Doppler radars can measure all three components of the vehicle's velocity relative to the Earth. Doppler radars are widely used on military helicopters.

The most accurate dead-reckoning system is an inertial navigator in which accelerometers measure the vehicle's acceleration while gyroscopes measure the orientation of the accelerometers. An on-board computer resolves the accelerations into navigation coordinates and integrates them to obtain velocity and position. The gyroscopes and accelerometers are mounted either directly to the airframe or on a servo-stabilized platform. When fastened directly to the airframe ("strap-down"), the sensors are exposed to the angular rates and angular accelerations of the vehicle. In the 2000s, virtually all inertial navigators were strap-down. Attitude is computed by a quaternion algorithm (Kayton and Fried, 1997, pg. 352–356) that integrates measured angular increments in three dimensions at a faster rate than the navigation coordinates are calculated.

When accelerometers and gyros are mounted on a servo-stabilized platform, gimbals angularly isolate them from rotations of the vehicle. The earliest inertial navigators used gimbals. In 2006, gimballed

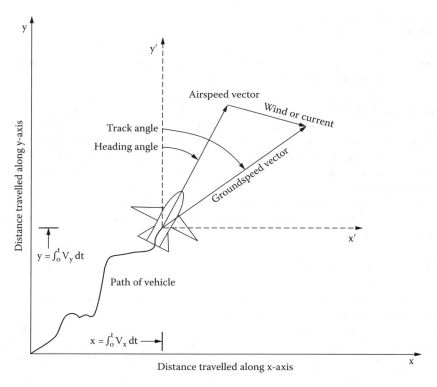

FIGURE 18.2 Geometry of Dead-Reckoning.

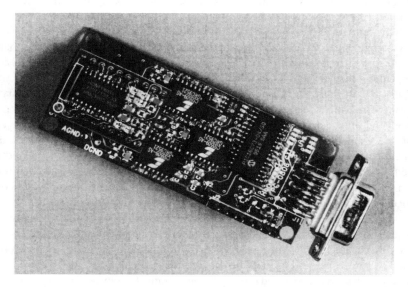

FIGURE 18.3 Circuit Board from 3-Axis Digital Magnetometer. A single-axis sensor chip and a 2-axis sensor chip are mounted orthogonally at the end opposite the connector. The sensor chips are magneto-resistive bridges with analog outputs that are digitized on the board. Photo courtesy of Honeywell, Copyright 2004.

FIGURE 18.4 GPS-Inertial Navigator. The inertial instruments are mounted at the rear with two laser gyroscopes and electrical connectors visible. The input-output board is next from the rear; it excites and reads the inertial sensors. The computer board is next closest to the observer and includes MIL-STD-1553 and RS-422 external interfaces. The power supply is in front. Between them is the single-board shielded GPS receiver. Round connectors on the front are for signals and electric power. A battery is in the case behind the handle. Weight 10 kg, power consumption 40 watts. This navigation set is used in the F-22 and many other military aircraft and helicopters. Photo courtesy of Northrop-Grumman Corporation.

navigators were only used on specialized high-accuracy military aircraft. The gimbal angles measure attitude directly, without computation. The instruments are in a benign angular environment and held at a constant orientation relative to gravity, which simplifies the computations and reduces the error in mechanical instruments.

Inertial systems measure vehicle orientation within 0.1 degree for steering and pointing. Most accelerometers consist of a gram-sized proof-mass that is mounted on a flexure pivot. The least expensive ones measure the deflection of the proof mass; the most precise accelerometers restore the proof mass to null, thus not relying on the structural properties of the flexure. The newest accelerometers are etched into silicon chips. The oldest gyroscopes contained metal wheels rotating in ball bearings or gas bearings that measured angular velocity or angular-increments relative to inertial space; more recently, gyroscopes contained rotating, vibrating rings whose frequency of oscillation measured the instrument's angular rates. The newest gyroscopes are evacuated cavities or optical fibers in which counter-rotating laser beams are compared in phase to measure the sensor's angular velocity relative to inertial space about an axis normal to the plane of the beams. Vibrating hemispheres and rotating, vibrating bars are the basis of some navigation-quality gyroscopes (drift rates less than 0.1 deg/hr).

Fault-tolerant configurations of cleverly-oriented redundant gyroscopes and accelerometers (typically four to six of each) detect and correct sensor failures. Inertial navigators are used in long-range airliners, business jets, most military fixed-wing aircraft, space boosters, entry vehicles, and manned spacecraft.

TABLE 18.1 Worldwide Radio Navigation Aids for Aviation

System	Frequency		Number of Stations	Number of Aeronautical Users
	Hz	Band		
Loran-C/Chaika	100 kHz	LF	50	130,000
Beacon*	200–1600 kHz	MF	4000	130,000
Instrument Landing System (ILS)*	{ 108–112 MHz	VHF	1500	200,000
	329–335 MHz	UHF		
VOR*	108–118 MHz	VHF	1500	200,000
SARSAT/COSPAS	{ 121.5 MHz	VHF	5 satellites	200,000
	243,406 MHz	UHF		
JTIDS	960–1213 MHz	L	Network	500
DME*	962–1213 MHz	L	1500	150,000
Tacan*	962–1213 MHz	L	1000	15,000
Secondary Surveillance Radar (SSR)*	1030, 1090 MHz	L	1000	250,000
GPS-GLONASS	1227, 1575 MHz	L	24 satellites	150,000
Radar Altimeter	4200 MHz	C	None	40,000
MLS*	5031–5091 MHz	C	25	100
Weather/map radar	10 GHz	X	None	10,000
Airborne Doppler radar	13–16 GHz	Ku	None	40,000
SPN-41 carrier-landing monitor	15 GHz	Ku	25	1600
SPN-42/46 carrier-landing radar	33 GHz	Ka	25	1600

* Standardized by International Civil Aviation Organization.

18.5   Radio Navigation

Scores of radio navigation aids have been invented, and many of them have been widely deployed, as summarized in Table 18.1. The most precise is the Global Positioning System (GPS), a network of 24 satellites and 16 ground stations (in 2006) for monitoring and control. An aircraft derives its three-dimensional position and velocity from one-way ranging signals at 1.575 GHz received from four or more satellites (military users also receive 1.227 GHz) [U.S. Air Force (2000)]. The one-way ranging measurements depend on precise atomic clocks on the spacecraft (one part in 10^{13}) and on precise clocks on the aircraft (one part in 10^8) that can be calibrated to behave briefly like atomic clocks by taking redundant range measurements from the GPS satellites. GPS offers better than 30-meter ranging errors to civil users and 5-meter ranging errors to military users. Simple receivers were available for less than $100 in 2006, with larger errors and without the ability to track at aircraft speeds. GPS provides continuous worldwide navigation for the first time in history. It made dead reckoning unnecessary on many vehicles and reduced the cost of most navigation systems. Figure 18.5 is an artist's drawing of a GPS Block 2F spacecraft, scheduled to be launched in 2007. During the 1990s, Russia deployed a satellite navigation system called GLONASS, incompatible with GPS, which they casually maintain. In 2006, the European Union was in the final stages of defining its own navigation satellite system, called Galileo, which will offer free and paid services [Hein (2003)]. The United States began a minor upgrade of GPS in 2006 and plans a major upgrade by 2015 to reduce vulnerability to jamming [Enge (2004)]. After 2006, civil users will also receive at 1.176 GHz, thus allowing them to estimate ionospheric errors.

Differential GPS (DGPS) employs ground stations at known locations that receive GPS signals and transmit measured ranging errors on a radio link to nearby vehicles. DGPS improves accuracy (centimeters for fixed observers) and detects faults in GPS satellites immediately. In 2003, the United States created a nationwide aeronautical DGPS system consisting of about 50 stations and monitoring sites. This Wide-Area Augmentation System (WAAS), which transmits its corrections via geosynchronous communication satellites, will eventually replace VORTAC on less-used airways and Category I ILS (page 18-8). In 2006, the United States experimented with a dense network of DGPS sites at airports called the Local-Area Augmentation System (LAAS). The intent is to replace several ILS and MLS (Microwave

FIGURE 18.5 Global Positioning Satellite, Block 2F, Courtesy of Rockwell.

Landing System) landing aids with a small number of DGPS stations at each airport. The experiments show that to achieve Category II and III landing (all-weather), inertial aiding will be needed. Error detection for WAAS and LAAS occurs in about one second.

Loran is used by general aviation aircraft for *en route* navigation and for nonprecision approaches to airports (in which the cloud bottoms are more than 400 feet above the runway). Loran's 100 KHz signals are usable within 1000 nmi (a nautical mile is 1852 meters exactly) of a chain consisting of three or four stations (Kayton and Fried, Chapter 4.5). Chains cover the United States, parts of western Europe, Japan, Saudi Arabia, and a few other areas. Russia has a compatible system called Chaika. The vehicle-borne receiver measures the difference in time of arrival of pulses emitted by two ground stations, thus locating the vehicle on one branch of a hyperbola whose foci are at the stations. Two or more station pairs give a two-dimensional position fix at the intersection of the hyperbolas, whose typical accuracy is 0.25 nmi, limited by propagation uncertainties over the terrain between the transmitting station and the aircraft. The measurement of hundred-microsecond time differences is made with a low-quality clock (one part in 10^4) in the vehicle. Loran stations were upgraded in the late 1990s so service is assured for the first two decades of the third millennium. Loran will be a coarse monitor of GPS and a stand-alone navigation aid whenever GPS is deliberately taken out of service by the U.S. military, its sole operator. GPS monitor functions might alternatively be provided by European or Russian navigation satellites or by private nav-com satellites. These satellite-based monitors are more accurate than Loran but are subject to the same outages as GPS: solar flares and jammers, for example.

The most widely used aircraft radio aid at the start of the third millennium is VORTAC, whose stations offer three services:

1. Analog bearing measurements at 108–118 MHz (called very high frequency omni range (VOR)); the vehicle compares the phases of a rotating cardioid pattern and an omni-directional sinusoid emitted by the ground station
2. Pulse distance measurements (called distance measuring equipment (DME)) at 1 GHz. DME measures the time delay for an aircraft to interrogate a VORTAC station and receive a reply
3. Tacan bearing information, conveyed in the amplitude modulation of the DME—replies from the VORTAC stations

On short flights over-ocean, the inertially derived state vector drifts one to two nmi per hour. When an aircraft approaches shore, it acquires a VOR station and updates the inertial state vector. Navigation then continues to the destination using inertia and VOR and perhaps ILS for landing. On long over-ocean flights (e.g., trans-Pacific), GPS is usually used with one or more inertial navigators to protect against failures in GPS satellites or in airborne receivers.

Throughout the western world, civil aircraft use VOR/DME, whereas military aircraft use Tacan/DME for *en route* navigation. In the 1990s, China and the successor states to the Soviet Union began to replace their direction-finding stations with International Civil Aviation Organization (ICAO) standard navigation aids (VOR, DME, and ILS) at their international airports and along the corridors that lead to them from their borders. DGPS sites will eventually replace most VORTACs; fifty DGPS sites could replace a thousand VORTACs, thus saving an immense sum of money for maintenance. Nevertheless, VORTAC stations are likely to be retained on important routes through 2025 though decommissioning will begin in 2010 (Fed. Radionav Plan, 2005).

Omega was a worldwide radio aid that consisted of eight ground stations, each of which emitted continuous sine waves at 10–13 KHz. Most vehicles measured the range differences between two stations by observing the phase differences between the received sinusoids. Ambiguous lanes were created when phase differences exceeded 360 degrees. Errors were about 2 nmi due to radio propagation irregularities. Omega, used by submarines, over-ocean general-aviation aircraft, and a few international air carriers, was decommissioned in 1997.

ILS is used in landing guidance throughout the world (in the 2000s, even in China, India, and the former Soviet Union). Transmitters adjacent to the runway create a horizontal guidance signal near 110 MHz and a vertical guidance signal near 330 MHz. Both signals are modulated such that the nulls intersect along a line in space that leads an aircraft from a distance of about 15 nmi to within 50 feet above the runway. ILS gives no information about where the aircraft is located along the beam, except at two or three vertical marker beacons. Most ILS installations are certified to the ICAO's Category I, in which the pilot must abort the landing if the runway is not visible at an altitude of 200 feet while descending. About two hundred ILSs are certified to Category II, which allows the aircraft to descend to 100 feet above the runway before aborting for lack of visibility. Category III allows an aircraft to land at still lower weather ceilings. About fifty ILSs are certified to Category III, mostly in Western Europe, which has the worst flying weather in the developed world. Category III ILSs detect their own failures and switch to a redundant channel within one second to protect aircraft that are flaring out (within 50 feet above the runway) and can no longer execute a missed approach. Once above the runway, the aircraft's bottom-mounted radio altimeter measures altitude and either the electronics or the pilot guides the flare maneuver. Landing aids are described by Kayton and Fried (1997) in Chapter 13. Category I ILS will begin to be decommissioned after 2010 (Fed. Radionav Plan, 2005).

U.S. Navy aircraft find their seaborne carriers with Tacan and use a microwave scanning system at 15.6 GHz to land; NASA's Space Shuttle used the Navy system to land at its spaceports but an inertially aided DGPS system replaced it. Another microwave landing system (MLS) at 5 GHz was supposed to replace the ILS in civil operations, especially for Categories II and III. However, experiments during the 1990s showed that DGPS with a coarse inertial supplement could achieve an in-flight accuracy of better than three meters as a landing aid and could detect satellite errors within a second. Hence, it is likely that LAAS will replace or supplement ILS, which has been guaranteed to remain in service at least until the year 2010 (Federal Radionavigation Plan), more likely past 2020. NATO may use portable MLS or LAAS for flights into tactical airstrips.

Position-location systems monitor the state vectors of many vehicles simultaneously and usually display the data in a control room or dispatch center. Some vehicles derive their state vector from the ranging modulations (e.g., DME); others merely report an independently derived position (e.g., GPS). The aeronautical bureaucracy calls the reporting of independently derived position Automatic Dependent Surveillance. The continuous broadcast of on-board-derived position (probably GPS-based) may be the basis of over-ocean air traffic control systems of the early twenty-first century. Table 18.1 lists Secondary Surveillance Radars that receive coded replies from aircraft so they can be identified by human air-traffic controllers and by collision-avoidance algorithms.

Low-altitude commercial communication satellites will offer digital-ranging services worldwide for a fee. The intermittent nature of these commercial fixes would require that vehicles dead-reckon between fixes, perhaps using solid-state inertial instruments. Therefore, if taxpayers insist on collecting fees for navigation services, private comm-nav networks (such as Galileo's commercial service) may replace the government-funded GPS and air-traffic communication networks in the mid-twenty-first century.

Military communication-navigation systems measure the position of air, land, and naval vehicles on battlefields and report to headquarters; examples are the American Joint Tactical Information Distribution System (JTIDS) and the Position Location Reporting System (PLRS). Their terminals were said to cost more than $100,000 each in 2000.

A worldwide network of approximately 40 SARSAT-COSPAS stations monitors signals from Emergency Location Transmitters (on aircraft, ships, and land users) on the three international distress frequencies (121.5, 243, and 406 MHz), relayed via low-orbit satellite-based transponders. Software at the listening stations calculates the position of the Emergency Location Transmitters within 5–15 km at 406 MHz and 15–30 km for the others, based on the Doppler-shift history observed by the satellites, so that rescue vehicles can be dispatched. Some 406-MHz Emergency Location Transmitters contain GPS sets; they transmit their position to geostationary satellites. SARSAT-COSPAS has saved more than 12,000 lives worldwide, from arctic bush-pilots to tropical fishermen, since 1982 [NASA/NOAA].

18.6 Celestial Navigation

Human navigators use sextants to measure the elevation angle of celestial bodies above the visible horizon. The peak elevation angle occurs at local noon or midnight:

$$\text{elev angle (degrees)} = 90 - \text{latitude} + \text{declination}$$

Thus, at local noon or midnight, latitude can be calculated by simple arithmetic from a table of declination (the angle of the sun or star above the Earth's equatorial plane). When time began to be broadcast to vehicles in the 1930s, off-meridian observations of the elevation angles of two or more celestial bodies became possible at any known time of night (cloud cover permitting). These fixes were hand-calculated using logarithms, then plotted on charts by a navigator. In the 1930s, hand-held bubble-level sextants were built to measure the elevation of celestial bodies from an aircraft without the need to see the horizon. The human navigator observed sun and stars through an astrodome on top of the aircraft. The accuracy of celestial fixes was 5–50 miles in the air, limited by the uncertainty in the horizon and the inability to make precise angular measurements on a pitching, rolling vehicle. Kayton (1990) reviews the history of celestial navigation at sea and in the air.

The first automatic star trackers, built in the late 1950s, measured the azimuth and elevation angles of stars relative to a gyroscopically stabilized platform. Approximate position measurements by dead reckoning allow the telescope to point within a fraction of a degree of the desired star. Thus, a narrow field-of-view is possible, permitting the telescope and photodetector to track stars in the daytime through a window on top of the aircraft. An on-board computer stores the right ascension and declination of 20–100 stars and computes the vehicle's position. Automatic star trackers, used in long-range military aircraft and on Space Shuttles, are physically mounted on the stable element of a gimballed inertial navigator. Clever design of the optics and of stellar-inertial signal-processing filters achieves accuracies better than 500 feet [Kayton and Fried, Chapter 12]. Future lower-cost systems may mount the star tracker directly to the vehicle.

18.7   Map-Matching Navigation

On aircraft, mapping radars and optical sensors present an image of the terrain to the crew, whereas on unmanned aircraft, navigation must be autonomous. Automatic map-matchers have been built since the 1960s that correlate the observed image to stored images of patches of distinctive terrain, choosing the closest match to update the dead-reckoned state vector. Since 1980, aircraft and cruise missiles measure the vertical profile of distinctive patches of terrain below the aircraft and match it to a stored profile. Updating with the matched profile, perhaps hourly, reduces the long-term drift of the inertial navigator. The profile of the terrain is measured by subtracting the readings of a baro-inertial altimeter (calibrated for altitude above sea level) and a radio altimeter (measuring terrain clearance). An on-board computer calculates the cross-correlation function between the measured profile and each of many stored profiles on possible parallel paths of the vehicle. The on-board inertial navigator usually contains a digital filter that corrects the drift of the azimuth gyroscope as a sequence of fixes is obtained. Hence, the direction of flight through the stored map is known, saving the considerable computation time that would be needed to correlate for an unknown azimuth of the flight path.

The most complex mapping systems observe their surroundings by digitized video (often stereo) and create their own map of the navigated space. In the 2000s, optical mappers are being developed that allow landings at fields that are not equipped with electronic aids.

18.8   Navigation Software

Navigation software is sometimes embedded in a central processor with other avionic software or confined to one or more navigation computers. The navigation software contains algorithms and data that process the measurements made by each sensor (e.g., GPS, inertial, or air data). It contains calibration constants, initialization sequences, self-test algorithms, reasonability tests, and alternative algorithms for periods when sensors have failed or are not receiving information. In the simplest systems, a state vector is calculated independently from each sensor while the navigation software calculates the best estimate of position and velocity. Prior to 1970, the best estimate was calculated from a least-squares algorithm with constant weighting functions or from a frequency-domain filter with constant coefficients. Now, a Kalman filter calculates the best estimate from mathematical models of the dynamics of each sensor (Kayton and Fried, 1997, Chapter 3).

Digital maps, often stored on compact disc, are carried on an increasing number of aircraft; position can be visually displayed to the crew and terrain warnings issued. Military aircraft superimpose their navigated position on a stored map of terrain and cultural features to aid in the penetration of and escape from enemy territory. Algorithms for waypoint steering and for control of the vehicle's attitude are contained in the software of the flight management and flight control subsystems.

Specially equipped aircraft are often used for the routine calibration of radio navigation aids, speed and velocity sensors, heading sensors, and new algorithms. Airborne test beds and hardware-software integration laboratories are routinely used to develop algorithms and sensor-software interfaces.

18.9   Design Trade-Offs

The navigation-system designer conducts trade-offs for each vehicle to determine which navigation systems to use and how to interface them. Trade-offs consider the following attributes:

FIGURE 18.6 Navigation displays in the U.S. Air Force C-5 transport showing flat-panel displays in front of each pilot; vertical situation display outboard and horizontal situation display inboard. Waypoints are entered on the horizontally-mounted Control-Display Unit just visible aft of the throttles. In the center of the instrument panel are status and engine displays and backup analog instruments. Photo courtesy of Honeywell, Copyright 2004.

1. Cost, including the construction and maintenance of transmitter stations and the purchase of on-board electronics and software. Users are concerned only with the costs of on-board hardware and software.
2. Accuracy of position and velocity, which is specified as a Circular Error Probable (CEP, in meters or nautical miles). The maximum allowable CEP is often based on the calculated risk of collision on a typical mission.
3. Autonomy, the extent to which the vehicle determines its own position and velocity without external aids. Autonomy is important to certain military vehicles and to civil vehicles operating in areas of inadequate radio-navigation coverage. Degrees of autonomy are described in Kayton and Fried, 1997, page 10.
4. Time delay in calculating position and velocity, caused by computational and sensor delays.
5. Geographic coverage. Radio systems operating below 100 KHz can be received beyond line-of-sight on Earth; those operating above 100 MHz are confined to line-of-sight.
6. Automation. The vehicle's operator (on-board crew or ground controller) receives a direct reading of position, velocity, and equipment status, usually without human intervention. The navigator's crew station disappeared in civil aircraft in the 1970s, because electronic equipment automatically selects stations, calculates waypoint-steering, and accommodates failures (Figure 18.6).

References

Books

Battin, R.H., *An Introduction to the Mathematics and Methods of Astrodynamics*, AIAA Press, Washington, D.C., 1987.

Bowditch, N., *The American Practical Navigator*, U.S. Government Printing Office, Washington, D.C., 1995. Re-issued approximately every five years.

Enge, P., Global Positioning System, *Scientific American*, May 2004, pp. 91–97.

Hein, G.W., et al., Galileo frequency and signal design, GPS World, Jan. 2003, p. 30–45.

Kayton, M., 1990. NAVIGATION: LAND, SEA, AIR, AND SPACE. IEEE Press, New York. 461 pgs.

Kayton, M. and Fried, W.R., *Avionics Navigation Systems*, 2nd ed., John Wiley, New York, 1997.

Minzner, R.A., The U.S. Standard Atmosphere 1976, NOAA report 76-1562, NASA SP-390, 1976 or latest edition.

Misra, P. and Enge P., *Global Positioning System*, Ganga-Jamuna Press, 2001.

Parkinson, B.W. and Spilker, J.J., Eds., *Global Positioning System, Theory and Applications*, American Institute of Aeronautics and Astronautics, Washington, D.C., 1996. 2 volumes.

Quinn, J., 1995 Revision of joint US/UK geomagnetic field models, *J. Geomagnetism Geo-electricity,* Fall 1996.

U.S. Air Force, *Navstar-GPS Space Segment/Navigation User Interfaces*, IRN-200c-004, ARINC Research, Annapolis, MD, 2000.

U.S. Government, Federal Radionavigation Plan, Departments of Defense and Transportation, Washington, D.C., issued biennially, latest edition at www.navcen.uscg.gov/frp/.

U.S. Government, Federal Radionavigation Systems, Departments of Defense and Transportation, Washington, D.C., issued biennially.

U.S. Government, WGS-84 World Geodetic System. U.S. Defense Mapping Agency, Washington, D.C., 1991.

Zhao, Y., *Vehicle Location and Navigation Systems*, Artech House, Boston, 1997.

Journals

AIAA J. Guidance, Control, Dynamics; bimonthly.

Commercial aeronautical standards produced by International Civil Aviation Organization (ICAO, Montreal), Aeronautical Radio Inc. (ARINC, Annapolis, MD), RTCA Inc., Washington, D.C., and European Commission for Aviation Electronics (EUROCAE, Paris).

IEEE Trans. Aerosp. Electron. Syst.; bimonthly through 1991, now quarterly.

J. Navigation (Royal Institute of Navigation, UK); quarterly.

Navigation (Journal of the U.S. Institute of Navigation); quarterly.

Proc. IEEE Position Location Navigation Symp. (PLANS); biennially.

Web Sites

Federal Aviation Administration www.faa.gov

U.S. Coast Guard Navigation Center www.navcen.uscg.gov (GPS, Maritime DGPS, Loran)

Navstar Global Positioning System Joint Program Office http://gps.losangeles.af.mil

Inmarsat www.inmarsat.org

NOAA Satellite and Information Service www.sarsat.noaa.gov

NASA www.nasa.gov

ARINC www.arinc.com

International Loran Association www.loran.org

Glonass www.glonass-ianc.RSA.RU (in Russian)

19

Navigation and Tracking

James L. Farrell
VIGIL, Inc.

19.1 Introduction

The task of navigation ("Nav") interacts with multiple avionics functions. To clarify the focus here, this chapter will not discuss tight formations, guidance, steering, minimization of fuel/noise/pollution, or managing time of arrival. The accent instead is on determining position and velocity (plus, where applicable, other variables such as acceleration, verticality, heading) with maximum accuracy reachable from whatever combination of sensor outputs are available at any time. Position can be expressed as a vector displacement from a designated point or in terms of latitude/longitude/altitude above mean sea level, above the geoid—or both. Velocity can be expressed in a locally level coordinate frame with various choices for an azimuth reference (e.g., geodetic North, Universal Transverse Mercator [UTM] grid North, runway centerline, wander azimuth with or without Earth sidereal rate torquing). In principle any set of axes could be used — such as an Earth Centered Earth Fixed (ECEF) frame for defining position by a Cartesian vector; or velocity in Cartesian coordinates or in terms of groundspeed, flight path angle, and ground track angle — in either case it is advisable to use accepted conventions.

Realization of near-optimal accuracy with any configuration under changing conditions is now routinely achievable. The method uses a means of dead-reckoning — preferably an Inertial Navigation System (INS) — which can provide essentially continuous position, velocity, and attitude in three dimensions by performing a running accumulation from derivative data. Whenever a full or partial fix is available from a nav sensor, a discrete update is performed on the entire set of variables representing the state of the nav system; the amount of reset for each state variable is determined by a weighting computation based on modern estimation. In this way, "initial" conditions applicable to the dead-reckoning device in effect are reinitialized as the "zero" time is advanced (and thus kept current) with each update. Computer-directed operations easily accommodate conditions that may arise in practice (incomplete fixes, inconsistent data rates, intermittent availability, changing measurement geometry, varying accuracies) while providing complete flexibility for backup with graceful degradation. The approach inherently combines short-term accuracy of the dead-reckoning data with the navaids' long-term accuracy. A commonly cited example of synergy offered by the scheme is a tightly coupled GPS/INS wherein the inertial information provides short-term aiding that vastly improves responsiveness of narrowband code and/or carrier tracking, while GPS information counteracts the long-term accumulation of INS error.

The goal of navigation has progressed far beyond mere determination of geographic location. Efforts to obtain double and triple "mileage" from inertial instruments, by integrating nav with sensor stabilization and flight control, are over a decade old. Older yet are additional tasks such as target designation, precision pointing, tracking, antenna stabilization, imaging sensor stabilization (and therefore transfer alignment). Digital beamforming (DBF) for array antennas (including graceful degradation to recover when some elements fail), needs repetitive data for instantaneous relative position of those elements; on deformable structures this can require multiple low-cost transfer-aligned Inertial Measuring Units (IMUs) and/or the fitting of spatial data to an aeroelastic model. The multiplicity of demands underlines the importance of integrating the operations; the rest of this chapter describes how integration should be done.

19.2   Fundamentals

To accomplish the goals just described, the best available balance is obtained between old and new information — avoiding both extremes of undue clinging to old data and jumping to conclusions at each latest input. What provides this balance is a modern estimation algorithm that accepts each data fragment as it appears from a nav sensor, immediately weighing it in accordance with its ability to shed light on every variable to be estimated. That ability is determined by accounting for all factors that influence how much or how little the data can reveal about each of those variables: Those factors include

- Instantaneous geometry (e.g., distance along a skewed line carries implications about more than one coordinate direction),
- Timing of each measurement (e.g., distance measurements separated by known time intervals carry implications about velocity as well as position), and
- Data accuracy, compared with the accuracy of estimates existing before measurement.

Only when all these factors are taken into account are accuracy and flexibility as well as versatility maximized. To approach the ramifications gradually, consider a helicopter hovering at constant altitude, which is to be determined on the basis of repeated altimeter observations. After setting the initial *a posteriori* estimate to the first measurement $\hat{Y}$, an *a priori* estimate $\hat{x}_2^{(-)}$ is predicted for the second measurement and that estimate is refined by a second observation,

$$\hat{x}_2^{(-)} = \hat{x}_1^{(+)}; \qquad \hat{x}_2^{(+)} = \hat{x}_2^{(-)} + \frac{1}{2}z_2, \quad z_2 \triangleq \hat{Y}_2 - \hat{x}_2^{(-)} \tag{19.1}$$

and a third observation,

$$\hat{x}_3^{(-)} = \hat{x}_2^{(+)}; \qquad \hat{x}_3^{(+)} = \hat{x}_3^{(-)} + \frac{1}{3}z_3, \quad z_3 \triangleq \hat{Y}_3 - \hat{x}_3^{(-)} \tag{19.2}$$

and then a fourth observation,

$$\hat{x}_4^{(-)} = \hat{x}_3^{(+)}; \qquad \hat{x}_4^{(+)} = \hat{x}_4^{(-)} + \frac{1}{4}z_4, \quad z_4 \triangleq \hat{Y}_4 - \hat{x}_4^{(-)} \tag{19.3}$$

which now clarifies the general expression for the mth observation,

$$\hat{x}_m^{(-)} = \hat{x}_{m-1}^{(+)}; \qquad \hat{x}_m^{(+)} = \hat{x}_m^{(-)} + \frac{1}{m}z_m, \qquad z_m \triangleq \hat{Y}_m - \hat{x}_m^{(-)} \tag{19.4}$$

which can be rewritten as

$$\hat{x}_m^{(+)} = \frac{m-1}{m}\hat{x}_m^{(-)} + \frac{1}{m}\hat{Y}_m, \qquad m > 0 \tag{19.5}$$

Substitution of $m = 1$ into this equation produces the previously mentioned condition that the first *a posteriori* estimate is equal to the first measurement; substitution of $m = 2$, combined with that condition, yields a second *a posteriori* estimate equal to the average of the first two measurements. Continuation with $m = 3, 4, \dots$ yields the general result that, after m measurements, estimated altitude is simply the average of all measurements.

This establishes an equivalence between the *recursive* estimation formulation expressed in (19.1)–(19.5) and the *block* estimate that would have resulted from averaging all data together in one step. Since that average is widely known to be optimum when all observations are statistically equally accurate, the recursion shown here must then be optimum under that condition. For measurement errors that are sequentially independent random samples with zero mean and variance R, it is well known that the mean squared estimation error $P_m^{(+)}$ after averaging m measurements is just R/m. That is the variance of the *a posteriori* estimate (just *after* inclusion of the last observation); for the *a priori* estimate the variance $P_m^{(-)}$ is $R/(m-1)$. It is instructive to express the last equation above as a blended sum of old and new data, weighted by factors

$$\frac{R}{P_m^{(-)} + R} \equiv \frac{R/P_m^{(-)}}{1 + R/P_m^{(-)}} = \frac{m-1}{m} \tag{19.6}$$

and

$$\frac{P_m^{(-)}}{P_{\dots}^{(-)} + R} = \frac{1}{m} \tag{19.7}$$

respectively; weights depend on variances, giving primary influence to information having lower mean squared error. This concept, signified by the left-hand sides of the last two equations, is extendable to more general conditions than the restrictive (uniform variance) case considered thus far. We are now prepared to address more challenging tasks.

As a first extension, let the sequence of altimeter measurements provide repetitive refinements of estimates for both altitude x_1 and vertical velocity x_2. The general expression for the mth observation now takes a more inclusive form

$$\hat{\mathbf{x}}_m^{(-)} = \Phi_m \hat{\mathbf{x}}_{m-1}^{(+)}; \qquad \hat{\mathbf{x}}_m^{(+)} = \hat{\mathbf{x}}_m^{(-)} + \mathbf{W}_m z_m, \qquad z_m \triangleq \hat{Y}_m - \hat{x}_{1,m}^{(-)}, \qquad \mathbf{x}_m \triangleq \begin{bmatrix} x_{m,1} \\ x_{m,2} \end{bmatrix} \tag{19.8}$$

The method accommodates estimation of multiple unknowns, wherein the status of a system is expressed in terms of a *state vector* ("*state*") $\mathbf{x}$, in this case a 2×1 vector containing two *state variables* ("*states*"); superscripts and subscripts continue to have the same meaning as in the introductory example, but for these states the conventions $m,1$ and $m,2$ are used for altitude and vertical velocity, respectively, at time t_m. For this dynamic case the *a priori* estimate at time t_m is not simply the previous *a posteriori* estimate; that previous state must be premultiplied by the *transition matrix*,

$$\Phi_m = \begin{bmatrix} 1 & t_m - t_{m-1} \\ 0 & 1 \end{bmatrix} \tag{19.9}$$

which performs a time extrapolation. Unlike the static situation, elapsed time now matters since imperfectly perceived velocity enlarges altitude uncertainty between observations — and position measurements separated by known time intervals carry implicit velocity information (thus enabling vector estimates to be obtained from scalar data in this case). Weighting applied to each measurement is influenced by three factors:

- A sensitivity matrix $\mathbf{H}_m$ whose (i, j) element is the partial derivative of the ith component of the mth measured data vector to the jth state variable. In this scalar measurement case $\mathbf{H}_m$ is a 1×2 matrix [1 0] for all values of m.
- A covariance matrix $\mathbf{P}_m$ of error in state estimate at time t_m [the ith diagonal element = mean squared error in estimating the ith state variable and, off the diagonal, $P_{ij} = P_{ji} = \sqrt{P_{ii}P_{jj}} \times$ (correlation coefficient between i^{th} and j^{th} state variable uncertainty)].
- A covariance matrix $\mathbf{R}_m$ of measurement errors at time t_m (in this scalar measurement case $\mathbf{R}_m$ is a 1×1 "matrix," i.e., a scalar variance R_m).

Although formation of $\mathbf{H}_m$ and $\mathbf{R}_m$ follows directly from their definitions, $\mathbf{P}_m$ changes with time (e.g., recall the effect of velocity error on position error) and with measurement events (because estimation errors fall when information is added). In this "continuous-discrete" approach, uncertainty is decremented at the discrete measurement events

$$\mathbf{P}_m^{(+)} = \mathbf{P}_m^{(-)} - \mathbf{W}_m \mathbf{H}_m \mathbf{P}_m^{(-)} \tag{19.10}$$

and, between events, dynamic behavior follows a continuous model of the form

$$\dot{\mathbf{P}} = \mathbf{AP} + \mathbf{PA}^T + \mathbf{E} \tag{19.11}$$

where $\mathbf{E}$ acts as a forcing function to maintain positive definiteness of $\mathbf{P}$ (thereby providing stability and effectively controlling the remembrance duration—the "data window" denoted herein by T— for the estimator) while $\mathbf{A}$ defines dynamic behavior of the state to be estimated ($\dot{\mathbf{x}} = \mathbf{Ax}$ and $\dot{\Phi} = \mathbf{A}\Phi$). In the example at hand,

$$\mathbf{A} = \begin{bmatrix} 0 & 1 \\ 0 & 0 \end{bmatrix}; \qquad \begin{bmatrix} \dot{x}_1 \\ \dot{x}_2 \end{bmatrix} = \mathbf{A} \begin{bmatrix} x_1 \\ x_2 \end{bmatrix} \tag{19.12}$$

Given $\mathbf{H}_m$, $\mathbf{R}_m$, and $\mathbf{P}_m^{(-)}$ the optimal (Kalman) weighting matrix is

$$\mathbf{W}_m = \mathbf{P}_m^{(-)} \mathbf{H}_m^T (\mathbf{H}_m \mathbf{P}_m^{(-)} \mathbf{H}_m^T + \mathbf{R}_m)^{-1} \tag{19.13}$$

which for a scalar measurement produces a vector $\mathbf{W}_m$ as the above inversion simplifies to division by a scalar (which becomes the variance R_m added to P_{11} in this example):

$$\mathbf{W}_m = \mathbf{P}_m^{(-)} \mathbf{H}_m^T / (\mathbf{H}_m \mathbf{P}_m^{(-)} \mathbf{H}_m^T + R_m) \tag{19.14}$$

The preceding (hovering helicopter) example is now recognized as a special case of this vertical nav formulation. To progress further, horizontal navigation addresses another matter, i.e., location

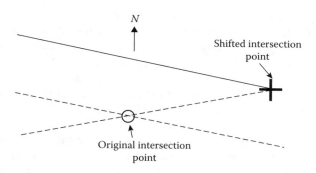

FIGURE 19.1 Nonorthogonal LOPs.

uncertainty in more than one direction—with measurements affected by more than one of the unknowns (e.g., lines of position [LOPs] skewed off a cardinal direction such as North or East; Figure 19.1). In the classic "compass-and-dividers" approach, dead reckoning would be used to plot a running accumulation of position increments until the advent of a fix from two intersecting straight or curved LOPs. The position would then be reinitialized at that fixed position, from whence dead reckoning would continue until the next fix. For integrated nav we fundamentally alter that procedure as follows:

- In the reintialization, data imperfections are taken into account. As already discussed, Kalman weighting (Equations 19.13 and 19.14) is based on accuracy of the dead-reckoning extrapolation as well as the variance of each measurement *and* its sensitivity to each state variable. An optimal balance is provided between old and new information, and the optimality inherently applies to updating of every state variable (e.g., to velocity estimates as well as position, even when only position is observed directly).
- Fixes can be incomplete. In this example, one of the intersecting LOPs may be lost. An optimal update is still provided by the partial fix data, weighted by $\mathbf{W}_m$ of (19.14).

Implications of these two alterations can be exemplified by Figure 19.1, depicting a pair of LOPs representing partial fixes, not necessarily synchronous. Each scalar measurement allows the entire state vector to be optimally updated with weighting from (19.14) in the relation,

$$\hat{\mathbf{x}}_{\mathbf{m}}^{(+)} = \hat{\mathbf{x}}_{\mathbf{m}}^{(-)} + \mathbf{W}_m z_m \tag{19.15}$$

where z_m is the *predicted residual* formed by subtracting the predicted measurement from the value observed at time t_m and acceptance-tested to edit out wild data points;

$$z_m = y_m + \varepsilon = Y_m - \hat{Y}_m^{(-)} + \varepsilon = \hat{Y}_m - \hat{Y}_m^{(-)}; \quad \hat{Y}_m^{(-)} = Y(\hat{\mathbf{x}}_m^{(-)}) \tag{19.16}$$

The measurement function $Y(\mathbf{x})$ is typically a simple analytical expression (such as that for distance from a designated point, the difference between distances from two specified station locations, GPS pseudorange or carrier phase difference, etc.). Its partial derivative with respect to each position state is obtained by simple calculus; other components of $\mathbf{H}_m$ (e.g., sensitivity to velocity states) are zero, in which case updating of those states occurs due to dynamics from off-diagonal elements of $\mathbf{P}$ in the product $\mathbf{P}_m^{(-)}\mathbf{H}_m^T$. R_m — whether constant or varying (e.g., with signal strength) — is treated as a known quantity; if not accurately known, a conservative upper bound can be used. The same is true for the covariance matrix $\mathbf{P}_0$ of error in state estimate at the time of initiating the estimation process — after which the changes are tracked by (19.10) at each measurement event, and by (19.11) between measurements — thus $\mathbf{P}$ is always available for Equations 19.13 and 19.14.

It is crucial to note that the updates are *not* obtained in the form of newly measured coordinates, as they would have been for the classical "compass-and-dividers" approach. Just as old navigators knew how to use partial information, a properly implemented modern estimator would not forfeit that capability. The example just shown provides the best updates, even with no dependably precise way of obtaining a point of intersection when motion occurs between measurements. Furthermore, even with a valid intersection from synchronized observations, the North coordinate of the intersection in Figure 19.1 would be more credible than the East. To show this, consider the consequence of a measurement error effectively raising the dashed LOP to the solid curve as shown; the North coordinate of the new intersection point "+" exceeds that of point "**O**" — but by less than the East-West coordinate shift.

Unequal sensitivity to different directions is automatically taken into account via $\mathbf{H}_m$ —just as the dynamics of $\mathbf{P}$ will automatically provide velocity updating without explicitly forming velocity in terms of sequential changes in measurements — and just as individual values of R_m inherently account for measurement accuracy variations.

Theoretically then, usage of Kalman weighting unburdens the designer while ensuring optimum performance; no other weighing could provide lower mean squared error in the estimated value of any state. Practically, the fulfillment of this promise is realized by observing additional guidelines, some of which apply "across the board" (e.g., usage of algorithms that preserve numerical stability) while others are application dependent.

Now that a highly versatile foundation has been defined for general usage, the way is prepared for describing some specific applications. The versatility just mentioned is exhibited in the examples that follow. Attention is purposely drawn to the standard process cycle; models of dynamics and measurements are sufficient to define the operation.

19.3   Applications

Various operations will now be described, using the unified form to represent the state dynamics* with repetitive instantaneous refresh via discrete or discretized observations (fixes, whether full or partial). Finite space necessitates some limitations in scope here. First, all updates will be from position-dependent measurements (e.g., Doppler can be used as a source of continuous dead-reckoning data but is not considered herein for the discrete fixes). In addition, all nav reference coordinate frames under consideration will be locally level. In addition to the familiar North-East-Down (NED) and East-North-Up (ENU) frames, this includes any wander azimuth frame (which deviates from the geographic by only an azimuth rotation about the local vertical). Although these reference frames are not inertial (thus the velocity vector is not exactly the time integral of total acceleration as expressed in a nav frame), known kinematical adjustments will not be described in any depth here. This necessitates restricting the aforementioned data window T to intervals no greater than a tenth of the 84-minute Schuler period. The limitation is not very severe when considering the amount of measured data used by most modern avionics applications within a few minutes duration.

Farrell[1] is cited here for expansion of conditions addressed, INS characterization, broader error modeling, increased analytical development, the physical basis for that analysis, and myriad practical "*do* s and *don't* s" for applying estimation in each individual operation.

19.3.1   Position and Velocity along a Line

The vertical nav case shown earlier can be extended to the case of time-varying velocity; with accurately (not necessarily exactly) known vertical acceleration Z_V,

$$\begin{bmatrix} \dot{x}_1 \\ \dot{x}_2 \end{bmatrix} = \begin{bmatrix} 0 & 1 \\ 0 & 0 \end{bmatrix} \begin{bmatrix} x_1 \\ x_2 \end{bmatrix} + \begin{bmatrix} 0 \\ Z_V \end{bmatrix} \tag{19.17}$$

---

*A word of explanation is in order: For classical physics the term *dynamics* is reserved for the relation between forces and translational acceleration, or torques and rotational acceleration — while *kinematics* describes the relation between acceleration, velocity, and position. In the estimation field, all continuous time-variation of the state is lumped together in the term *dynamics*.

which allows interpretation in various ways. With a positive upward convention (as in the ENU reference, for example), x_1 can represent altitude above any datum while x_2 is upward velocity; a positive downward convention (NED reference) is also accommodated by simple reinterpretation. In any case, the above equation correctly characterizes actual vertical position and velocity (with true values for Z_V and all xs), and likewise characterizes *estimated* vertical position and velocity (denoted by circumflexes over Z_V and all xs). Therefore, by subtraction, it also characterizes *uncertainty in* vertical position and velocity (i.e., error in the estimate, with each circumflex replaced by a tilde ~). That explains the role of this expression in two separate operations:

- Extrapolation of the *a posteriori* estimate (just after inclusion of the last observation) to the time of the next measurement, to obtain an *a priori* estimate of the state vector — which is used to predict the measurement's value. If a transition matrix can readily be formed (e.g., Equation 19.9 in the example at hand), it is sometimes, but not always, used for that extrapolation.

- Propagation of the covariance matrix from time t_{m-1} to t_m via (19.11) initialized at the *a posterori* value $\mathbf{P}_{m-1}^{(+)}$ and ending with the *a priori* value $\mathbf{P}_m^{(-)}$. Again, an alternate form using (19.9) is an option.

After these two steps, the cycle at time t_m is completed by forming gain from (19.14), predicted residual from (19.16), update via (19.15), and decrement by (19.10).

The operation just described can be visualized in a generic pictorial representation. Velocity data in a dead-reckoning (DR) accumulation of position increments predicts the value of each measurement. The difference z between the prediction and the observed fix (symbolically shown as a discrete event depicted by the momentary closing of a switch) is weighted by position gain W_{pos} and velocity gain W_{vel} for the update. Corrected values, used for operation thereafter, constitute the basis for further subsequent corrections.

For determination of altitude and vertical velocity, the measurement prediction block in Figure 19.2 is replaced by a direct connection; altimeter fixes are compared vs. the repeatedly reinitialized accumulation of products (time increment) $\times$ (vertical velocity). In a proper implementation of Figure 19.2 time history of *a posteriori* position tracks the truth; RMS position error remains near $\sqrt{P_{11}}$. At the first measurement, arbitrarily large initial uncertainty falls toward sensor tolerance — and promptly begins rising at a rate dictated by $\sqrt{P_{22}}$. A second measurement produces another descent followed by another climb, but now at gentler slope, due to implicit velocity information gained from repeated position observations within a known time interval. With enough fix data the process approaches a quasi-static condition with $\sqrt{P_{11}}$ maintained at levels near RMS sensor error.

Extensive caveats, ramifications, etc. could be raised at this point; some of the more obvious ones will be mentioned here.

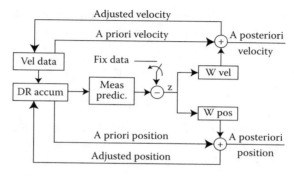

FIGURE 19.2 Position and velocity estimation.

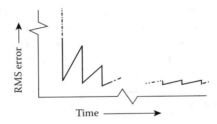

FIGURE 19.3　Time history of accuracy.

- In analogy with the static example, the *left* side of (19.7), with $P_{m11}^{(-)}$ substituted for $P_m^{(-)}$, implies high initial weighting followed by lighter weights as measurements accumulate. If fixes are from sensors with varying tolerance, the entire approach remains applicable; only parameter values change. The effect in Figure 19.3 would be a smaller step decrement, and less reduction in slope, when RMS fix error is larger.

- Vertical velocity can be an accumulation of products, involving instantaneous vertical acceleration which comes from data containing an accelerometer offset driven by a randomly varying error, e.g., having spectral density in conformance to **E** of (19.11). With this offset represented as a third state, another branch would be added to Figure 19.2 and an augmented form of (19.17) could define dynamics in instantaneous altitude, vertical velocity, and vertical acceleration (instead of a constant bias component, extension to exponential correlation is another common alternative);

$$
\begin{bmatrix} \dot{x}_1 \\ \dot{x}_2 \\ \dot{x}_3 \end{bmatrix} = \begin{bmatrix} 0 & 1 & 0 \\ 0 & 0 & 1 \\ 0 & 0 & 0 \end{bmatrix} \begin{bmatrix} x_1 \\ x_2 \\ x_3 \end{bmatrix} + \begin{bmatrix} 0 \\ 0 \\ e \end{bmatrix}
\tag{19.18}
$$

Rather than ramping between fixes, position uncertainty then curves upward faster than the linear rate in Figure 19.3; curvature starts to decrease after the *third* fix. It takes longer to reach quasi-static condition, and closeness of "steady-state" $\sqrt{P_{11}}$ to RMS sensor error depends on measurement scheduling density within a data window.

- (19.12) and Figure 19.2 can also represent position and velocity estimation along another direction, e.g., North or East — or both, as developed in the next section.

19.3.2　Position and Velocity in Three-Dimensional Space

For brevity, only a succinct description is given here. First consider excursion over a meridian with position x_1 expressed as a product [latitude (*Lat*) increment] × [total radius of curvature (R_M + altitude)],

$$
R_M = \frac{a_E(1 - e_E^2)}{[1 - e_E^2 \sin^2(Lat)]^{3/2}}; \quad a_E = 6378137 \,\text{m.}; \quad e_E^2 = (2 - f)f, f = \frac{1}{298.25722}
\tag{19.19}
$$

so that, for usage of **A** in (19.12), x_2 is associated with North component V_N of velocity. North position fixes could be obtained by observing the altitude angle of Polaris (appropriately corrected for slight deviation off the North Pole). To use the formulation for travel in the East direction, the curvature radius is (R_P + h),

$$R_p = a_E / \sqrt{1 - e_E^2 \sin^2(\textbf{\textit{Lat}})}; \quad h = \textbf{altitude} \tag{19.20}$$

and, while the latitude rate is $V_N / (R_M + h)$, the longitude rate is $V_E \sec(\textit{Lat}) / (R_p + h)$.

Even for limited distance excursions within a data window, these spheroidal expressions would be used in kinematic state extrapolation, while our short-term ground rule allows a simplified ("flat-Earth" Cartesian) model to be used as the basis for matrix extrapolation in (19.11). The reason lies with *very* different sensitivities in Equations 19.15 and 19.16. The former is significantly less critical; a change $\delta \textbf{W}$ would modify the *a posteriori* estimate by only the second-order product $z_m \delta \textbf{W}_m$. By way of contrast, small variations in an anticipated measurement (from seemingly minor model approximations) can produce an unduly large deviation in the residual — a small difference of large quantities.

Thus, for accuracy of *additive* state *vector* adjustments (such as *velocity* $\times \Delta time$ products in dynamic propagation), Equations 19.19 and 19.20 properly account for path curvature and for changes in direction of the nav axes as the path progresses. At the poles, the well-known singularity in {sec(*Lat*)} of course necessitates a modified expression (e.g., Earth-centered vector).

In applying (19.12) to all three directions, a basic decision must be made at the outset. Where practical, it is desirable for axes to remain separated, which produces three uncoupled two-state estimators. An example of this form is radar tracking at long range — long enough so that, within a data window duration, the line-of-sight (LOS) direction remains substantially fixed (i.e., nonrotating). If all three axes are monitored at similar data rates and accuracies, experience has shown that even a fully coupled six-state estimator has position error ellipsoid axes aligned near the sensor's range/azimuth/elevation directions. In that case, little is lost by ignoring coupling across sensor reference axes — hence the triad of uncoupled two-state estimators, all in conformance to (19.12). To resolve vectors along cardinal directions at any time, all that is needed is the direction cosine matrix transformation between nav and sensor axes, which is always available.

When the conditions mentioned above do not hold, the reasoning needs to be revisited. If LOS direction rotates (which happens at short range), or if all three axes are *not* monitored at similar data rates, decoupling may or may not be acceptable; in any case it is suboptimal. If one axis (or a pair of axes) is *un*monitored a fully coupled six-state estimator can dramatically outperform the uncoupled triad. In that case, although (19.12) represents uncoupled dynamics for each axis, coupling comes from multiple changing projections in measurement sensitivity $\textbf{H}$ as the sensor sight-line direction rotates.

Even the coupled formulation has a simple dynamic model in partitioned form; for a relative position vector $\textbf{R}$ and velocity $\textbf{V}$ driven by perturbing acceleration $\textbf{e}$,

$$\begin{bmatrix} \dot{\textbf{R}} \\ \dot{\textbf{V}} \end{bmatrix} = \begin{bmatrix} \textbf{0} & \textbf{I} \\ \textbf{0} & \textbf{0} \end{bmatrix} \begin{bmatrix} \textbf{R} \\ \textbf{V} \end{bmatrix} + \begin{bmatrix} \textbf{0} \\ \textbf{e} \end{bmatrix} \tag{19.21}$$

where $\textbf{I}$ and $\textbf{0}$ are null and identity partitions. The next section extends these concepts.

19.3.3 Position, Velocity, and Acceleration of a Tracked Object

In this chapter it has been repeatedly observed that velocity can be inferred from position-dependent measurements separated by known time intervals. In fact, a velocity *history* can be inferred. As a further generalization of methods just shown, the position reference need not be stationary. In the example now to be described, the origin will move with a supersonic jet carrying a radar and INS. Furthermore, the object whose state is being estimated can be external, with motions that are independent of the platform carrying the sensors that provide all the measurements.

For tracking, first consider the uncoupled case already described, wherein each of three separate estimator channels corresponds to a sensor reference axis direction and each channel has three kinematically related states, representing that directional component of relative (sensor-to-tracked-object) position, relative velocity, and total (not relative) acceleration of the tracked object.* The expression used to propagate state estimates between measurements in a channel conforms to standard kinematics, i.e.,

$$
\begin{bmatrix} \hat{x}_{m1}^{(-)} \\ \hat{x}_{m2}^{(-)} \\ \hat{x}_{m3}^{(-)} \end{bmatrix} = \begin{bmatrix} 1 & t_m - t_{m-1} & \frac{1}{2}(t_m - t_{m-1})^2 \\ 0 & 1 & t_m - t_{m-1} \\ 0 & 0 & 1 \end{bmatrix} \begin{bmatrix} \hat{x}_{m-1,1}^{(+)} \\ \hat{x}_{m-1,2}^{(+)} \\ \hat{x}_{m-1,3}^{(+)} \end{bmatrix} - \begin{bmatrix} \frac{1}{2}(t_m - t_{m-1})q_m \\ q_m \\ 0 \end{bmatrix} \tag{19.22}
$$

where q_m denotes the component, along the sensor channel direction, of the change in INS velocity during $(t_m - t_{m-1})$. In each channel, $\mathbf{E}$ of (19.11) has only one nonzero value, a spectral density related to data window and measurement error variance σ^2 by

$$
E_{33} = (20\sigma^2/T^5)/g^2 \qquad (g/\mathbf{sec})^2/\mathbf{Hz} \tag{19.23}
$$

To change this to a fully coupled 9-state formulation, partition the 9×1 state vector into three 3×1 vectors $\mathbf{R}$ for relative position, $\mathbf{V}_r$ for relative velocity, and $\mathbf{Z}_T$ for the tracked object's total acceleration — all expressed in the INS reference coordinate frame. The partitioned state transition matrix is then constructed by replacing each diagonal element in (19.22) by a 3×3 identity matrix $\mathbf{I}_{33}$, each zero by a 3×3 null matrix, and multiplying each above-diagonal element by $\mathbf{I}_{33}$. Consider this transition matrix to propagate covariances as expressed in *sensor reference axes*, so that parameters applicable to a sensing channel are used in (19.23) for each measurement.

Usage of different coordinate frames for states (e.g., geographic in the example used here) and $\mathbf{P}$ (sensor axes) must of course be taken into account in characterizing the estimation process. An orthogonal triad $\mathbf{I}_b\mathbf{J}_b\mathbf{K}_b$ conforms to directions of sensor sight-line $\mathbf{I}_b$, its elevation axis $\mathbf{J}_b$ in the normal plane, and the aximuth axis $\mathbf{I}_b \times \mathbf{J}_b$ normal to both. The instantaneous direction cosine matrix $\mathbf{T}_{b/A}$ will be known (from the sensor pointing control subsystem) at each measurement time. By combination with the transformation $\mathbf{T}_{A/G}$ from geographic to airframe coordinates (obtained from INS data), the transformation from geographic to sensor coordinates is

$$
\mathbf{T}_{b/G} = \mathbf{T}_{b/A}\mathbf{T}_{A/G} \tag{19.24}
$$

which is used to resolve position states along $\mathbf{I}_b\mathbf{J}_b\mathbf{K}_b$:

$$
\frac{1}{|\mathbf{R}|}\mathbf{T}_{b/G}\mathbf{R} = \begin{bmatrix} 1 \\ p_A \\ -p_E \end{bmatrix} \tag{19.25}
$$

where p_A and p_E — small fractions of a radian — are departures above and to the right, respectively, of the *a priori* estimated position from the sensor sight-line (which due to imperfect control does not look exactly where the tracked object is anticipated at t_m).

For application of (19.16), p_A and p_E are recognized in the role of *a priori* estimated measurements

---

*Usage of relative acceleration states would have sacrificed detailed knowledge of INS velocity history, characterizing ownship acceleration instead with the random model used for the tracked object. To avoid that unnecessary performance degradation the dynamic model used here, in contrast to (19.18), has a forcing function with nonzero mean.

— adjusting the "dot-off-the-crosshairs" azimuth ("AZ") and elevation ("EL") observations so that a full three-dimensional fix (range, AZ, EL) in this operation would be

$$
\begin{bmatrix} y_R \\ y_{AZ} \\ y_{EL} \end{bmatrix} = \begin{bmatrix} 1 & 0 & 0 \\ 0 & \dfrac{1}{|\mathbf{R}|} & 0 \\ 0 & 0 & -\dfrac{1}{|\mathbf{R}|} \end{bmatrix} \mathbf{T}_{b/G} \mathbf{R} - \begin{bmatrix} 0 \\ p_A \\ -p_E \end{bmatrix}
\tag{19.26}
$$

Since $\mathbf{R}$ contains the first three states, its matrix coefficient in (19.26) provides the three nonzero elements of $\mathbf{H}$; e.g., for scalar position observables, these are comprised of:

- The top row of $\mathbf{T}_{b/G}$ for range measurements,
- The middle row of $\mathbf{T}_{b/G}$ divided by scalar range for azimuth measurements,
- The bottom row of $\mathbf{T}_{b/G}$ divided by scalar range $\times$ (-1), for elevation measurements.

By treating scalar range coefficients as well as the direction cosines as known quantities in this approach, *both* the dynamics *and* the observables are essentially linear in the state. This has produced success in nearly all applications within the experience of this writer. The sole need for extension arose when distances and accuracies of range data were extreme (the cosine of the angle between the sensor sight-line and range vector could not be set at unity). Other than that case, the top row of $\mathbf{T}_{b/G}$ suffices for relative position states, and also for relative velocity states when credible Doppler measurements are available.

A more thorough discourse would include a host of additional material, including radar and optical sensing considerations, sensor stabilization — with its imperfections isolated from tracking, error budgets, kinematical correction for gradual rotation of the acceleration vector, extension to multiple track files, sensor fusion, myriad disadvantages of alternative tracking estimator formulations, etc. The ramifications are too vast for inclusion here.

19.3.4 Position, Velocity, and Attitude in Three-Dimensional Space (INS Aiding)

In the preceding section, involving determination of velocity history from position measurement sequences, dynamic velocity variations were expressed in terms of an acceleration vector. For nav (as opposed to tracking of an external object) with high dynamics, the history of velocity is often tied to the angular orientation of an INS. In straight-and-level Northbound flight, for example, an unknown tilt ψ_N about the North axis would produce a fictitious ramping in the indicated East velocity V_E; in the short-term this effect will be indistinguishable from a bias n_{aE} in the indicated lateral component (here, East) of accelerometer output. More generally, velocity *vector* error will have a rate

$$
\dot{\mathbf{v}} = \boldsymbol{\psi} \times \mathbf{A} + \mathbf{n}_a = -\mathbf{A} \times \boldsymbol{\psi} + \mathbf{n}_a
\tag{19.27}
$$

where bold symbols ($\mathbf{v}$, $\mathbf{n}$) contain the geographic components equal to corresponding scalars denoted by italicized quantities (v,n) and $\mathbf{A}$ represents the vector, also expressed in geographic coordinates, of the total nongravitational acceleration experienced by the IMU. Combined with the intrinsic kinematical relation between $\mathbf{v}$ and a position vector error $\mathbf{r}$, in a nav frame rotating at $\tilde{\boldsymbol{\omega}}$ rad/sec, the 9-state dynamics with a time-invariant misorientation $\boldsymbol{\psi}$ can be expressed via 3×3 matrix partitions,

$$
\begin{bmatrix} \dot{\mathbf{r}} \\ \dot{\mathbf{v}} \\ \dot{\boldsymbol{\psi}} \end{bmatrix} = \begin{bmatrix} -\tilde{\boldsymbol{\omega}}\times & \mathbf{I} & 0 \\ 0 & 0 & (-\mathbf{A}\times) \\ 0 & 0 & 0 \end{bmatrix} \begin{bmatrix} \mathbf{r} \\ \mathbf{v} \\ \boldsymbol{\psi} \end{bmatrix} + \begin{bmatrix} 0 \\ \mathbf{n}_a \\ \mathbf{e} \end{bmatrix} \tag{19.28}
$$

which lends itself to numerous straighforward interpretations. For brevity, these will simply be listed here:

- For strapdown systems, it is appropriate to replace vectors such as $\mathbf{A}$ and $\mathbf{n}_a$ by vectors initially expressed in vehicle coordinates and transformed into geographic coordinates, so that parameters and coefficients will appear in the form received.
- Although both $\mathbf{n}_a$ and $\mathbf{e}$ appear as forcing functions, the latter drives the highest-order state and thus exercises dominant control over the data window.
- If $\mathbf{n}_a$ and $\mathbf{e}$ contain both bias and time-varying random (noisy) components, (19.28) is easily reexpressible in augmented form, wherein the biases can be estimated along with the corrections for estimated position, velocity, and orientation. Especially for accelerometer bias elements, however, observability is often limited; therefore the usage of augmented formulations should be adopted judiciously. In fact, the number of states should in many cases be *reduced*, as in the next two examples:
  - In the absence of appreciable sustained horizontal acceleration, the azimuth element of misorientation is significantly less observable than the tilt components. In some operations this suggests replacing (19.28) with an eight-state version obtained by omitting the ninth state and deleting the last row and column of the matrix.
  - When the last *three* states are omitted — while the last three rows and columns of the matrix are deleted — the result is the fully coupled three-dimensional position and velocity estimator (19.21).

The options just described can be regarded as different modes of the standard cyclic process already described, with operations defined by dynamics and measurement models. Any discrete observation could be used with (19.28) or an alternate form just named, constituting a mode subject to restrictions that were adopted here for brevity (position-dependent observables only, with distances much smaller than Earth radius).

At this point, expressions could be given for measurements as functions of the states and their sensitivities to those state variables: (19.26) provides this for range and angle data; it is now appropriate to discuss GPS information in an integrated nav context.

19.3.5 Individual GPS Measurements as Observables

The explosive growth of navigation applications within the past two decades has been largely attributed to GPS. Never before has there been a nav data source of such high accuracy, reachable from any location on the Earth's surface at any time. Elsewhere in this book the reader has been shown how GPS data can be used to:

- Solve for 3D position and user clock offset with pseudorange observations received simultaneously from each of four space vehicles (SVs),
- Use local differential GPS corrections that combine, for each individual SV, compensation for propagation delays plus SV clock and ephemeris error,
- Compensate via wide-area augmentation which, though not as accurate as local, is valid for much greater separation distances between the user and reference station,
- Use differencing techniques with multiple SVs as well as multiple receivers to counteract the effects of the errors mentioned and of user clock offsets,

- Apply these methods to carrier phase as well as to pseudorange so that, once the cycle count ambiguities are resolved, results can be accurate to within a fraction of the L-band wavelength.

Immediately we make a definite departure from custom here; each scalar GPS observable will call for direct application of (19.15). To emphasize this, results will first be described for instances of sparse measurement scheduling. Initial runs were made with real SV data, taken before the first activation of selective availability (SA) degradations, collected from a receiver at a known stationary location but spanning intervals of several hours. Even with that duration consumed for the minimum required measurements, accuracies of 1 or 2 m were obtained: not surprising for GPS with good geometry and no SA.

The results just mentioned, while not considered remarkable, affirm the point that full fixes are not at all necessary with GPS. They also open the door for drawing dependable conclusions when the same algorithms are driven by simulated data containing errors from random number generators. A high-speed aircraft simulation was run under various conditions, always with no more than one pseudo-range observation every 6 sec. (and furthermore with some gaps even in that slow data rate). Since the results are again unremarkable, only a brief synopsis suffices here:

- Estimates converged as soon as the measurements accumulated were sufficient to produce a nav solution (e.g., two asynchronous measurements from each of three noncoplanar SVs for a vehicle moving in three dimensions, with known clock state, or four SVs with all states initially unknown).
- Initial errors tended to wash out; accuracies of the estimates just mentioned were determined by measurement error levels amplified by geometry.
- Velocity errors tended toward levels proportional to a ratio (RMS measurement error)/(T), where T here represents time elapsed since the first measurement on a course leg, or the data window — whichever is smaller. The former definition of the denominator produced a transient at the onset and when speed or direction changed.
- Doppler data reduced the transient, and INS velocity aiding minimized or removed it.
- Extreme initial errors interfered with these behavioral patterns somewhat — readily traceable to usage of imprecise direction cosines — but the effects could be countered by reinitialization of estimates with *a posteriori* values and recycling the measurements.

These results mirror familiar real-world experience (including actual measurement processing by this author); they are used here to emphasize adequacy of partial fixes at low rates, which many operational systems to this day still fail to exploit.[6]

Although the approach just described is well known (i.e., in complete conformance to the usual Kalman filter updating cycle) and the performance unsurprising, the last comment is significant. There are numerous applications wherein SV sight-lines are often obscured by terrain, foliage, buildings, or structure of the vehicle carrying the GPS receiver. In addition, there can be SV outages (whether from planned maintenance or unexpected failures), intermittently strong interference or weak signals, unfavorable multipath geometry in certain SV sight-line directions, etc., and these problems can arise in critical situations.

At the time of this writing there are widespread opportunities, prompted by genuine need, to replace loose (cascaded) configurations by tightly coupled (integrated) configurations. Accentuating the benefit is the bilateral nature of tight integration. As the tracking loops (code loop and, where activated, carrier phase track) contribute to the estimator, the updated state enhances ability to maintain stable loop operation. For a properly integrated GPS/INS, this enhancement occurs even with narrow bandwidth in the presence of rapid change. Loop response need not follow the dynamics, only the error in perceived dynamics.

It is also noted that the results just described are achievable under various conditions and can be scaled over a wide accuracy range. Sensitivity **H** of an individual SV observation contains the SV-to-receiver unit vector; when satellite observations are differenced, the sensitivity **H** contains the difference of two SV-to-receiver unit vectors. Measurements may be pseudoranges ($\sigma < 10$ m, typically), differentially corrected pseudoranges ($\sigma = 1$ or 2 m), or carrier phase (with ambiguities resolved, σ at 1 cm or less).

In all cases, attainable performance is determined by σ and the span of **H** for each course leg. An analogous situation holds for other navaids when used with the standard updating procedure presented herein.

19.4   Conclusion

Principles of nav system integration have been described, within the limits of space. Inevitably, some restrictions in scope were adopted; those wishing to pursue the topic in greater depth may consult the sources which follow.

References

1. Farrell, J. L., *Integrated Aircraft Navigation,* Academic Press, New York, 1976. (Now available in paperback only; 800/628-0885 or 410/647-6165.)
2. Bierman, *Factorized Methods for Discrete Sequential Estimation,* Academic Press, New York, 1977.
3. Institute of Navigation Redbooks (reprints of selected GPS papers); Alexandria, VA, 703/683-7101.
4. Brown and Hwang, *Introduction to Random Signals and Applied Kalman Filtering,* John Wiley & Sons, New York, 1996.
5. Kayton and Fried (Eds.), *Avionics Navigation Systems,* John Wiley & Sons, New York, 1997.
6. Farrell, Stephens, and McConkey, "Send Measurements, Not Coordinates," *Navigation* (Journal of the Institute of Navigation), v. 46, n. 3, Fall 1999, pp. 203–15.

Further Information

1. Journal and Conference Proceedings from the Institute of Navigation, Alexandria, VA.
2. Tutorials from Conferences sponsored by the Institute of Navigation (Alexandria, VA) and the Position Location And Navigation Symposium (PLANS) of the Institute of Electrical and Elecronic Engineers (IEEE).
3. Transactions of the Institute of Electrical and Electronic Engineers (IEEE) Aerospace and Electronic Systems Society (AES).

20

Flight Management Systems

Randy Walter

Smiths Industries

20.1 Introduction

The flight management system typically consists of two units, a computer unit and a control display unit. The computer unit can be a standalone unit providing both the computing platform and various interfaces to other avionics or it can be integrated as a function on a hardware platform such as an Integrated Modular Avionics (IMA) cabinet. The Control Display Unit (CDU or MCDU) provides the primary human/machine interface for data entry and information display. Since hardware and interface implementations of flight management systems can vary substantially, this discussion will focus on the functional aspects of the flight management system.

The flight management system provides the primary navigation, flight planning, and optimized route determination and en route guidance for the aircraft and is typically comprised of the following interrelated functions: navigation, flight planning, trajectory prediction, performance computations, and guidance.

To accomplish these functions the flight management system must interface with several other avionics systems. As mentioned above, the implementations of these interfaces can vary widely depending upon the vintage of equipment on the aircraft but generally will fall into the following generic categories.

- Navigation sensors and radios
  - Inertial/attitude reference systems
  - Navigation radios
  - Air data systems
- Displays
  - Primary flight and navigation
  - Multifunction
  - Engine
- Flight control system
- Engine and fuel system
- Data link system
- Surveillance systems

Figure 20.1 depicts a typical interface block diagram.

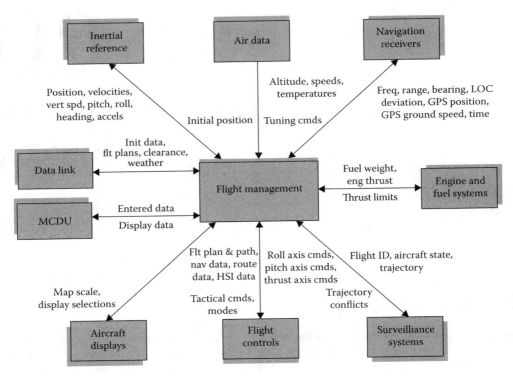

FIGURE 20.1 Typical interface block diagram.

Today, flight management systems can vary significantly in levels of capability because of the various aviation markets they are intended to serve. These range from simple point to point lateral navigators to the more sophisticated multisensor navigation, optimized four-dimensional flight planning/guidance systems. The flight management system in its simplest form will slowly diminish as reduced separation airspace standards place more demands on the aircraft's ability to manage its trajectory more accurately, even though lateral-only navigators will continue to have a place in recreational general aviation.

With its current role in the aircraft, the flight management system becomes a primary player in the current and future communications navigation surveillance for air traffic management (CNS/ATM) environment. Navigation within required navigation performance (RNP) airspace, data-linked clearances and weather, aircraft trajectory-based traffic management, time navigation for aircraft flow control, and seamless low-visibility approach guidance all are enabled through advanced flight management functionality.

20.2 Fundamentals

At the center of the FMS functionality is the flight plan construction and subsequent construction of the four-dimensional aircraft trajectory defined by the specified flight plan legs and constraints and the aircraft performance. Flight plan and trajectory prediction work together to produce the four-dimensional trajectory and consolidate all the relevant trajectory information into a flight plan/profile buffer. The navigation function provides the dynamic current aircraft state to the other functions. The vertical, lateral steering, and performance advisory functions use the current aircraft state from navigation and the information in the flight plan/profile buffer to provide guidance, reference, and advisory information relative to the defined trajectory and aircraft state.

- The navigation function — responsible for determining the best estimate of the current state of the aircraft.
- The flight planning function — allows the crew to establish a specific routing for the aircraft.

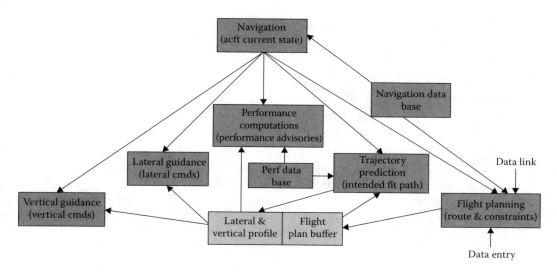

FIGURE 20.2 Flight management functional block diagram.

- The trajectory prediction function — responsible for computing the predicted aircraft profile along the entire specified routing.
- The performance function — provides the crew with aircraft unique performance information such as takeoff speeds, altitude capability, and profile optimization advisories.
- The guidance functions — responsible for producing commands to guide the aircraft along both the lateral and vertical computed profiles.

Depending on the particular implementation, the ancillary input/output (I/O), built-in test equipment (BITE), and control display functions may be included as well. Since the ancillary functions can vary significantly, this discussion will focus on the core flight management functions.

There are typically two loadable databases that support the core flight management functions. These are the navigation database which must be updated on a 28-day cycle and the performance database that only gets updated if there's been a change in the aircraft performance characteristics (i.e., engine variants or structural variants affecting the drag of the aircraft).

The navigation database contains published data relating to airports, navaids, named waypoints, airways and terminal area procedures along with RNP values specified for the associated airspace. The purpose of the navigation data base is twofold. It provides the navigation function location, frequency, elevation, and class information for the various ground-based radio navigation systems. This information is necessary to select, auto-tune, and process the data from the navigation radios (distance, bearing, or path deviation) into an aircraft position. It also provides the flight plan function with airport, airport-specific arrival, departure, and approach procedures (predefined strings of terminal area waypoints), airways (predefined enroute waypoint strings), and named waypoint information that allows for rapid route construction. A detailed description of the actual data content and format can be found in ARINC 424.

The performance database contains aircraft/engine model data consisting of drag, thrust, fuel flow, speed/altitude envelope, thrust limits, and a variety of optimized and tactical speed schedules that are unique to the aircraft. Figure 20.2 shows the interrelationships between the core functions and the databases.

20.2.1 Navigation

The navigation function within the FMS computes the aircraft's current state (generally WGS-84 geodetic coordinates) based on a statistical blending of multisensor position and velocity data. The aircraft's current state data usually consists of:

- Three-dimensional position (latitude, longitude, altitude)
- Velocity vector
- Altitude rate
- Track angle, heading, and drift angle
- Wind vector
- Estimated Position Uncertainty (EPU)
- Time

The navigation function is designed to operate with various combinations of autonomous sensors and navigation receivers. The position update information from the navigation receivers is used to calibrate the position and velocity data from the autonomous sensors, in effect providing an error model for the autonomous sensors. This error model allows for navigation coasting based on the autonomous sensors while maintaining a very slow growth in the EPU. If the updating from navigation aids such as distance measurement equipment (DME), very high frequency omni range (VOR), or global positioning system (GPS) is temporarily interrupted, navigation accuracy is reasonably maintained, resulting in seamless operations. This capability becomes very important for operational uses such as RNAV approach guidance where the coasting capability allows completion of the approach even if a primary updating source such as GPS is lost once the approach is commenced. A typical navigation sensor complement consists of:

- Autonomous sensors
  - Inertial reference
  - Air data
- Navigation receivers
  - DME receivers
  - VOR/LOC receivers
  - GPS receivers

The use of several navigation data sources also allows cross-checks of raw navigation data to be performed to ensure the integrity of the FMS position solution.

20.2.1.1 Navigation Performance

The navigation function, to be RNP airspace compliant per DO-236, must compute an Estimated Position Uncertainty (EPU) that represents the 95% accuracy performance of the navigation solution. The EPU is computed based on the error characteristics of the particular sensors being used and the variance of the individual sensors position with respect to other sensors. The RNP for the airspace is defined as the minimum navigation performance required for operation within that airspace. It is specified by default values based on the flight phase retrieved from the navigation data base for selected flight legs or crew-entered in response to ATC-determined airspace usage. A warning is issued to the crew if the EPU grows larger than the RNP required for operation within the airspace. The table below shows the current default RNP values for the various airspace categories.

Airspace Definition	Default RNP
Oceanic — no VHF navaids within 200 nm	12.0 nm
Enroute — above 15,000 ft	2.0 nm
Terminal	1.0 nm
Approach	0.5 nm

A pictorial depiction of the EPU computation is shown below for a VOR/VOR position solution. A similar representation could be drawn for other sensors.

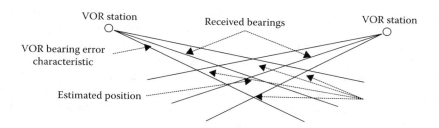

As can be seen from the diagram, the estimated position uncertainty (EPU) is dependent on the error characteristics of the particular navigation system being used as well as the geometric positioning of the navaids themselves. Other navigation sensors such as an inertial reference system have error characteristics that are time-dependent. More information pertaining to EPU and various navigation navaid system error characteristics can be found in RTCA DO-236.

20.2.1.2 Navigation Receiver Management

The various navigation receivers require different levels of FMS management to obtain a position update solution.

GPS — The GPS receiver is self-managing in that the FMS receives position, velocity, and time information without any particular FMS commands or processing. Typically, the FMS will provide an initial position interface to reduce the satellite acquire time of the receiver and some FMSs may provide an estimated time of arrival associated with a final approach fix waypoint to support the Predictive Receiver Autonomus Integrity Monitor (PRAIM) function in the GPS. More information on the GPS interface and function can be found in ARINC 743.

VHF navaids (DME/VOR/ILS) — The DME/VOR/ILS receivers must be tuned to an appropriate station to receive data. The crew may manually tune these receivers but the FMS navigation function will also auto-tune the receivers by selecting an appropriate set of stations from its stored navigation database and sending tuning commands to the receiver(s). The selection criteria for which stations to tune are:

- Navaids specified within a selected flight plan procedure, while the procedure is active.
- The closest DME navaids to the current aircraft position of the proper altitude class that are within range (typically 200 nm).
- Collocated DME/VORs within reasonable range (typically 25 nm).
- ILS facilities if an ILS or localizer (LOC) approach has been selected into the flight plan and is active.

Since DMEs receive ranging data and VORs receive bearing data from the fixed station location, the stations must be paired to determine a position solution as shown below:

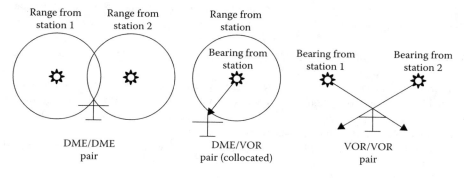

The pairing of navaids to obtain a position fix is based on the best geometry to minimize the position uncertainty (minimize the portion of EPU caused by geometric dilution of precision, GDOP). As can be seen from the figure above, the FMS navigation must process range data from DMEs and bearing data from VORs to compute an estimated aircraft position. Further, since the DME receives slant range data from ground station to aircraft, the FMS must first correct the slant range data for station elevation and aircraft altitude to compute the actual ground-projected range used to determine position. The station position, elevation, declination, and class are all stored as part of the FMS navigation data base. There are variations in the station-tuning capabilities of DME receivers. A standard DME can only accept one tuning command at a time, an agility-capable DME can accept two tuning commands at a time, and a scanning DME can accept up to five tuning commands at a time. VOR receivers can only accept one tuning command at a time.

An ILS or LOC receiver works somewhat differently in that it receives cross-track deviation information referenced to a known path into a ground station position. These facilities are utilized as landing aids and therefore are located near runways. The FMS navigation function processes the cross-track information to update the cross-track component of its estimated position. More information about DME/VOR/ILS can be found in ARINC 709, 711, and 710, respectively.

20.2.2 Flight Planning

The basis of the FMC flight profile is the route that the aircraft is to fly from the departure airport to the destination airport. The FMS flight planning function provides for the assembly, modification, and activation of this route data known as a flight plan. Route data are typically extracted from the FMC navigation data base and typically consists of a departure airport and runway, a standard instrument departure (SID) procedure, enroute waypoints and airways, a standard arrival (STAR) procedure, and an approach procedure with a specific destination runway. Often the destination arrival (or approach transition) and approach procedure are not selected until the destination terminal area control is contacted. Once the routing, along with any route constraints and performance selections, are established by the crew, the flight plan is assembled into a "buffer" that is used predominantly by the trajectory predictions in computing the lateral and vertical profile the aircraft is intended to fly from the departure airport to the destination airport.

The selection of flight planning data is done by the crew through menu selections either on the MCDU or navigation display or by data link from the airline's operational control. Facilities are also provided for the crew to define additional navigation/route data by means of a supplemental navigation data base. Some of the methods for the crew to create new fixes (waypoints) are listed below.

PBD Waypoints — Specified as bearing/distance off existing named waypoints, navaids, or airports.

PB/PB Waypoints — Specified as the intersections of bearings from two defined waypoints.

ATO Waypoints — Specified by an along-track offset (ATO) from an existing flight plan waypoint. The waypoint that is created is located at the distance entered and along the current flight plan path from the waypoint used as the fix. A positive distance results in a waypoint after the fix point in the flight plan while a negative distance results in a waypoint before the fix point.

Lat/Lon Waypoints — Specified by entering in the latitude/longitude coordinates of the desired waypoint.

Lat/Lon Crossing Waypoints — Created by specifying a latitude or longitude. A waypoint will be created where the active flight plan crosses that latitude or longitude. Latitude or longitude increments can also be specified, in which case several waypoints are created where the flight plan crosses the specified increments of latitude or longitude.

Intersection of Airways — Created by specifying two airways. A waypoint will be created at the first point where the airways cross.

Fix Waypoints — Created by specifying a "fix" reference. Reference information includes creation of abeam waypoints and creation of waypoints where the intersections of a specified radial or distance from the "fix" intersects the current flight plan.

Runway Extension Waypoints — Created by specifying a distance from a given runway. The new waypoint will be located that distance from the runway threshold along the runway heading.

Abeam Waypoints — If a direct-to is performed, selection of abeam points results in waypoints being created at their abeam position on the direct-to path. Any waypoint information associated with the original waypoint is transferred to the newly created waypoints.

FIR/SUA Intersection Waypoints — Creates waypoints where the current flight plan crosses flight information region (FIR) boundaries and special use areas (SUA) that are stored in the navigation data base.

The forward field of view display system shows a presentation of the selected segments of the flight plan as the flight plan is being constructed and flown.

The crew can modify the flight plan at any time. The flight plan modification can come from crew selections or via data link from the airline operational communications or air traffic control in response to a tactical situation. An edit to the flight plan creates a modified (or temporary) version of the flight plan that is a copy of the active flight plan plus any accrued changes made to it. Trajectory predictions are performed on the modified flight plan with each edit and periodically updated, which allows the crew to evaluate the impact of the flight plan changes prior to acceptance. When the desired changes have been made to the crew's satisfaction this modified flight plan is activated by the crew.

20.2.2.1 Flight Plan Construction

Flight plans are normally constructed by linking data stored in the navigation data base. The data may include any combination of the following items:

- SID/STAR/approach procedures
- Airways
- Prestored company routes
- Fixes (en route waypoints, navaids, nondirectional beacons, terminal waypoints, airport reference points, runway thresholds)
- Crew-defined fixes (as referenced above)

These selections may be strung together using clearance language, by menu selection from the navigation data base, by specific edit actions, or data link.

Terminal area procedures (SIDs, STARs, and approaches) consist of a variety of special procedure legs and waypoints. Procedure legs are generally defined by a leg heading, course or track, and a leg termination type. The termination type can be specified in many ways such as an altitude, a distance, or intercept of another leg. More detail on the path construction for these leg types and terminators will be discussed in the trajectory predictions section. Refer to ARINC 424 specification for further detail about what data and format are contained in the NDB to represent these leg types and terminations.

AF	DME Arc to a Fix
CA	Course to an Altitude
CD	Course to a Distance
CF*	Course to a Fix
CI	Course to an Intercept
CR	Course to Intercept a Radial
DF*	Direct to a Fix

---

*These leg types are recommended in DO-236 as the set that produces consistent ground tracks and the only types that should be used within RNP airspace.

FA*	Course from Fix to Altitude
FC	Course from Fix to Distance
FD	Course from Fix to DME Distance
FM	Course from Fix to Manual Term
HA*	Hold to an Altitude
HF*	Hold, Terminate at Fix after 1 Circuit
HM*	Hold, Manual Termination
IF*	Initial Fix
PI	Procedure Turn
RF*	Constant Radius to a Fix
TF*	Track to Fix
VA	Heading to Altitude
VD	Heading to Distance
VI	Heading to Intercept next leg
VM	Heading to Manual Termination
VR	Heading to Intercept Radial

Many of these leg types and terminations have appeared because of the evolution of equipment and instrumentation available on the aircraft and do not lend themselves to producing repeatable, deterministic ground tracks. For example, the ground track for a heading to an altitude will not only be dependent on the current wind conditions but also the climb performance of each individual aircraft. One can readily see that to fly this sort of leg without an FMS, the crew would follow the specified heading using the compass until the specified altitude is achieved, as determined by the aircraft's altimeter. Unfortunately, every aircraft will fly a different ground track and in some cases be unable to make a reasonable maneuver to capture the following leg. For the FMS, the termination of the leg is "floating" in that the lat/lon associated with the leg termination must be computed. These nondeterministic-type legs present problems for the air traffic separation concept of RNP airspace and for this reason RTCA DO-236 does not recommend the use of these legs in terminal area airspace, where they are frequently used today. These leg types also present added complexity in the FMS path construction algorithms since the path computation becomes a function of aircraft performance. With the advent of FMS and RNAV systems, in general, the need for non-deterministic legs simply disappears along with the problems and complexities associated with them.

Waypoints may also be specified as either "flyover" or nonflyover". A flyover waypoint is a waypoint whose lat/lon position must be flown over before the turn onto the next leg can be initiated whereas a nonflyover waypoint does not need to be overflown before beginning the turn onto the next leg.

20.2.2.2 Lateral Flight Planning

To meet the tactical and strategic flight planning requirements of today's airspace, the flight planning function provides various ways to modify the flight plan at the crew's discretion.

Direct-to — The crew can perform a direct-to to any fix. If the selected fix is a downtrack fix in the flight plan, then prior flight plan fixes are deleted from the flight plan. If the selected fix is not a downtrack fix in the flight plan, then a discontinuity is inserted after the fix and existing flight plan data are preserved.

Direct/intercept — The direct/intercept facility allows the crew to select any fixed waypoint as the active waypoint and to select the desired course into this waypoint. This function is equivalent to a direct-to except the inbound course to the specified fix which may be specified by the crew. The inbound course may be specified by entering a course angle, or if the specified fix is a flight plan fix, the crew may also select the prior flight plan-specified course to the fix.

Holding pattern — Holding patterns may be created at any fix or at current position. All parameters for the holding pattern are editable including entry course, leg time/length, etc.

Fixes — Fixes may be inserted or deleted as desired. A duplicate waypoint page will automatically be displayed if there is more than one occurrence of the fix identifier in the navigation database. Duplicate fixes are arranged starting from the closest waypoint to the previous waypoint in the flight plan.

Procedures — Procedures (SIDs, STARs, and approaches including missed approach procedures) may be inserted or replaced as desired. If a procedure is selected to replace a procedure that is in the flight plan, the existing procedure is removed and replaced with the new selection.

Airway segments — Airway segments may be inserted as desired.

Missed approach procedures — The flight planning function also allows missed approach procedures to be included in the flight plan. These missed approach procedures can either come from the navigation database where the missed approach is part of a published procedure, in which case they will be automatically included in the flight plan, or they can be manually constructed by entry through the MCDU. In either case, automatic guidance will be available upon activation of the missed approach.

Lateral offset — The crew can create a parallel flight plan by specifying a direction (left or right of path) and distance (up to 99 nm) and optionally selecting a start and/or end waypoint for the offset flight plan. The flight planning function constructs an offset flight plan, which may include transition legs to and from the offset path.

20.2.2.3   Vertical Flight Planning

Waypoints can have associated speed, altitude, and time constraints. A waypoint speed constraint is interpreted as a "cannot exceed" speed limit, which applies at the waypoint and all waypoints preceding the waypoint if the waypoint is in the climb phase, or all waypoints after it if the waypoint is in the descent phase. A waypoint altitude constraint can be of four types — "at," "at or above," "at or below," or "between." A waypoint time constraint can be of three types — "at," "after," "before," "after" and "before" types are used for en route track-crossings and the "at" type is planned to be used for terminal area flow control.

Vertical flight planning consists of selection of speed, altitude, time constraints at waypoints (if required or desired), cruise altitude selection, aircraft weight, forecast winds, temperatures, and destination barometric pressure as well as altitude bands for planned use of aircraft anti-icing. A variety of optimized speed schedules for the various flight phases are typically available. Several aircraft performance-related crew selections may also be provided. All these selections affect the predicted aircraft trajectory and guidance.

20.2.2.4   Atmospheric Models

Part of the flight planning process is to specify forecast conditions for temperatures and winds that will be encountered during the flight. These forecast conditions help the FMS to refine the trajectory predictions to provide more accurate determination of estimated times of arrival (ETAs), fuel burn, rates of climb/descent, and leg transition construction.

The wind model for the climb segment is typically based on an entered wind magnitude and direction at specified altitudes. The value at any altitude is interpolated between the specified altitudes to zero on the ground and merged with the current sensed wind. Wind models for use in the cruise segment usually allow for the entry of wind (magnitude and direction) for multiple altitudes at en route waypoints. Future implementation of en route winds may be via a data link of a geographical current wind grid database maintained on the ground. The method of computing winds between waypoints is accomplished by interpolating between entries or by propagating an entry forward until the next waypoint entry is encountered. Forecast winds are merged with current winds obtained from sensor data in a method which gives a heavier weighting to sensed winds close to the aircraft and converges to sensed winds as each waypoint-related forecast wind is sequenced. The wind model used for the descent segment is a set of altitudes with associated wind vector entered for different altitudes. The value at any altitude is interpolated from these values, and blended with the current sensed wind.

Forecast temperature used for extrapolating the temperature profile is based on the International Standard Atmosphere (ISA) with an offset (ISA deviation) obtained from pilot entries and/or the actual sensed temperature.

Forecast temperature = 15 + *ISA dev* − 0.00198 × *altitude*   *altitude* < 36,089
Forecast temperature = −56.5   *altitude* > 36,089

Air pressure is also used in converting speed between calibrated airspeed, mach, and true airspeed.

δ (*Pressure ratio*) = $(1 - 0.0000068753 * altitude)^{5.2561}$ *altitude* < 36,089

δ (*Pressure ratio*) = $0.22336 * e^{(4.8063 * (36089 - altitude)/100,000)}$

20.2.3 Trajectory Predictions

Given the flight plan, the trajectory prediction function computes the predicted four-dimensional flight profile (both lateral and vertical) of the aircraft within the specified flight plan constraints and aircraft performance limitations based on entered atmospheric data and the crew-selected modes of operation. The lateral path and predicted fuel, time, distance, altitude, and speed are obtained for each point in the flight plan (waypoints as well as inserted vertical breakpoints such as speed change, cross-over, level off, top of climb (T/C), top of descent (T/D) points). The flight profile is continuously updated to account for nonforecasted conditions and tactical diversions from the specified flight plan.

To simplify this discussion, the flight path trajectory is broken into two parts — the lateral profile (the flight profile as seen from overhead) and the vertical profile (the flight profile as seen from the side). However, the lateral path and vertical path are interdependent in that they are coupled to each other through the ground speed parameter. Since the speed schedules that are flown are typically constant CAS/mach speeds for climb and descent phases, the TAS (or ground speed) increases with altitude for the constant CAS portion and mildly decreases with altitude for the constant mach portion, as shown in the following equations.

$$Mach = sqrt\ [(1/\delta\{[1 + 0.2(CAS/661.5)^2]^{3.5} - 1\} + 1)^{0.286} - 1]$$

$$TAS = 661.5 \times mach \times sqrt[\theta]$$

$$CAS = calibrated\ airspeed\ in\ knots$$

$$TAS = true\ airspeed\ in\ knots$$

$$\delta = atmospheric\ pressure\ ratio\ (actual\ temperature\ /S.L.\ std.\ temperature)$$

$$\theta = atmospheric\ temperature\ ratio\ (actual\ temperature\ /S.L.\ std.\ temperature)$$

The significance of the change in airspeed with altitude will become apparent in the construction of the lateral and vertical profile during ascending and descending flights as described in the next section. Further, since the basic energy balance equations used to compute the vertical profile use TAS, these speed conversion formulas are utilized to convert selected speed schedule values to true airspeed values.

20.2.3.1 Lateral Profile

Fundamentally, the lateral flight profile is the specified route (composed of procedure legs, way-points, hold patterns, etc.), with all the turns and leg termination points computed by the FMS according to how the aircraft should fly them. The entire lateral path is defined in terms of straight segments and turn segments which begin and end at either fixed or floating geographical points. Computing these segments can be difficult because the turn transition distance and certain leg termination points are a function of predicted aircraft speed (as noted in the equations below), wind, and altitude, which, unfortunately, are dependent on how much distance is available to climb and descend. For example, the turn transition at a waypoint requires a different turn radius and therefore a different distance when computed with different speeds. The altitude (and therefore speed of the aircraft) that can be obtained at a waypoint is dependent upon how much distance is available to climb or desend. So, the interdependency between speed and leg distance presents a special problem in formulating a deterministic set of algorithms for computing the trajectory. This effect becomes significant for course changes greater than 45°, with the largest effect for legs such as procedure turns which require a 180° turn maneuver.

Lateral turn construction is based on the required course change and the aircraft's predicted ground speed during the turn. If the maximum ground speed that the aircraft will acquire during the required course change is known, a turn can be constructed as follows:

$$Turn\ Radius\ (ft) = (GS^2)/(g \times tan\phi)$$
$$Turn\ Arclength\ (ft) = \Delta\ Course \times Turn\ Radius$$
$$GS = maximum\ aircraft\ ground\ speed\ during\ the\ turn$$
$$g = acceleration\ due\ to\ gravity$$
$$\phi = nominal\ aircraft\ bank\ angle\ used\ to\ compute\ a\ turn.$$

For legs such as constant radius to a fix (RF) where the turn radius is specified, a different form of the equation is used to compute the nominal bank angle that must be used to perform the maneuver.

$$\phi = arctan[GS^2/(\textbf{turn radius} \times g)]$$

To determine the maximum aircraft ground speed during the turn the FMC must first compute the altitude at which the turn will take place, and then the aircraft's planned speed based on the selected speed schedule and any applicable wind at that altitude. The desired bank angle required for a turn is predetermined based on a trade-off between passenger comfort and airspace required to perform a lateral maneuver.

The basis for the lateral profile construction is the leg and termination types mentioned in the flight plan section. There are four general leg types:

- Heading (V) — aircraft heading
- Course (C) — fixed magnetic course
- Track (T) — computed great circle path (slowly changing course)
- Arc (A or R) — an arc defined by a center (fix) and a radius

There are six leg terminator types:

- Fix (F) — terminates at geographic location
- Altitude (A) — terminates at a specific altitude
- Intercept next leg (I) — terminates where leg intercepts the next leg
- Intercept radial (R) — terminates where leg intercepts a specific VOR radial
- Intercept distance (D or C) — terminates where leg intercepts a specific DME distance or distance from a fix
- Manual (M) — leg terminates with crew action

Not all terminator types can be used with all leg types. For example, a track leg can only be terminated by a fix since the definition of a track is the great circle path between two geographic locations (fixes). Likewise, arc legs are only terminated by a fix. In a general sense, heading and course legs can be graphically depicted in the same manner understanding that the difference in the computation is the drift angle (or aircraft yaw).

Figure 20.3 depicts a graphical construction for the various leg and terminator types. The basic construction is straightforward. The complexity arises from the possible leg combinations and formulating proper curved transition paths between them. For example, if a TF leg is followed by a CF

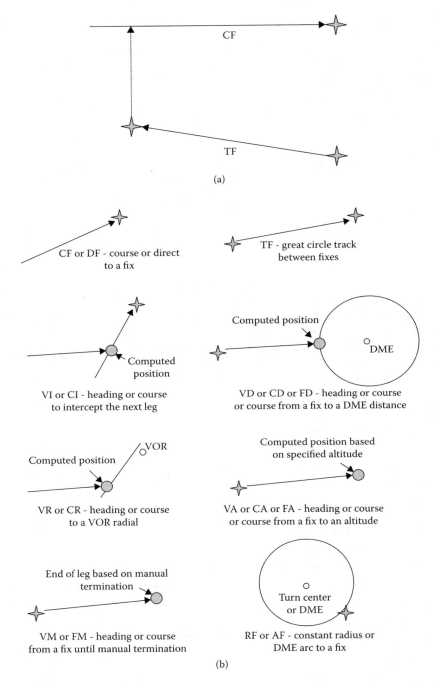

FIGURE 20.3   Basic lateral leg construction.

leg where the specified course to the fix does not pass through the terminating fix for the prior TF leg, then a transition path must be constructed to complete a continuous path between the legs.

In summary, the lateral flight path computed by the FMC contains much more data than straight lines connecting fixed waypoints. It is a complete prediction of the actual lateral path that the aircraft will fly under FMS control. The constructed lateral path is critical because the FMC will actually control the aircraft to it by monitoring cross-track error and track angle error, and issuing roll commands to the autopilot as appropriate.

20.2.3.2 Vertical Profile

The fundamental basis for the trajectory predictor is the numerical integration of the aircraft energy balance equations including variable weight, speed, and altitude. Several forms of the energy balance equation are used to accommodate unrestricted climb/descent, fixed gradient climb/descent, speed change, and level flight. The integration steps are constrained by flight plan-imposed altitude and speed restrictions as well as aircraft performance limitations such as speed and buffet limits, maximum altitudes, and thrust limits. The data that drives the energy balance equations come from the airframe/engine-dependent thrust, fuel flow, drag, and speed schedule models stored in the performance data base. Special construction problems are encountered for certain leg types such as an altitude-terminated leg because the terminator has a floating location. The location is dependent upon where the trajectory integration computes the termination of the leg. This also determines the starting point for the next leg.

The trajectory is predicted based on profile integration steps — the smaller the integration step the more accurate the computed trajectory. For each step the aircraft's vertical speed, horizontal speed, distance traveled, time used, altitude change, and fuel burned is determined based on the projected aircraft target speed, wind, drag, and engine thrust for the required maneuver. The aircraft's vertical state is computed for the end of the step and the next step is initialized with those values. Termination of an integration step can occur when a new maneuver type must be used due to encountering an altitude or speed constraint, flight phase change, or special segments such as turn transitions where finer integration steps may be prudent. The vertical profile is comprised of the following maneuver types: unrestricted ascending and descending segments, restricted ascending and descending segments, level segments, and speed change segments. Several forms of the energy balance equation are used depending on the maneuver type for a given segment of the vertical profile. Assumptions for the thrust parameter are maneuver type and flight phase dependent.

20.2.3.3 Maneuver Types

Unrestricted ascending and descending segments — The following form of the equation is typically used to compute the average vertical speed for fixed altitude steps (*dh* is set integration step). Using fixed altitude steps for this type of segment allows for deterministic step termination at altitude constraints. For ascending flight the thrust is generally assumed to be the take-off, go-around, or climb thrust limit. For descending flight the thrust is generally assumed to be at or a little above flight idle.

$$V/S = \frac{\dfrac{(T - D)V_{ave}}{GW}}{\dfrac{T_{act}}{T_{std}} + \dfrac{V_{ave}}{g}\dfrac{dV_{true}}{dh}}$$

where:

T	=	Avg. thrust (lb)
D	=	Avg. drag (lb)
GW	=	A/C gross wt (lb)
T_{act}	=	Ambient temp (K)
T_{std}	=	Std. day temp (K)

V_{ave} = Average true airspeed (ft/sec)

g = 32.174 ft/sec^2

dV_{true} = Delta V_{true} (ft/sec)

dh = Desired altitude step (ft)

The projected aircraft true airspeed is derived from the pilot-selected speed schedules and any applicable airport or waypoint-related speed restrictions. Drag is computed as a function of aircraft configuration, speed, and bank angle. Fuel flow and therefore weight change is a function of the engine thrust. Once V/S is computed for the step the other prediction parameters can be computed for the step.

$$dt = \frac{dh}{V/S}, \quad \text{where} \quad dt = \text{delta time for step}$$

$ds = dt(V_{true} + \text{average along track wind for segment}), \quad \text{where} \quad ds = \text{delta distance for step}$

$dw = dt \times \text{fuel flow}(T), \quad \text{where} \quad dw = \text{delta weight for step}$

Restricted ascending and descending segments — The following form of the equation is typically used to compute the average thrust for fixed altitude steps (dh and V/S are predetermined). Using fixed altitude steps for this type of segment allows for deterministic step termination at altitude constraints. The average V/S is either specified or computed based on a fixed flight path angle (FPA).

$$V/S_{ave} = GS_{ave} \tan FPA, \quad \text{where} \quad GS_{ave} = \text{segment ground speed (ft/sec)}.$$

The fixed *FPA* can in turn be computed based on a point to point vertical flight path determined by altitude constraints, which is known as a geometric path. With a specified *V/S* or *FPA* segment the thrust required to fly this profile is computed.

$$T = \frac{W \times V/S_{ave}}{V_{ave}} - \left(1 + \frac{V_{ave}}{g} \frac{dV_{true}}{dh}\right) + D$$

The other predicted parameters are computed as stated for the unrestricted ascending and descending segment.

Level segments — Constant-speed-level segments are a special case of the above equation. Since dV_{true} and V/S_{ave} are by definition zero for level segments, the equation simplifies to $T = D$. Level segments are typically integrated based on fixed time or distance steps so the other predicted parameters are computed as follows:

dt = set integration step

       and

$ds = dt(V_{true} + \text{average along track wind for segment}) \quad\quad ds = \text{delta distance for step}$

       or

ds = set integration step

       and

$dt = ds/(V_{true} + \text{average along track wind for segment}) \quad\quad dt = \text{delta time for step}$

$dw = dt \times \text{fuel flow}(T) \quad\quad\quad\quad\quad\quad\quad\quad\quad\quad\quad\quad dw = \text{delta weight for step}$

Speed change segments — The following form of the equation is typically used for speed change segments to compute the average time for a fixed dV_{true} step. The V/S_{ave} used is predetermined based on ascending, descending, or level flight along with the operational characteristics of the flight controls or as for the case of geometric paths computed based on the required *FPA*. The thrust is assumed to be flight idle for descending flight, take-off or climb thrust limit for ascending flight, or cruise thrust limit for level flight.

$$dt = dV_{true}/g \left\{ \frac{(T - D)}{GW} - \left(\frac{T_{act}}{T_{std}} \frac{V/S_{ave}}{V_{ave}} \right) \right\}$$

$$dh = V/S_{ave} \times dt$$

For all maneuver types the altitude rate, speed change, or thrust must be corrected for bank angle effects if the maneuver is performed during a turn transition. The vertical flight profile that the FMC computes along the lateral path is divided into three phases of flight: climb, cruise, and descent.

The climb phase — The climb phase vertical path, computed along the lateral path, is typically composed of the segments shown in Figure 20.4. In addition to these climb segments there can also be altitude level-off segments created by altitude restrictions at climb waypoints and additional target speed acceleration segments created by speed restrictions at climb waypoints.

The cruise phase — The cruise phase vertical path, computed along the lateral path, is very simple. It's typically composed of a climb speed to cruise speed acceleration or deceleration segment followed by a segment going to the FMC-computed top of descent. The cruise phase typically is predicted level at cruise altitude via several distance- or time-based integration steps. Unlike the climb and descent phase, the optimal cruise speeds slowly change with the changing weight of the aircraft, caused by fuel burn. If step climb or descents are required during the cruise phase, these are treated as unrestricted ascending flight and fixed V/S or FPA descents. At each step the FMC computes the aircraft's along-path speed, along-path distance traveled, and fuel burned based on the projected aircraft target speed, wind, drag, and engine thrust. The projected aircraft true airspeed is derived from the pilot-selected cruise speed schedule and applicable airport-related speed restrictions. Drag is computed as a function of aircraft speed and bank angle. For level flight, thrust must be equal to drag. Given the required thrust, the engine power setting can be computed, which becomes the basis for computing fuel burn and throttle control guidance.

The descent phase — The descent phase vertical path, computed along the lateral path, can be composed of several vertical leg types as shown in Figure 20.6.

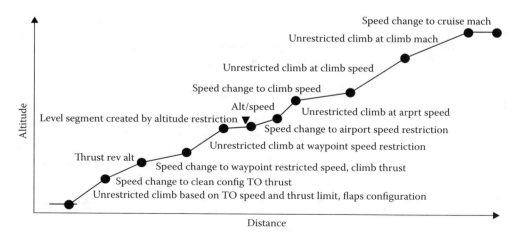

FIGURE 20.4 Typical climb profile.

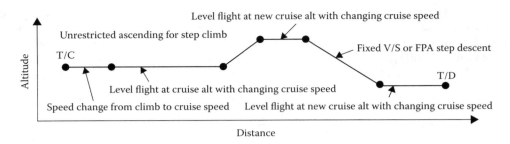

FIGURE 20.5   Typical cruise profile.

In addition to these descent segments, there can also be altitude level-off segments created by altitude restrictions at descent waypoints and additional targets speed deceleration segments created by speed restrictions at descent waypoints as well as eventual deceleration to the landing speed for the selected flaps configuration.

20.2.3.4   NDB Vertical Angles

These leg types are generally used in the approach. The desired approach glide slope angle that assures obstacle clearance is packed as part of the waypoint record for the approach in the Navigation Data Base (NDB). The angle is used to compute the descent path between the waypoint associated with the angle and the first of the following to be encountered (looking backwards):

1. Next lateral waypoint with an NDB vertical angle record
2. Next "at" constraint
3. First approach waypoint

A new NDB gradient can be specified on any waypoint. This allows the flexibility to specify multiple FPAs for the approach if desired. The integration basis for this leg assumes a thrust level compatible with maintaining the selected speed schedule at the descent rate specified by the NDB angle. Decelerations that can occur along these legs because of various restrictions (both regulatory and airframe) assume performing the speed change at idle thrust at the vertical speed specified by the NDB angle. If within the region where flaps are anticipated, then the deceleration model is based on a flaps configuration performance model.

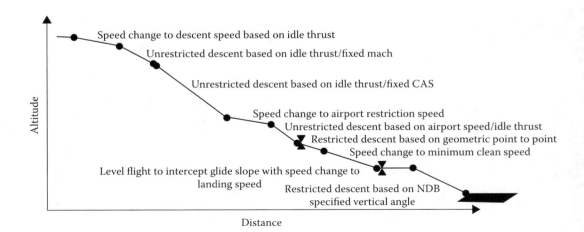

FIGURE 20.6   Typical descent profile.

Default approach vertical angle — Generally, this leg is used in lieu of a specified NDB angle to construct a stabilized nominal glide slope between the glide slope intercept altitude (typically 1500 ft above the runway) to the selected runway. The integration basis for this leg is the same as the NDB angle.

Direct to vertical angle — This leg type provides a vertical "Dir to" capability for use in tactical situations. The path constructed is the angle defined by the current 3-D position of the aircraft and the next appropriate reference point (usually the next altitude constraint). For a pending vertical "direct to" the direct to angle is updated on a periodic basis to account for the movement of the aircraft. In determining the direct to angle the aircraft 3-D position is extrapolated to account for the amount of time required to compute the trajectory for VNAV guidance to avoid path overshoots when the trajectory is available. The integration basis for this leg assumes a thrust level compatible with maintaining the selected speed schedule at the descent rate specified by the direct to gradient. Decelerations that can occur along these descent legs because of various restrictions (both regulatory and aircraft) assume performing the speed change at idle thrust for the anticipated flaps/landing gear configuration.

Computed vertical angle — This leg type provides constant angle vertical paths between constraints that are part of the vertical flight plan. These geometric paths provide for repeatable, stabilized, partial power descent paths at lower altitudes in the terminal area. The general rules for proper construction of these paths are:

- Vertical maneuvering should be minimized. This implies that a single angle to satisfy a string of altitude constraints is preferred. This can occur when "At or above" and "At or below" altitude constraints are contained in the flight plan.
- If a string of "At or above" and/or "At or below" constraints can be satisfied with an unrestricted, idle power path, then that path is preferred.
- Computed gradient paths should be checked for flyability (steeper than idle). The computation of the idle path (for the anticipated and idle with drag devices deployed) should account for a minimum deceleration rate if one is contained within the computed gradient leg.

The integration basis for this leg assumes a thrust level compatible with maintaining the selected speed schedule at the descent rate specified by the computed vertical angle. Decelerations that can occur along these descent legs because of various restrictions (both regulatory and airframe) assume performing the speed change at idle thrust limited for a maximum deceleration rate.

Constant V/S — This leg type provides a strategic, shallower-than-idle initial descent path if desired. The construction of this path is dependent on a vertical speed and intercept altitude being requested. The integration basis for this leg assumes a thrust level compatible with maintaining the selected speed schedule at the descent rate specified by the commanded v/s. Decelerations that can occur along these descent legs because of various restrictions (both regulatory and airframe) assume performing the speed change at idle thrust limited for a maximum deceleration rate.

Unrestricted descent — The unrestricted descent uses performance data to construct an energy-balanced idle descent path when not constrained by altitude constraints. The integration basis for this leg assumes maintaining the selected speed schedule at idle thrust. This results in a changing vertical speed profile. Decelerations that can occur along these descent legs because of various restrictions (both regulatory and aircraft) assume performing the speed change at a minimum vertical speed rate and idle thrust limited for a maximum deceleration rate. The minimum vertical speed can be based on energy sharing or a precomputed model. An idle thrust factor allows the operator to create some margin (shallower or steeper) in the idle path construction.

20.2.4 Performance Computations

The performance function provides the crew information to help optimize the flight or provide performance information that would otherwise have to be ascertained from the aircraft performance manual. FMSs implement a variety of these workload reduction features; only the most common functions are discussed here.

20.2.4.1 Speed Schedule Computation

Part of the vertical flight planning process is the crew selection of performance modes for each flight phase based on specific mission requirements. These performance modes provide flight profile optimization through computation of flight phase-dependent, optimized speed schedules that are used as a basis for both the trajectory prediction, generation of guidance speed targets, and other performance advisories.

The selection of a specific performance mode for each flight phase results in the computation of an optimized speed schedule, which is a constant CAS, constant mach pair, which becomes the planned speed profile for each flight phase. The altitude where the CAS and mach are equivalent is known as the crossover altitude. Below the crossover altitude the CAS portion of the speed schedule is the controlling speed parameter and above the crossover altitude the mach portion is the controlling speed. The performance parameter that is optimized is different for each performance mode selection.

Climb

- Economy (based on Cost Index) — speed that optimizes overall cost of operation (lowest cost).
- Maximum angle of climb — speed that produces maximum climb rate with respect to distance.
- Maximum rate of climb — speed that produces maximum climb rate with respect to time.
- Required time of arrival speed (RTA) — speed that optimizes overall cost of operation, while still achieving a required time of arrival at a specific waypoint.

Cruise

- Economy (based on Cost Index) — speed that optimizes overall cost of operation (lowest cost).
- Maximum endurance — speed that produces lowest fuel burn rate, maximizing endurance time.
- Long range cruise — speed that produces best fuel mileage, maximizing range.
- Required time of arrival (RTA) — speed that optimizes overall cost of operation, while still achieving a required time of arrival at a specific waypoint.

Descent

- Economy (based on Cost Index) — speed that optimizes overall cost of operation (lowest cost).
- Maximum descent rate — speed that produces maximum descent rate with respect to time.
- Required time of arrival (RTA) — speed that optimizes overall cost of operation, while still achieving a required time of arrival at a specific waypoint.

All flight phases allow a manually entered CAS/mach pair as well.

It may be noted that one performance mode that is common to all flight phases is the "economy" speed mode which minimizes the total cost of operating the airplane on a given flight. This performance mode uses a Cost Index, which is the ratio of time-related costs (crew salaries, maintenance, etc.) to fuel cost as one of the independent variables in the speed schedule computation.

Cost Index (CI) = flight time-related cost/fuel cost

The cost index allows airlines to weight time and fuel costs based on their daily operations.

20.2.4.2 Maximum and Optimum Altitudes

An important parameter for the flight crew is the optimum and maximum altitude for the aircraft/engine type, weight, atmospheric conditions, bleed air settings, and the other vertical flight planning parameters. The optimum altitude algorithm computes the most cost-effective operational altitude based solely on aircraft performance and forecasted environmental conditions. Fundamentally, the algorithm searches for the altitude that provides the best fuel mileage.

Altitude that maximizes the ratio: ground speed/fuel burn rate

The maximum altitude algorithm computes the highest attainable altitude based solely on aircraft performance and forecasted environmental conditions, while allowing for a specified rate of climb margin.

Altitude that satisfies the equality: min climb rate $=$ TAS $\times$ (thrust $-$ drag)/weight

Optimum altitude is always limited by maximum altitude. The algorithms for these parameters account for the weight reduction caused by the fuel burn in achieving the altitudes. The speeds assumed are the selected performance modes.

Trip altitude — Another important computation that allows the crew to request an altitude clearance to optimize the flight is the recommended cruise altitude for a specified route known as trip altitude. This altitude may be different from the optimum altitude in that for short trips the optimum altitude may not be achievable because of the trip distance. This algorithm searches for the altitude that satisfies the climb and descent while preserving a minimum cruise time.

Alternate destinations — To help reduce crew workload during flight diversion operations the FMS typically provides alternate destination information. This computation provides the crew with distance, fuel, and ETA for selected alternate destinations. The best trip cruise altitude may be computed as well. The computations are based either on a direct route from the current position to the alternate or continuing to the current destination, execution of a missed approach at the destination, and then direct to the alternate. Also computed for these alternate destinations are available holding times at the present position and current fuel state vs. fuel required to alternates. Usually included for the crew convenience is the CDU/MCDU retrieval of suitable airports that are nearest the aircraft.

Step climb/descent — For longer-range flights often the achievable cruise altitude is initially lower than the optimum because of the heavy weight of the aircraft. As fuel is burned off and the aircraft weight reduced, it becomes advantageous to step climb to a higher altitude for more efficient operation. The FMS typically provides a prediction of the optimum point(s) at which a step climb/descent maneuver may be initiated to provide for more cost-effective operation. This algorithm considers all the vertical flight planning parameters, particularly the downstream weight of the aircraft, as well as entered wind data. The time and distance to the optimum step point for the specified step altitude is displayed to the crew, as well as the percent savings/penalty for the step climb/descent vs. the current flight plan. For transoceanic aircraft it is typical for the trajectory prediction function to assume that these steps will be performed as part of the vertical profile, so that the fuel predictions are more aligned with what the aircraft will fly.

Thrust limit data — To prevent premature engine maintenance/failure and continued validation of engine manufacturer's warrantees, it becomes important not to overboost the aircraft engines. The engine manufacturers specify flight phase-dependent thrust limits that the engines are designed to operate reliably within. These engine limits allow higher thrust levels when required (take-off, go-around, engine out) but lower limits for non-emergency sustained operation (climb and cruise). The thrust limits for take-off, climb, cruise, go around, and continuous modes of operation are computed based on the current temperature, altitude, speed, and type of engine/aircraft and engine bleed settings. Thrust limit data are usually represented by "curve sets" in terms of either engine RPM (N1) or engine pressure ratio (EPR), depending on the preferred engine instrumentation package used to display the actual engine thrust. The "curve sets" typically have a temperature-dependent curve and an altitude-dependent curve along with several correction curves for various engine bleed conditions. The algorithms used to compute the thrust limits vary among engine manufacturers.

Take-off reference data — The performance function provides for the computation, or entry, of V_1, V_R and V_2 take-off speeds for selected flap settings and runway, atmospheric, and weight/CG conditions. These speeds are made available for crew selection for display on the flight instruments. In addition, take-off configuration speeds are typically computed. The take-off speeds and configuration speeds are stored as data sets or supporting data sets in the performance database.

Approach reference data — Landing configuration selection is usually provided for each configuration appropriate for the operation of the specific aircraft. The crew can select the desired approach configuration and the state of that selection is made available for other systems. Selection of an approach configuration also results in the computation of a landing speed based on a manually entered wind correction for the destination runway. In addition, approach configuration speeds are computed and displayed for reference and selection for display on the flight instruments. The approach and landing speeds are stored as data sets in the performance database.

Engine-out performance — The performance function usually provides engine-out performance predictions for the loss of at least one engine. These predictions typically include:

- Climb at engine-out climb speed
- Cruise at engine-out cruise speed
- Driftdown to engine-out maximum altitude at driftdown speed
- Use of maximum continuous thrust

The engine out speed schedules are retrieved from the performance data base and the trajectory predictions are computed based on the thrust available from the remaining engines and the increased aircraft drag created by engine windmilling and aircraft yaw caused by asymmetrical thrust.

20.2.5 Guidance

The FMS typically computes roll axis, pitch axis, and thrust axis commands to guide the aircraft to the computed lateral and vertical profiles as discussed in the trajectory predictions section. These commands may change forms depending on the particular flight controls equipment installed on a given aircraft. Other guidance information is sent to the forward field of view displays in the form of lateral and vertical path information, path deviations, target speeds, thrust limits and targets, and command mode information.

20.2.5.1 Lateral Guidance

The lateral guidance function typically computes dynamic guidance data based on the predicted lateral profile described in the trajectory predictions section. The data are comprised of the classic horizontal situation information:

- Distance to go to the active lateral waypoint (DTG)
- Desired track (DTRK)
- Track angle error (TRKERR)
- Cross-track error (XTRK)
- Drift angle (DA)
- Bearing to the go to waypoint (BRG)
- Lateral track change alert (LNAV alert)

A common mathematical method to compute the above data is to convert the lateral path lat/lon point representation and aircraft current position to earth-centered unit vectors using the following relationships:

P = *earth centered unit position vector with x, y, z components*

$X = COS\ (lat)\ COS\ (lon)$

$Y = COS\ (lat)\ SIN\ (lon)$

$Z = SIN\ (lat)$

For the following vector expressions $\times$ is the vector cross product and $\cdot$ is the vector dot product. For any two position vectors that define a lateral path segment:

$N = Pst \times Pgt$ $\qquad\qquad$ $N = $ unit vector normal to **Pst and Pgt**

$Pap = N \times (Ppos \times N)$ $\qquad$ $Pgt = $ go to point unit position vector

$\qquad\qquad\qquad\qquad\qquad\qquad\qquad$ $Pst = $ start point unit position vector

$DTGap = $ earth radius $*$ arcCOS $(Pgt \cdot Pap)$ $\qquad$ $Pap = $ along path position unit vector

$DTGpos = $ earth radius $*$ arcCOS $(Pgt \cdot Ppos)$ $\qquad$ $Ppos = $ current position unit vector

$XTRK = -$earth radius $*$ arcCOS $(Pap \cdot Ppos)$ $\quad$ (full expression)

$XTRK = -$earth radius $* N \cdot Ppos$ $\qquad\qquad$ (good approximation)

$\qquad\qquad Est = Z \times P$ $\qquad\qquad$ $E_{st} = $ East-pointing local level unit vector

$\qquad\qquad Nth = P \times Est$ $\qquad\qquad$ $Nth = $ North-pointing local level unit vector

$$Z = \begin{vmatrix} 0 \\ 0 \\ 1 \end{vmatrix} \quad Z \text{ axis unit vector}$$

$DTRK = $ arcTAN $[(-N \cdot Nth_{ap})/(-N \cdot Est_{ap})]$

$BRG = $ arcTAN $[(-N \cdot Nth_{pos})/(-N \cdot Est_{pos})]$

$TRKERR = DTRK - $ Current Track

$DA = $ Current Track $- $ Current Heading

LNAV Alert is set when the DTG/ground speed < 10 sec from turn initiation

The above expressions can also be used to compute the distance and course information between points that are displayed to the crew for the flight plan presentation. The course information is generally displayed as magnetic courses, due to the fact that for many years a magnetic compass was the primary heading sensor and therefore all navigation information was published as magnetic courses. This historical-based standard requires the installation of a worldwide magnetic variation model in the FMS since most of the internal computations are performed in a true course reference frame. Conversion to magnetic is typically performed just prior to crew presentation.

The lateral function also supplies data for a graphical representation of the lateral path to the navigation display, if the aircraft is so equipped, such that the entire lateral path can be displayed in an aircraft-centered reference format or a selected waypoint center reference format. The data for this display are typically formatted as lat/lon points with identifiers and lat/lon points with straight and curved vector data connecting the points. Refer to ARINC 702A for format details. In the future the FMS may construct a bit map image of the lateral path to transmit to the navigation display instead of the above format.

Lateral leg switching and waypoint sequencing — As can be seen in the lateral profile section, the lateral path is composed of several segments. Most lateral course changes are performed as "flyby" transitions. Therefore anticipation of the activation of the next vertical leg is required, such that a smooth capture of that segment is performed without path overshoot. The turn initiation criteria are based on the extent of the course change, the planned bank angle for the turn maneuver, and the ground speed of the aircraft.

$$Turn\ Radius = Ground\ Speed^2/[g * TAN\ (\phi_{nominal})]$$

$$Turn\ initiation\ Distance = Turn\ Radius/TAN\ (Course\ Change/2) + roll\ in\ distance$$

The roll in distance is selected based on how quickly the aircraft responds to a change in the aileron position. Transitions that are flyby but require a large course change (>135°) typically are constructed for a planned overshoot because of airspace considerations. Turn initiation and waypoint sequence follow the same algorithms except the course change used in the above equations is reduced from the actual course change to delay the leg transition and create the overshoot. The amount of course change reduction is determined by a balance in the airspace utilized to perform the overall maneuver. For "flyover" transitions, the activation of the next leg occurs at the time the "flyover" waypoint is sequenced.

The initiation of the turn transition and the actual sequence point for the waypoint are not the same for "flyby" transitions. The waypoint is usually sequenced at the turn bisector point during the leg transition.

Roll control — Based on the aircraft current state provided by the navigation function and the stored lateral profile provided by the trajectory prediction function, lateral guidance produces a roll steering command that can be engaged by the flight controls. This command is both magnitude and rate limited based on aircraft limitations, passenger comfort, and airspace considerations. The roll command is computed to track the straight and curved path segments that comprise the lateral profile. The roll control is typically a simple control law driven by the lateral cross-track error and track error as discussed in the prior subsection as well as a nominal roll angle for the planned turn transitions. The nominal roll angle is zero for straight segments but corresponds to the planned roll angle used to compute lateral transition paths to follow the curved segments.

$$Roll = xtrk\ gain \times xtrk + trk\ gain \times trk\ error + \phi_{nominal}$$

where

$$\phi_{nominal} = nominal\ planned\ roll\ angle.$$

The gain values used in this control loop are characteristic of the desired aircraft performance for a given airframe and flight controls system.

Lateral capture path construction — At the time LNAV engagement with the flight controls occurs, a capture path is typically constructed that guides the airplane to the active lateral leg. This capture path is usually constructed based on the current position and track of the aircraft if it intersects the active lateral leg. If the current aircraft track does not intersect the active lateral leg, then LNAV typically goes into an armed state waiting for the crew to steer the aircraft into a capture geometry before fully engaging to automatically steer the aircraft. Capture of the active guidance leg, is usually anticipated to prevent overshoot of the lateral path.

20.2.5.2 Vertical Guidance

The vertical guidance function provides commands of pitch, pitch rate, and thrust control to the parameters of target speeds, target thrusts, target altitudes, and target vertical speeds (some FMS provide only the targets depending on the flight management/flight control architecture of the particular aircraft). Much like the lateral guidance function, the vertical guidance function provides dynamic guidance parameters for the active vertical leg to provide the crew with vertical situation awareness. Unlike the lateral guidance parameters, the vertical guidance parameters are somewhat flight phase dependent.

Flight Phase	Vertical Guidance Data
Takeoff	Take-off speeds V1, V2, VR
	Take-off thrust limit
Climb	Target speed based on selected climb speed schedule,
	flight plan speed restriction, and airframe limitations
	Target altitude intercept
	Alt constraint violation message
	Distance to top of climb
	Climb thrust limits
Cruise	Target speed based on selected cruise speed schedule,
	flight plan speed restriction, and airframe limitations
	Maximum and optimum altitude
	distance to step climb point
	Distance to top of descent
	Cruise thrust limit
	Cruise thrust target
Descent	Target speed based on selected descent speed schedule,
	flight plan speed restriction, and airframe limitations
	Target altitude intercept
	Vertical deviation
	Desired V/S
	Energy bleed-off message
Approach	Target speed based on dynamic flap configuration
	Vertical deviation
	Desired V/S
Missed Approach	Target speed based on selected climb speed schedule,
	flight plan speed restriction, and airframe limitations
	Target altitude intercept
	Alt constraint violation msg
	Distance to top of climb
	Go-around thrust limit

Vertical guidance is based on the vertical profile computed by the trajectory prediction function as described in a previous section as well as performance algorithms driven by data from the performance data base.

The mathematical representation of the vertical profile is the point type identifier, distance between points, which includes both lateral and vertical points, speed, altitude, and time at the point. Given this information, data for any position along the computed vertical profile can be computed.

$$Path\ gradient = (alt_{start} - alt_{end})/distance\ between\ points$$

Therefore the path reference altitude and desired *V/S* at any point is given by:

$$Path\ altitude = alt_{end} + path\ gradient * DTGap$$

$$Vertical\ deviation = current\ altitude - path\ altitude$$

$$Desired\ V/S = path\ gradient * current\ ground\ speed$$

In the same manner time and distance data to any point or altitude can be computed as well. The target speed data are usually not interpolated from the predicted vertical profile since it is only valid for on-path flight conditions. Instead, it is computed based on the current flight phase, aircraft altitude, relative position with respect to flight plan speed restrictions, flaps configuration, and airframe speed envelope limitations. This applies to thrust limit computations as well.

Auto flight phase transitions — The vertical guidance function controls switching of the flight phase during flight based on specific criteria. The active flight phase becomes the basis for selecting the controlling parameters to guide the aircraft along the vertical profile. The selected altitude is used as a limiter in that the vertical guidance will not allow the aircraft to fly through that altitude (except during approach operations where the selected altitude may be pre-set for a missed approach if required). When on the ground with the flight plan and performance parameters initialized, the flight phase is set to take-off. After liftoff, the phase will switch to climb when the thrust revision altitude is achieved. The switch from climb to cruise (level flight) phase usually occurs when the aircraft is within an altitude acquire band of the target altitude.

$$|Cruise\ altitude\ -\ current\ altitude|\ <\ capture\ gain\ *\ current\ vertical\ speed$$

The capture gain is selected for aircraft performance characteristics and passenger comfort. The switch from cruise to descent can occur in various ways. If the crew has armed the descent phase by lowering the preselected altitude below cruise altitude, then descent will automatically initiate at an appropriate distance before the computed T/D to allow for sufficient time for the engine to spool down to descent thrust levels so that the aircraft speed is coordinated with the initial pitch-over maneuvers. If the crew has not armed the descent by setting the selected altitude to a lower level, then cruise is continued past the computed T/D until the selected altitude is lowered to initiate the descent. Facilities are usually provided for the crew to initiate a descent before the computed T/D in response to ATC instructions to start descending.

Vertical leg switching — As can be seen in the vertical profile section, the vertical path is composed of several segments. Just as in the lateral domain it is desirable to anticipate the activation of the next vertical leg such that a smooth capture of that segment is performed without path overshoot. It therefore becomes necessary to have an appropriate criteria for vertical leg activation. This criteria is typically in the form of an inequality involving path altitude difference and path altitude rate difference.

$$|Path\ altitude\ (n)\ -\ path\ altitude\ (n\ +\ 1)|\ *\ capture\ gain\ <\ |desired\ V/S\ (n)\ -\ desired\ V/S\ (n\ +\ 1)|$$

The capture gain is determined based on airframe performance and passenger comfort.

Pitch axis and thrust axis control — The pitch command produced by vertical guidance is based on tracking the speed target, FMS path, or acquiring and holding a target altitude depending on the flight phase and situation. If VNAV is engaged to the flight controls an annunciation of the parameter controlling pitch is usually displayed in the crew's forward field of view.

Control strategy may vary with specific implementations of FMSs. Based on the logic in the above table, the following outer loop control algorithms are typically used to compute the desired control parameters.

Pitch axis control — The control algorithms below are representative of control loop equations that could be used and are by no means the only forms that apply. Both simpler and more complex variations may be used.

Vspd

> **Capture**
> Delta pitch = speed rate gain * (airspeed rate − capture rate)
> **Track**
> Delta pitch = (airspeed gain * airspeed error + speed rate gain * airspeed rate)/V_{true}

Flight Phase	Pitch Axis Control	Thrust Axis Control	Pitch/Thrust Mode Annunciation
Take-off	None until safely off ground then same as climb	Take-off thrust limit	Vspd/TO limit
Climb and cruise climb	Capture and track speed target	Climb thrust limit	Vspd/CLB limit
Level flight	Capture and maintain altitude	Maintain speed target	Valt/CRZ limit
Unrestricted descent	Capture and track vertical path	Set to flight idle	Vpath/CRZ limit
Restricted descent and approach	Capture and track vertical path	Set to computed thrust required, then maintain speed	Vpath/CRZ limit
Descent path capture from below and cruise descent	Capture and track fixed V/S capture path	Set to computed thrust required, then maintain speed	Vpath/CRZ limit
Descent path capture from above	Capture and track upper speed limit	Set to flight idle	Vspd/CRZ limit
Missed approach	Capture and track speed target	Go-around thrust limit	Vspd/GA limit

Vpath

> ***Capture***
> *V/S error = fixed capture V/S − current V/S*
> *Delta pitch = path capture gain * arcSIN (V/S error/V_{true})*
> ***Track***
> *Delta pitch = (VS gain * V/S error + alt error gain * alt error)/V_{true}*

Valt

> ***Capture***
> *Capture V/S = alt capture gain * alt error*
> *V/S error = capture V/S − current V/S*
> *Delta pitch = V/S gain * * arcSIN (V/S error/V_{true})*
> ***Track***
> *Delta pitch = (VS gain * current V/S + alt error gain * alt error)/V_{true}*

Proper aircraft pitch rates and limits are typically applied before final formulation of the pitch command. Once again, the various gain values are selected based on the aircraft performance and passenger comfort.

Thrust axis control — The algorithms below are representative of those that could be utilized to determine thrust settings. Quite often the thrust setting for maintaining a speed is only used for an initial throttle setting. Thereafter the speed error is used to control the throttles.

Thrust Limit

> *Thrust limit = f (temp, alt, spd, engine bleed air): stored as data sets in the performance DB*

Flight Idle

> *Idle thrust = f (temp, alt, spd, engine bleed air): stored as data sets in the performance DB*

Thrust Required

$$T = \frac{W \times V/S_{ave}}{V_{ave}}\left(1 + \frac{V_{ave}}{g}\frac{dV_{true}}{dh}\right) + D$$

RTA (required time of arrival) — The required time of arrival or time navigation is generally treated as a dynamic phase-dependent speed schedule selection (refer to the performance section). From this standpoint the only unique guidance requirements are the determination of when to recompute the phase-dependent speed schedules based on time error at the specified point and perhaps the computation of the earliest and latest times achievable at the specified point.

RNAV approach with VNAV guidance — The only unique guidance requirement is the increased scale in the display of vertical deviation when the approach is initiated. The vertical profile for the approach is constructed as part of the vertical path by trajectory predictions, complete with deceleration segments to the selected landing speed (refer to the performance section).

20.3   Summary

This chapter is an introduction to the functions that comprise a flight management system and has focused on the basic functionality and relationships that are fundamental to understanding the flight management system and its role in the operations of the aircraft. Clearly, there is a myriad of complexity in implementing each function that is beyond the scope of this publication.

The future evolution of the flight management system is expected to focus not on the core functions as described herein, but on the use within the aircraft and on the ground of the fundamental information produced by the flight management system today. The use of the FMS aircraft state and trajectory intent within the aircraft and on the ground to provide strategic conflict awareness is a significant step toward better management of the airspace. Communication of the optimized user-preferred trajectories will lead to more efficient aircraft operation. The full utilization of RNP-based navigation will increase the capacity of the airspace. Innovative methods to communicate FMS information and specify flight plan construction with the crew to make flight management easier to use are expected as well. Clearly, the FMS is a key system in moving toward the concepts embodied in CNS future airspace.

21

TCAS II

Steve Henely
Rockwell Collins

21.1   Introduction

The Traffic Alert and Collision Avoidance System (TCAS) provides a solution to the problem of reducing the risk of midair collisions between aircraft. TCAS is a family of airborne systems that function independently of ground-based air traffic control (ATC) to provide collision avoidance protection. The TCAS concept makes use of the radar beacon transponders carried by aircraft for ground ATC purposes and provides no protection against aircraft that do not have an operating transponder.

TCAS I provides proximity warning only, to aid the pilot in the visual acquisition of potential threat aircraft. TCAS II provides traffic advisories and resolution advisories (recommended evasive maneuvers) in a vertical direction to avoid conflicting traffic. Development of TCAS III, which was to provide traffic advisories and resolution advisories in the horizontal as well as the vertical direction, was discontinued in favor of emerging systems such as the ADS-B system discussed elsewhere in this book. This chapter will focus on TCAS II.

Based on a congressional mandate (Public Law 100-223), the Federal Aviation Administration (FAA) issued a rule effective February 9, 1989 that required the equipage of TCAS II on airline aircraft with more than 30 seats by December 30, 1991. Public Law 100-223 was later amended (Public Law 101-236) to permit the FAA to extend the deadline for TCAS II fleetwide implementation to December 30, 1993. In December of 1998 the FAA released a Technical Standard Order (TSO) that approved Change 7, resulting in the DO-185A TCAS II requirement. Change 7 incorporates software enhancements to reduce the number of false alerts.

21.2   Components

TCAS II consists of the Mode S/TCAS Control Panel, the Mode S transponder, the TCAS computer, antennas, traffic and resolution advisory displays, and an aural annunciator. Figure 21.1 is a block diagram of TCAS II. Control information from the Mode S/TCAS Control Panel is provided to the TCAS computer via the Mode S Transponder. TCAS II uses a directional antenna, mounted on top of the aircraft. In addition to receiving range and altitude data on targets above the aircraft, this directional antenna is used to transmit interrogations at varying power levels in each of four 90° azimuth segments. An omnidirectional transmitting and receiving antenna is mounted at the bottom of the aircraft to provide

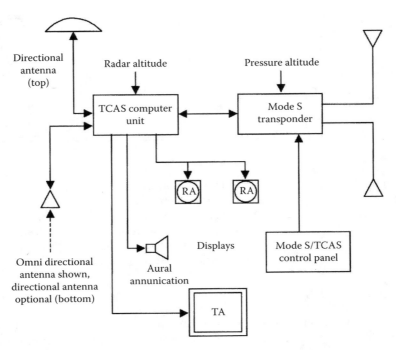

FIGURE 21.1   TCAS II block diagram.

TCAS with range and altitude data from traffic that is below the aircraft. TCAS II transmits transponder interrogations on 1030 MHz and receives transponder replies on 1090 MHz.

The Traffic Advisory (TA) display depicts the position of the traffic relative to the TCAS aircraft to assist the pilot in visually acquiring threatening aircraft. The Resolution Advisory (RA) can be displayed on a standard Vertical Speed Indicator (VSI), modified to indicate the vertical rate that must be achieved to maintain safe separation from threatening aircraft. When an RA is generated, the TCAS II computer lights up the appropriate display segments and RA compliance is accomplished by flying to keep the VSI needle out of the red segments. On newer aircraft, the RA display function is integrated into the Primary Flight Display (PFD). Displayed traffic and resolution advisories are supplemented by synthetic voice advisories generated by the TCAS II computer.

21.3   Surveillance

TCAS listens for the broadcast transmission (squitters) which is generated once per second by the Mode S transponder and contains the discrete Mode S address of the sending aircraft. Upon receipt of a valid squitter message the transmitting aircraft identification is added to a list of aircraft the TCAS aircraft will interrogate. Figure 21.2 shows the interrogation/reply communications between TCAS systems. TCAS sends an interrogation to the Mode S transponder with the discrete Mode S address contained in the squitter message. From the reply, TCAS can determine the range and the altitude of the interrogated aircraft.

There is no selective addressing capability with Mode A/C transponders, so TCAS uses the Mode C only all-call message to interrogate these types of Mode A/C transponders at a nominal rate of once per second. Mode C transponders reply with altitude data while Mode A transponders reply with no data in the altitude field. All Mode A/C transponders that receive a Mode C all-call interrogation from TCAS will reply. Since the length of the reply is 21 μs, Mode A/C-equipped aircraft within a range difference of 1.7 nmi from the TCAS will generate replies that overlap each other, as shown in Figure 21.3. These overlapping Mode A/C replies are known as synchronous garble.

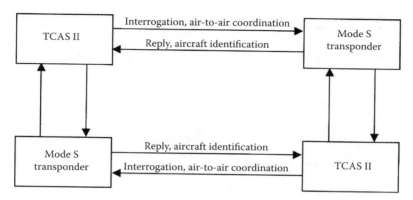

FIGURE 21.2   Interrogation/Reply between TCAS systems.

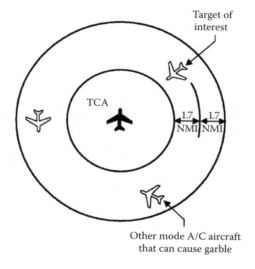

FIGURE 21.3   Synchronous garble area.

Hardware degarblers can reliably decode up to three overlapping replies. The whisper-shout technique and directional transmissions can be used to reduce the number of transponders that reply to a single interrogation. A low power level is used for the first interrogation step in a whisper-shout sequence. In the second whisper-shout step, a suppression pulse is first transmitted at a slightly lower level than the first interrogation, followed 2 μs later by an interrogation at a slightly higher power level than the first interrogation. The whisper-shout procedure shown in Figure 21.4 reduces the possibility of garble by suppressing most of the transponders that had replied to the previous interrogation, but eliciting replies from an additional group of transponders that did not reply to the previous interrogation. Directional interrogation transmissions further reduce the number of potential overlapping replies.

21.4   Protected Airspace

One of the most important milestones in the quest for an effective collision avoidance system is the development of the range/range rate (tau). This concept is based on time-to-go, rather than distance-to-go, to the closest point of approach. Effective collision avoidance logic involves a trade-off between providing the necessary protection with the detection of valid threats while at the same time avoiding

TABLE 21.1 Sensitivity Level Selection Based on Altitude

| Altitude (in Feet) | Sensitivity Level | Tau Values (in Seconds) | |
		TA	RA
0–1,000 AGL	2	20	N.A.
1,000–2,350 AGL	3	25	15
2,350–5,000 MSL	4	30	20
5,000–10,000 MSL	5	40	25
10,000–20,000 MSL	6	45	30
20,000–42,000 MSL	7	48	35
Greater than 42,000 MSL	7	48	35

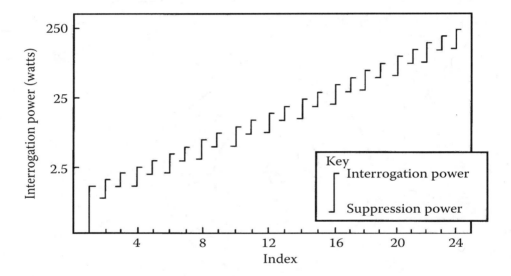

FIGURE 21.4 Whisper-Shout interrogation.

false alarms. This trade-off is accomplished by controlling the sensitivity level, which determines the tau, and therefore the dimensions of the protected airspace around each TCAS-equipped aircraft.

The pilot can select three modes of TCAS operation: STANDBY, TA-ONLY, and AUTOMATIC. These modes are used by the TCAS logic to determine the sensitivity level. When the STANDBY mode is selected, the TCAS equipment does not transmit interrogations. Normally, the STANDBY mode is used when the aircraft is on the ground. In TA-ONLY mode, the equipment performs all of the surveillance functions and provides TAs but not RAs. The TA-ONLY mode is used to avoid unnecessary distractions while at low altitudes and on final approach to an airport. When the pilot selects AUTOMATIC mode, the TCAS logic selects the sensitivity level based on the current altitude of the aircraft. Table 21.1 shows the altitude thresholds at which TCAS automatically changes its sensitivity level selection and the associated tau values for altitude-reporting aircraft.

The boundary lines depicted in Figure 21.5 show the combinations of range and range rate that would trigger a TA with a 40s tau and an RA with a 25s tau. These TA and RA values correspond to sensitivity level 5 from Table 21.1. As shown in Figure 21.5, the boundary lines are modified at close range to provide added protection against slow closure encounters.

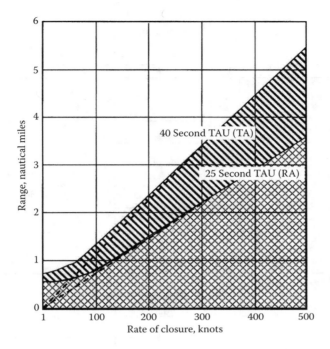

FIGURE 21.5   TA/RA Tau values for sensitivity level 5.

21.5   Collision Avoidance Logic

The collision avoidance logic functions are shown in Figure 21.6. This description of the collision avoidance logic is meant to provide a general overview. There are many special conditions relating to particular geometry, thresholds, and equipment configurations that are not covered in this description. Using surveillance reports, the collision avoidance logic tracks the slant range and closing speed of each target to determine the time in seconds until the closest point of approach. If the target is equipped with an altitude-encoding transponder, collision avoidance logic can project the altitude of the target at the closest point of approach.

A range test must be met and the vertical separation at the closest point of approach must be within 850 ft for an altitude-reporting target to be declared a potential threat and a traffic advisory to be generated. The range test is based on the RA tau plus approximately 15 s. A non-altitude-reporting target is declared a potential threat if the range test alone shows that the calculated tau is within the RA tau threshold associated with the sensitivity level being used.

A two-step process is used to determine the type of resolution advisory to be selected when a threat is declared. The first step is to select the sense (upward or downward) of the resolution advisory. Based on the range and altitude tracks of the potential threat, the collision avoidance logic models the potential threat's path to the closest point of approach and selects the resolution advisory sense that provides the greater vertical separation. The second resolution advisory step is to select the strength of the resolution advisory. The least disruptive vertical rate maneuver that will achieve safe separation is selected. Possible resolution advisories are listed in Table 21.2.

In a TCAS/TCAS encounter, each aircraft transmits Mode S coordination interrogations to the other to ensure the selection of complementary resolution advisories. Coordination interrogations contain information about an aircraft's intended vertical maneuver.

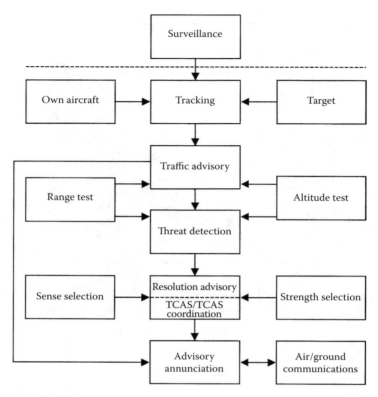

FIGURE 21.6   CAS logic functions.

TABLE 21.2   Resolution Advisories

Upward Sense	Type	Downward Sense
Increase Climb to 2500 fpm	Positive	Increase Descent to 2500 fpm
Reversal to Climb	Positive	Reversal to Descend
Maintain Climb	Positive	Maintain Descent
Crossover Climb	Positive	Crossover Descend
Climb	Positive	Descend
Don't Descend	Negative vsl	Don't Climb
Don't Descend >500 fpm	Negative vsl	Don't Climb >500 fpm
Don't Descend >1000 fpm	Negative vsl	Don't Climb >1000 fpm
Don't Descend >2000 fpm	Negative vsl	Don't Climb >2000 fpm

Note: Any combination of climb and descent restrictions may be given simultaneously (normally in multi-aircraft encounters); fpm = feet per minute; vsl = vertical speed limit.

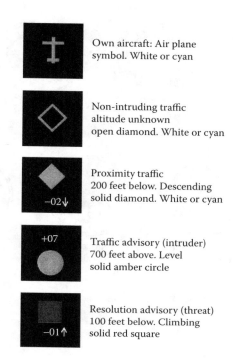

Own aircraft: Air plane
symbol. White or cyan

Non-intruding traffic
altitude unknown
open diamond. White or cyan

Proximity traffic
200 feet below. Descending
solid diamond. White or cyan

Traffic advisory (intruder)
700 feet above. Level
solid amber circle

Resolution advisory (threat)
100 feet below. Climbing
solid red square

FIGURE 21.7 Standardized symbology for TA display.

21.6 Cockpit Presentation

The traffic advisory display can either be a dedicated TCAS display or a joint-use weather radar and traffic display (see Figure 21.10). In some aircraft, the traffic advisory display will be an electronic flight instrument system (EFIS) or flat panel display that combines traffic and resolution advisory information on the same display. Targets of interest on the traffic advisory display are depicted in various shapes and colors as shown in Figure 21.7.

The pilot uses the resolution advisory display to determine whether an adjustment in aircraft vertical rate is necessary to comply with the resolution advisory determined by TCAS. This determination is based on the position of the vertical speed indicator needle with respect to the lighted segments. If the needle is in the red segments, the pilot should change the aircraft vertical rate until the needle falls within the green "fly-to" segment. This type of indication is called a corrective resolution advisory. A preventive resolution advisory is when the needle is outside the red segments and the pilot should simply maintain the current vertical rate. The green segment is lit only for corrective resolution advisories. Resolution advisory display indications corresponding to typical encounters are shown in Figure 21.8.

Figure 21.9 shows a combined traffic advisory/resolution advisory display indicating a traffic advisory (potential threat 200 ft below), resolution advisory (threat 100 ft above) and nonthreatening traffic (1200 ft above). The airplane symbol on the lower middle section of the display indicates the location of the aircraft relative to traffic. Figure 21.10 shows an example of a joint-use weather radar and traffic display.

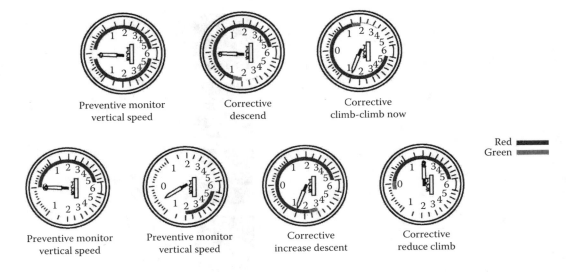

FIGURE 21.8 Typical resolution advisory indications.

FIGURE 21.9 Combined traffic advisory/resolution advisory display.

FIGURE 21.10 Joint use weather radar and traffic display.

22

Vehicle Health Management Systems

Philip A. Scandura, Jr.
Honeywell International

22.1   Introduction

The notion of vehicle health management (VHM) should be familiar to anyone who has ever operated or ridden in a vehicle, whether it was an automobile, truck, boat, or aircraft. Even the Wright Brothers performed VHM in their bicycle shop over 100 years ago. VHM includes the set of *activities that are performed to identify, mitigate, and resolve faults with the vehicle* [1]. These activities can be grouped into four phases, as illustrated in Figure 22.1.

The first phase, **health state determination**, is primarily concerned with determining the overall health state of the vehicle. By using diagnostic and prognostic algorithms, the vehicle and its systems are monitored to detect and isolate failures. This can be performed manually using procedures and observations, automatically using embedded hardware and software, or by some combination of these means.

The next phase, **mitigation**, involves the real-time assessment of the impact of the failures on the vehicle and its current mission. Once the impact is assessed, system redundancy management reconfigures the vehicle to maintain a safe operating condition and continue the mission, if possible. In those cases in which reconfiguration is not sufficient to continue the mission, the flight crew may modify the mission

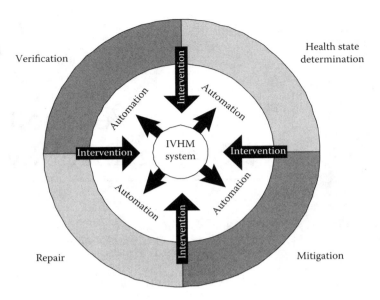

FIGURE 22.1 Health Management Activity Model [1].

to compensate. In the case of an aircraft, for example, an engine failure may require the crew to modify the original route to accommodate diversion to a closer airport.

Following mission completion, the **repair** phase conducts activities to return the vehicle to its nominal operating state. This typically involves the repair or replacement of the failed components. Depending upon the situation, such as a long-duration space mission to Mars, repairs may actually take place during the mission, as well as once the mission is completed.

The final phase, referred to as **verification**, consists of activities to ensure that all repairs have been performed correctly and that the system has been returned to full operational status. Depending upon the requirements of the governing regulatory authority, such as the Federal Aviation Administration (FAA), the verification phase may require the use of independent inspectors.

22.2 Definition of Integrated Vehicle Health Management

Although the foregoing discussion paints a reasonable picture of the activities involved in operating and maintaining a vehicle in an acceptable operating condition, there is actually more to VHM than just supporting vehicle operation. Integrated Vehicle Health Management (IVHM) spans the entire life cycle of the vehicle, including design, operation (as already discussed), and improvements. Not just focused on the vehicle itself, IVHM must also address the supporting infrastructure necessary to operate the vehicle. As illustrated in Figure 22.2, this requires taking an "enterprise-wide" approach to IVHM, addressing both the business cycle (vehicle development and continuous improvement) and the mission cycle (vehicle mission planning and execution).

22.2.1 A System Engineering Discipline

It is important to understand that IVHM is not just a stand-alone subsystem added on to an existing vehicle, nor should a group of sensors and related instrumentation system be considered IVHM. From a software perspective, IVHM is more that just fault models, algorithms, and sensor processing software. Although it is accurate to state that IVHM may use these components to perform its intended function, a true IVHM system is more than just a collection of "pieces and parts." In actual practice, IVHM must incorporate a philosophy, methodology, and process that focuses on design and development for safety, operability, maintainability, reliability, and testability. To be most effective, IVHM must be "designed into" the vehicle and its supporting

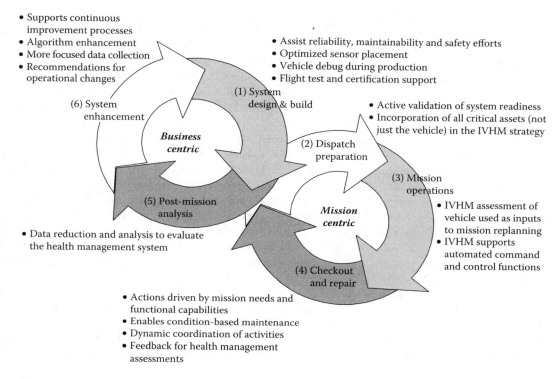

- Supports continuous improvement processes
- Algorithm enhancement
- More focused data collection
- Recommendations for operational changes

(6) System enhancement

- Assist reliability, maintainability and safety efforts
- Optimized sensor placement
- Vehicle debug during production
- Flight test and certification support

(1) System design & build

- Active validation of system readiness
- Incorporation of all critical assets (not just the vehicle) in the IVHM strategy

(2) Dispatch preparation

(3) Mission operations

- IVHM assessment of vehicle used as inputs to mission replanning
- IVHM supports automated command and control functions

Business centric

Mission centric

(5) Post-mission analysis

- Data reduction and analysis to evaluate the health management system

(4) Checkout and repair

- Actions driven by mission needs and functional capabilities
- Enables condition-based maintenance
- Dynamic coordination of activities
- Feedback for health management assessments

FIGURE 22.2 Enterprise-Wide Approach to IVHM [2].

infrastructure from the beginning of the program, not "added on" along the way.* IVHM principles must permeate the culture and mindset of the organization and be held in similar regard to safety. In summary, *IVHM must be elevated to the status of a system engineering discipline* [2].

22.2.2 Layered Approach

IVHM can be thought of as a distributed system, implemented as a series of layers, in which each layer performs a portion of the overall IVHM function, as illustrated in Figure 22.3. The first layer requires

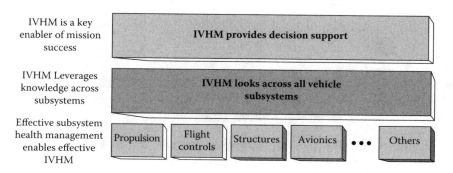

IVHM is a key enabler of mission success — **IVHM provides decision support**

IVHM Leverages knowledge across subsystems — **IVHM looks across all vehicle subsystems**

Effective subsystem health management enables effective IVHM — Propulsion | Flight controls | Structures | Avionics | ••• | Others

FIGURE 22.3 Layered Approach to IVHM [2].

---

*Although there are instances of successful add-on VHM systems, typically they are not as effective or efficient as those designed into the vehicle and its supporting infrastructure.

the establishment of a strong foundation of Subsystem Health Management (SHM), provided by embedded Built-In Test (BIT) and Fault Detection, Isolation, and Recovery (FDIR) capabilities that monitor the components within each subsystem boundary. A subsystem can be thought of as a component or collection of components that provide a higher-level vehicle function. The primary purpose of SHM is to ensure **safe operation** of the subsystem by providing the necessary subsystem monitors and functional tests as directed by the safety analysis for that subsystem. Often the functional design of a subsystem must be enhanced by BIT to mitigate latent faults or hazardous conditions. The secondary purpose of SHM is **economic** in that it helps to reduce the vehicle life cycle cost through improved maintainability, testability, and reliability. Commercial aviation experience has shown that the largest expense incurred over the life cycle of an aircraft (not including fuel and labor costs) is attributable to maintenance activities.* Without the establishment of accurate and reliable SHM, the effectiveness of the overall IVHM system will be severely limited.

In the middle layer, IVHM looks across all subsystems to assess overall vehicle health. It is important that SHM information from all subsystems be made visible to IVHM at the vehicle level, enabling IVHM to detect and isolate faults that may occur "between" subsystems or whose effects impact multiple subsystems. In this way, IVHM is able to determine overall vehicle health and use that information to annunciate possible vehiclewide mitigation strategies. Many modern aircraft employ some type of central maintenance system that fulfills the role of collecting faults from all subsystems, performing root-cause determination, and recommending repair actions.

Finally, at the highest layer, vehicle health and operations are integrated to maximize the benefits to the overall system. As illustrated in Figure 22.4, IVHM can provide decision support capabilities to all facets of the enterprise. It is important to note that the use of IVHM data depends upon the business case for the particular market segment and application. One must determine if the available IVHM data

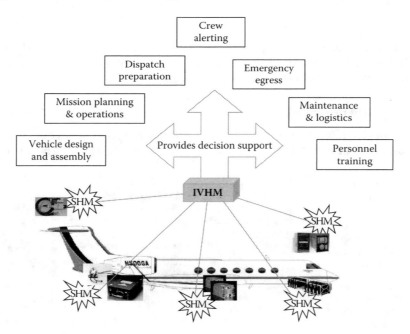

FIGURE 22.4 IVHM Provides Decision Support [2].

---

*In the year 2001, the average age of in-service commercial aircraft was approximately 12 years old [7]. Those readers who drive older model automobiles should agree that as a vehicle ages, it costs more to maintain it in a safe and operational condition.

creates the necessary value for the intended user community, be it maintenance crew, flight crew, airline operator, vehicle manufacturer, or even passenger. In addition to business needs, the IVHM system design (both the physical architecture and the processes used to develop the architecture) must support the safety and criticality needs of the consumers of IVHM data. Traditional maintenance systems have been used at the end of the flight to guide the repair of the aircraft, rather than during the flight to aid in the off-nominal operation of the aircraft, and have therefore been classified as noncritical systems. On the other hand, the Crew Alerting System (CAS) is used to determine the overall functional capability of the aircraft and plays a significant role in its operation; therefore, CAS has been classified as a critical system [3]. Satisfying such a broad spectrum of users emphasizes the need to determine the entire user community for IVHM data and ensure that the proper level of design assurance is applied to address their needs [2].

22.2.3 Health Management Integration

As previously discussed, IVHM must be designed into the system from the onset of the program, not added on as an afterthought. To do so, coordination and integration of IVHM efforts must occur across the program. The notion of **health management integration** helps to ensure that IVHM is properly designed into the system and involves the establishment of health management policies and processes that are enforced* across the design of the vehicle and supporting systems. Examples of health management policies and processes include the following:

- Fault detection and isolation philosophy
- Optimal sensor quantity and placement guidelines
- Standard designs and practices for developing BIT and FDIR
- IVHM/SHM metrics (e.g., fault coverage percentage, fault isolation accuracy percentage)
- IVHM/SHM test plans and procedures
- Fault modeling guidelines
- Interface standards between SHM and IVHM

Defining a comprehensive set of health management policies and processes requires the cooperation of the vehicle manufacturer and subsystem providers. Once established, periodic program and technical oversight helps to ensure consistent and correct application of the policies and processes across all subsystems. To provide such oversight, often a **health management integrator** is employed who is ultimately responsible for the coordination and integration activities that are crucial to the successful deployment of the IVHM system.

The ultimate goal of health management integration is to ensure an optimal IVHM balance across the system, resulting in improved system safety and reliability and reduced life cycle costs [2].

22.3 Evolution of VHM Standards

22.3.1 Commercial

Early commercial aircraft were composed primarily of mechanical and analog devices. Testing the functionality of a device typically employed nothing more than a simple push button that supplied current to the internal circuitry of the device. If continuity of the circuit was detected, a green light would illuminate, signifying a successful test. Referred to as Push-to-Test or Go/No-Go Test, these could loosely be considered as the beginning of BIT.

---

*Enforcement is typically implemented contractually in the form of requirements flowed down from the vehicle manufacturer to the various subsystem vendors, and associated design reviews of their designs against the requirements.

Starting in the early 1980s, commercial aircraft began to employ digital subsystems using hardware and software to perform functions previously performed by mechanical and analog means. These new digital subsystems, typically consisting of one or more Line Replaceable Units (LRUs), posed special challenges to aircraft mechanics as the ability to troubleshoot a "black box" was limited to the indications provided by the subsystem. The use of dedicated front panels with push buttons and simple display capability (e.g., lights, alpha-numeric readouts) provided the mechanic with the ability to test and query the subsystem.

As digital subsystems proliferated, it became apparent that standards were necessary, as mechanics were being overwhelmed with the varied and differing approaches taken by avionics manufactures. Working with the industry, Aeronautical Radio Inc. (ARINC) developed the first aviation industry standard for health management, entitled ARINC-604 "Guidance for Design and Use of Built-In Test Equipment."* Major document sections include Goals for BITE (Built-In Test Equipment), Maintenance Concept, BITE System Concepts, and Centralized Fault Display System Concepts. From this standard, the field of VHM in commercial aviation was born, although the acronym VHM would not come into usage until nearly 20 years later.

Following the release of ARINC-604, the advent of centralized display panels shared by several LRUs provided the mechanic with a single access point to several systems, theoretically reducing the amount of training required to learn the individual systems. It was not until the emergence of centralized maintenance computers in the late 1980s and early 1990s, however, that mechanics would truly benefit. These centralized systems gathered health and status data from several LRUs and performed fault consolidation and root-cause analysis, directing the mechanic to the offending system that required repair or replacement** and pointing to the applicable maintenance procedure. Referred to as the Central Maintenance Computer (CMC) or Onboard Maintenance System (OMS), these new systems were the result of further work by the aviation industry to produce updated standards, including ARINC-624, "Design Guidance for Onboard Maintenance System." Major document sections in ARINC-624 include Maintenance Concept, OMS Description, CMC Design Considerations, OMS Member System BITE, OMS Communications Protocol, Onboard Maintenance Documentation, and Airplane Condition Monitoring Function (ACMF) [4].

Several commercial aircraft use subsystems that follow ARINC-604-1. Newer aircraft include a CMC based upon ARINC-624. Table 22.1 provides a partial listing of these commercial aircraft.

TABLE 22.1 Commercial Aircraft Usage of ARINC-604-1 and ARINC-624 (Partial Listing)

Aircraft	Entry into Service (Approximate)	ARINC 604-1	ARINC 624	Simplified variant[a] of ARINC 604-1 and 624
Boeing 757/767	1982/1983	X		
Airbus A320	1988	X		
Boeing 747–400	1989	X		
McDonnell Douglas MD-11	1991	X		
Boeing 777	1995		X	
Boeing 717	1999	X		
Cessna Sovereign (Business Jet)	2004			X
Agusta AB-139 (helicopter)	2004			X
Gulfstream 450/500/550 (Business Jet)	2004			X
Embraer ERJ-170/190 (Regional Jet)	2004/2005			X

[a] Specialized standard developed for use with Honeywell International Primus Epic® systems.

---

*ARINC-604 was first published in 1985, soon after revised as ARINC-604-1 in 1988, as it still exists today.

**For more information regarding the emergence of the Central Maintenance Computer, see Maintaining Federated vs. Modular Avionics Systems.

MAINTAINING FEDERATED VS. MODULAR AVIONICS SYSTEMS

For many years, the traditional approach to avionics systems has been federated; that is, one or more LRUs have been dedicated to each aircraft function. For example, flight management hosted in its own flight management computer (usually dual); flight controls function hosted in a triplex configuration of flight control computers, cockpit displays with individual signal generator LRUs, and display control panels. In summary, each avionics subsystem resided in its own physical LRU with dedicated connections to control panels, sensors, and actuators.

In the early 1990s, the Boeing 777 broke with the federated tradition, using instead a modular avionics approach in which multiple functions were hosted on generic Line Replaceable Modules (LRMs) installed in two modular racks, designated left and right. Examples of LRMs include power supply modules, display processing modules, aircraft I/O modules, communication modules, and database modules. These first generation LRMs were not totally generic, however, as many contained unique hardware components restricting avionics functions to specific modules. Regardless of this limitation, the 777 approach boldly moved away from the one avionics function per LRU relationship and toward multiple avionics functions per LRM.

Subsequent generations of modular avionics systems further refined the generic nature of the LRMs, creating generic computing modules for processing both generic and custom I/O modules, network interface modules, data storage modules, and so on. In addition, the modular rack concept was expanded to allow the use of multiple racks networked together via an aircraftwide redundant databus. The key to the success of these newer systems was their scalability and extensibility to numerous aircraft and helicopter models produced by differing manufacturers for various markets (e.g., business aviation, general aviation, regional airlines, commercial airlines, search and rescue, military, etc.) rather than targeted to specific aircraft (such as the 777).

Although modular avionics systems provided many benefits, the challenge of maintaining the system had to be addressed. Line maintenance performed by the aircraft mechanic is relatively simple in a federated environment. Because most functions are hosted in their own dedicated LRU, failures in a particular system typically result in the removal and replacement of that LRU. "Shotgun troubleshooting" was common; that is, if a problem occurs in the left LRU, it can be swapped with the right LRU. If the problem follows the LRU, it is most likely being caused by that LRU.

In a modular system, however, functions no longer enjoy dedicated LRUs. Rather than a flight management function being hosted in a dedicated computer, it now lives in a processor module, receives aircraft data from an I/O module and uses navigation data stored on a database module; all three receive power from a power supply module. Failures of the flight management function can be attributed to faults in any of the modules or communication paths between them. It can be difficult, if not impossible, for the mechanic to determine which module is the cause of the problem without additional information from the system. The emergence of the CMC solved this problem by collecting health state data from all LRUs and LRMs, performing fault consolidation and root cause analysis to determine the faulty unit, and finally directing the line mechanic to the unit that required replacement. In this way the CMC provides an essential role in the troubleshooting and repair of aircraft with modular avionic systems [4].

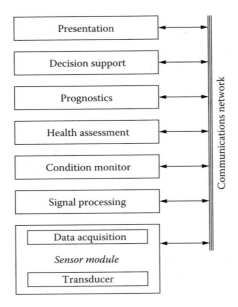

FIGURE 22.5 OSA-CBM layered approach.

22.3.2 Military

The ARINC standards discussed previously have been used by some military programs; however, many programs have transitioned to a more modern standard that traces its roots to network protocol layered architectures. The standard, known as Open Systems Architecture for Condition-Based Maintenance (OSA-CBM), was developed originally for use on Navy ships and is now steadily spreading to military ground vehicles and aircraft. A consortium including Boeing, Caterpillar, Mimosa, Oceana Sensor, Rockwell, Penn State, and the Office of Naval Research developed OSA-CBM. The consortium's mission statement declares the following:

> Condition-Based Maintenance (CBM) is becoming more widespread within US industry and military. A complete CBM system comprises a number of functional capabilities and the implementation of a CBM system requires the integration of a variety of hardware and software components. There exists a need for an Open System Architecture to facilitate the integration and interchangeability of these components from a variety of sources.[5]

Shown in Figure 22.5, OSA-CBM is a layered architecture approach to VHM in which each layer is viewed as a collection of similar tasks or functions. Following a hierarchical relationship, data flow through logical transitions from sensor outputs at the lowest layer to decision support at the highest layer through various intermediate layers. The intention of such an approach is to encourage standardization of product offerings for each layer.

A sample list of military programs currently using OSA-CBM or variants includes: On-Board Condition-Based Maintenance for the United States Navy; the Joint Strike Fighter (JSF F-35) for the United States Navy, Air Force and Marines; Future Combat Systems for the United States Army; and the Expeditionary Fighting Vehicle for the United States Marines.

22.4 Key Technologies

To this point, numerous process-related aspects of IVHM have been discussed, including design philosophy, health management polices, the role of the health management integrator, and various industry standards. Although each is essential to the success of an IVHM system, various key technologies enable

IVHM to accomplish its main objective of determining the health state of the vehicle. A brief overview of these technologies is provided in the sections that follow. For more in-depth information, consult the references and reading list at the end of this chapter.

22.4.1 Member System Concept

The definition of a subsystem requires the establishment of an arbitrary boundary in which components within the boundary are part of the subsystem, and those outside are part of other subsystems. Establishment of boundaries is important because each defines ownership of components for SHM purposes. Without clear boundaries, "orphan" components may result that are not monitored by any particular subsystem. This should be discouraged during the vehicle design phase, as these orphan components typically cause an unmaintainable and potentially unsafe condition.

A member system is defined as a subsystem that complies with the health management policies and processes established for the vehicle. Member systems are responsible for performing SHM upon the hardware and software components within the member system boundary and reporting faults and health status to the IVHM system, especially those that impact the intended function of the member system.

Member systems are categorized in terms of their level of compliance with the established health management policies and processes. This compliance falls into one of three categories:

- Fully compliant member system — those supporting all IVHM features and interfaces
- Partially compliant member system — those supporting a selected subset of IVHM features and interfaces
- Noncompliant (or nonmember) system — those that do not to interface to the IVHM system

While the latter category may seem counter-intuitive given the importance of IVHM, there business reasons (e.g., not cost effective) or technical reasons (e.g., increase in complexity) that dictate when an IVHM system interface is not added. In these cases, it is deemed acceptable to use the subsystem as is rather than incur the cost of updating and recertifying the system. It should be noted that existing or legacy subsystems chosen for use in a new aircraft design often fall into this category.

22.4.2 Diagnostics

A simple, straightforward definition of diagnostics is the detection of faults or anomalies in vehicle subsystems. The following sections address diagnostic methods to detect faults caused by hardware failures, software errors, and those occurring at the system level.

22.4.2.1 Hardware Diagnostic Methods

While not intended to be all-inclusive, many of the classical diagnostic methods for hardware have been listed in Table 22.2. In addition, detailed examples are provided for a selected subset of I/O methods, including output wraparound and input stimulus.

TABLE 22.2 Hardware Diagnostic Methods

Hardware Type	Diagnostic Method
Memory	ROM checksums, CRCs; RAM pattern tests (stuck address line, stuck data bit); parity, error detect and correct (EDAC)
CPU Core	Instruction sets; register/cache tests; PCI access; interrupts; watchdog timer
I/O Testing	Output wraparounds; input stimulus; A/D and D/A conversion; reference voltages;
Power Supply	Voltage monitors; current monitors; temperature monitors

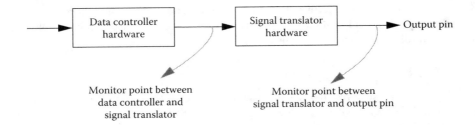

Example component technologies
Data controllers: ARINC-429 chip, discrete chip, A/D converter chip
Signal translators: Op amps, discrete drivers (ground/open or 28V/open)
Output pins: Card level pin, LRU pin

FIGURE 22.6   Output Wraparound Monitoring, General Approach.

22.4.2.1.1   *Output Wraparound Example*

The detection of output faults is an essential part of vehicle fault isolation. Output faults typically propagate downstream, appearing as input faults on other subsystems, which makes it very important for an LRU to detect the occurrence of an output failure, isolate the failed circuitry (if possible), and report the failure to the CMC. Figure 22.6 shows the general approach to output wraparound monitoring. The output signal can be monitored at two points: the first as it leaves the data controller, the second after it undergoes signal translation and is ready to leave the card or LRU. As suggested by Figure 22.6, placing the wraparound point closer to the external output pin maximizes the amount of circuitry tested by the monitor. Depending upon the specific circuit design, one monitor point may be more feasible to implement than the other. Regardless of the point chosen, designers must guard against adding more test circuitry than functional circuitry; that is, they must balance the amount of extra hardware added for testing versus the resulting reduction in circuit reliability due to the added components.

Figure 22.7 shows a simple example of wraparound circuitry for discrete outputs, and Figure 22.8 depicts a more complex example of wraparound circuitry used for digital bus outputs (in this case, ARINC-429 transceivers). The left-hand portion of the figure illustrates output transmitter wraparounds. The use of a multiplexed test receiver (Test MUX [Test Multiplexer] and Test Rx [Test Receiver]) allows

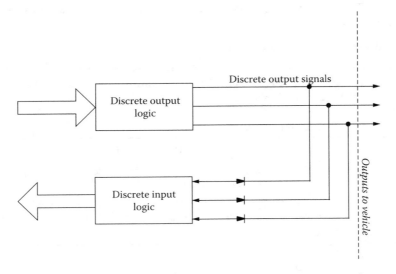

FIGURE 22.7   Simple Discrete Output Wraparound.

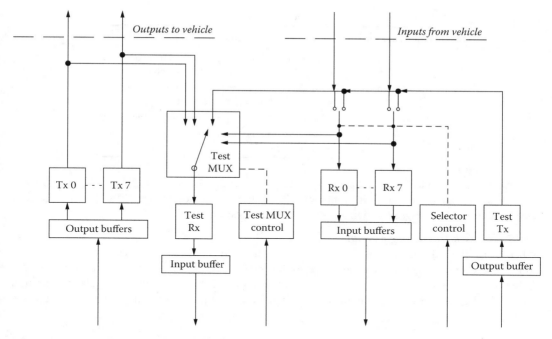

FIGURE 22.8 Complex Digital Bus Wraparound.

one receiver to listen to any selected transmitter, which permits the software to compare the input pattern at the test receiver with the intended output pattern of the transmitter. The remainder of Figure 22.8 illustrates the ability to stimulate the input receivers, as discussed in the next section.

22.4.2.1.2 *Input Stimulus Example*

Without the ability for an LRU to stimulate its own inputs, the only way to test them is to rely on the presence of external data from an upstream subsystem. If no data is present, the LRU cannot determine if its input circuitry is faulty or if the upstream subsystem is not sending data. By adding the ability to stimulate the inputs, the LRU can test them during the power-up sequence or following the detection of data loss from the external subsystem.

The design of most discrete input circuitry is simple enough that the addition of self-stimulating components violates the rule of thumb to "avoid adding more test hardware than functional hardware." Consequently, one rarely encounters self-stimulating discrete inputs. Digital bus inputs, on the other hand, are subject to various failure modes making it advisable to add self-stimulus circuitry. As illustrated in the right-hand portion of Figure 22.8, the addition of a test transmitter (Test Tx) can route test data to any selected input receiver. If the transmitted data does not match the received data, a hardware fault has occurred. In addition, the Test Tx is used to test the Test Rx prior to testing the other transmitters and receivers.

22.4.2.2 Software Diagnostic Methods

The topic of software diagnostic methods can be somewhat controversial, since classical design assurance processes (e.g., validation and verification) are intended to prevent software design errors from being released into the field. Regardless of the process employed, experience shows that escapes do occur, resulting in in-service problems due to software errors. The challenge here is that software errors typically result in transient or intermittent faults, which are very difficult to isolate to a root cause; therefore, it is important that software exception handlers are designed to capture the maximum amount of isolation and debug data as practical, allowing the design engineers to recreate the scenario and determine the

TABLE 22.3 Software Diagnostic Methods

Software Function	Diagnostic Method
Exception Handlers	Fatal exceptions; processor specific exceptions; access violation exceptions; floating point exceptions
Execution Monitors	Software heartbeat; software order; ticket punching; execution traceback and state capture
Threshold Monitors	Level sensing; rate of change; comparison/voting
Data Validation	Reasonableness checks; parity/checksums/CRCs; source selection/reversion

offending software code. Table 22.3 provides a sample listing of several diagnostic methods for software. What follows are detailed examples of methods this author has found to be very effective over the years, including ticket punching, execution traceback, and state capture.

22.4.2.2.1 Ticket Punching Example

When using the ticket punching method, software processes or threads "punch a ticket" when they complete execution during a given time slice, as illustrated in Figure 22.9. A ticket watcher process then gathers up all the required tickets and strobes the application heartbeat or enables the application's output process. In the case of a heartbeat implementation, missed ticket punches result in no heartbeat updates,

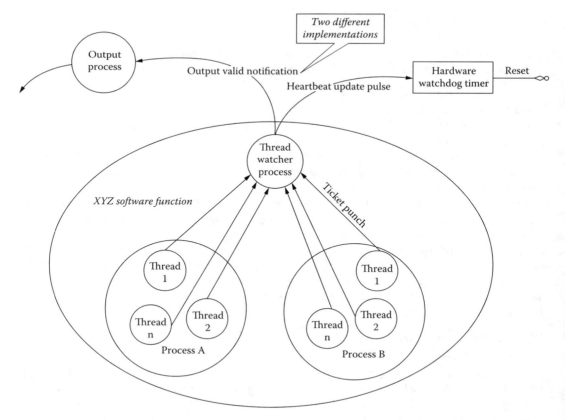

FIGURE 22.9 Software Ticket Punching.

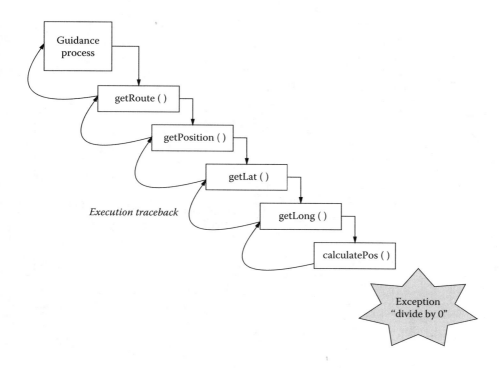

FIGURE 22.10 Software Execution Traceback.

typically causing a system reset. In an output process implementation, missed ticket punches result in the transmission of no data (the output goes silent), stale data (data is unchanged from the previous transmission), or data that has been flagged invalid, depending upon the desired output response. The ticket punching method protects against nonresponsive software by detecting it and responding accordingly.

22.4.2.2.2 *Execution Traceback and State Capture Examples*

The execution traceback method (Figure 22.10) is triggered by a software exception (e.g., a divide-by-zero violation), which causes the operating system to walk back through the execution stack, determining "who-called-who," and logging the resultant stack trace with the exception fault. Depending upon the operating system design, the stack trace may be captured in real-time using a sliding window technique, or it may be constructed on demand when the exception occurs. Not all commercial operating systems provide this useful feature; however, it can be incorporated into existing homegrown operating systems in use by many LRUs.

The state capture method (Figure 22.11) involves the collection of key state data on a periodic basis. When a software exception occurs, the latest data snapshot is logged with the exception fault. Examples of state data include vehicle-specific data (altitude, heading, position, etc.), software execution data (processor registers, system timers, stack pointers, etc.), or both.

When used together, the execution traceback and state capture methods are powerful allies in the struggle to debug in-service software errors.

22.4.2.3 System-Level Diagnostic Methods

System-level diagnostic methods are intended to detect faults occurring between two or more subsystems. Table 22.4 lists many of the methods currently in use, followed by examples explaining compatibility/configuration checking and redundancy management.

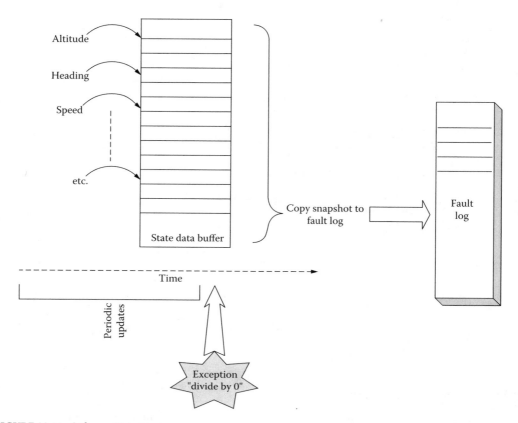

FIGURE 22.11   Software State Capture.

TABLE 22.4   System Level Diagnostic Methods

System Function	Diagnostic Method
Compatibility Checks	Hardware-to-hardware; hardware-to-software; software-to-software
Configuration Checks	Subsystem-to-vehicle; vehicle-to-mission
Redundancy Management	Comparison monitors (between two subsystems); voting monitors (between three or more subsystems)
Integrity Management	End-to-end protocol checks
Mode-Specific Tests	On-ground; safe-mode; power-up; background; power save

22.4.2.3.1   Compatibility/Configuration Checking Example

The goal of compatibility checking is to determine if two or more components play nicely together, that is, to determine if they function and interface together according to the established rules for those components. For example, a 10 mm metric socket is not compatible with a 1/2-inch bolt. Compatibility checking occurs at many levels, including hardware-to-hardware, hardware-to-software and software-to-software.

Configuration checking is a specialized form of compatibility checking in which the goal is to determine if the collection of systems is the correct set for the vehicle or if the vehicle is the correct choice for the mission. From the perspective of the certification authorities, such as the FAA, an aircraft is equipped with an approved set of systems and software (as documented in the Type Certificate). Although other versions of these systems and software may also be compatible with each other, they are not approved

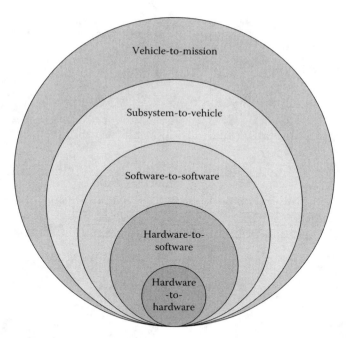

FIGURE 22.12 Compatibility/Configuration Checking.

as part of that aircraft's configuration. In other words, a successful compatibility check does not ensure a successful configuration check.

As summarized by Figure 22.12, the various levels of compatibility and configuration checks start at the innermost layer of the onion and build upon each other, working from the inside out.

22.4.2.3.2 *Redundancy Management Example*

Redundancy management is often used as an architectural method of increasing safety levels. Rather than depending upon a single instance of a subsystem to provide autopilot functionality, for example, modern aircraft often use multiple instances in which one instance is in control of the aircraft, while the others monitor its performance. In the event of a subsystem failure, the commanding instance is taken offline and control is transferred to another instance.

Redundancy management schemes can be implemented using simple comparison monitors between two instances of a subsystem, in which a miscompare event indicates a fault, although it cannot be determined which subsystem is in error. In theses cases, the subsystems may retry the operation, or take themselves offline and attempt recovery. More complex redundancy management schemes typically involve voting between three or more instances of a subsystem, in which case the majority rules.

Shown in Figures 22.13A, 22.13B, and 22.13C is a notional quad-redundant scheme (typically used in human-rated spacecraft control systems) in which three instances are active, with the fourth remaining as a hot spare (Figure 22.13A). Voting information is exchanged via a cross-channel data link bus. In the event of a voting failure, the minority system is taken off-line and the hot spare is made active (Figure 22.13B). If a second failure occurs, the system reverts to comparison monitoring between the remaining two instances (Figure 22.13C). This quad-redundant scheme is also referred to as "two-fault tolerant" in that it can experience two failures and still provide safe operation.

22.4.3 Prognostics

Unlike the field of diagnostics in which many of the methods discussed thus far are mature and in widespread use, the field of prognostics is still relatively young. In fact, some refer to prognostics as both "art" and "science." While there are differing definitions, this text has adopted the following definition

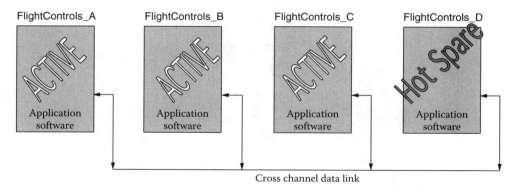

Voting occurs between active instances across the link.

# of failures	Instance A	Instance B	Instance C	Instance D
None	Voting	Voting	Voting	Hot spare

FIGURE 22.13A   Redundancy Management Example — No Failures (3-Way Voting).

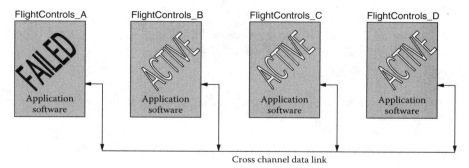

Voting occurs between active instances across the link.

# of failures	Instance A	Instance B	Instance C	Instance D
None	Voting	Voting	Voting	Hot spare
1	Shutdown	Voting	Voting	Voting

FIGURE 22.13B   Redundancy Management Example — One Failure (3-Way Voting).

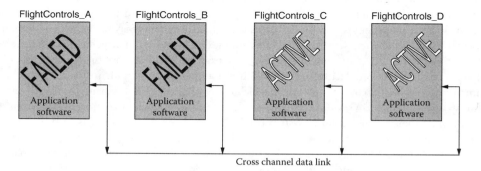

Voting occurs between active instances across the link.

# of failures	Instance A	Instance B	Instance C	Instance D
None	Voting	Voting	Voting	Hot spare
1	Shutdown	Voting	Voting	Voting
2	Shutdown	Shutdown	Comparison	Comparison

FIGURE 22.13C   Redundancy Management Example — Two Failures (2-Way Comparison).

from Engel: "The capability to provide early detection of the precursor and/or incipient fault condition (very 'small' fault) of a component, and to have the technology and means to manage and predict the progression of this fault condition to component failure" [6]. Common prognostic methods focus on the detection and diagnosis of events that mark the early stages of component failure, for example via parameter trending of vibrations, pressures, temperatures, etc. This knowledge can be used to accelerate vehicle maintenance, encouraging the correction of problems while they are still small rather than waiting for them to escalate into major failures.

There are those in the prognostic community that are also interested in solving the more difficult problem of estimating the remaining life of a component or system in an effort to extend the vehicle maintenance cycle. For example, while traditional maintenance methods may have required the replacement of component X every 500 hours, armed with a reliable estimate of remaining life the interval might be extended to 600 hours. It should be noted that beyond the technical challenges of estimating remaining life, there are certification challenges in proving to the authorities that it is safe to extend the maintenance interval.

Examples of prognostic methods include Probability Density Functions, Fuzzy Logic, Neural Networks, Genetic Algorithms, State-Based Feature Recognition and Dempster-Shafer Theory; all of which are beyond the scope of this text. For more information, the reader is encouraged to consult the references and reading list at the end of this chapter.

22.4.4 Intelligent Reasoners

The concept of "intelligent reasoning" refers to the collection of symptoms (e.g., vehicle status, diagnostic/prognostic results) from vehicle subsystems, the application of relationship-based rules or models to filter out cascaded effects, and the isolation of the incipient fault condition (i.e., the root cause) behind those symptoms. Before the advent of centralized maintenance computers (introduced previously as CMCs), the flight crew and/or aircraft mechanic manually performed these activities. Once CMCs came into existence, these activities were automated using intelligent reasoners to assist the aircraft mechanic. Functionally, these reasoners are categorized as either "rule-based" or "model-based."

22.4.4.1 Rule-Based Reasoners

Rule-based reasoners typically employed hard-coded logic in the form of "if then else" statements. System engineers serving as domain experts in each vehicle system would create logic rules to diagnose thousands of symptom scenarios. Software engineers would codify the logic rules, often resulting in highly complex logic. While the rule-based approach works for "small" systems, it doesn't scale up to handle the more complex systems and their interactions typically found on aircraft. In addition, updating rule-based systems has proven troublesome, as the impacts of changes are not always well understood. The seemingly simple change of a single rule could actually break many other rules making engineers very reluctant to update them.

22.4.4.2 Model-Based Reasoners

The advent of model-based reasoners attempted to address many of the limitations of the rule-based approach. While there are many types of models, most aircraft CMCs use fault propagation models, also referred to as "cause and effect" models. The basic premise is that internal LRU failures manifest themselves as output faults, the effects of which propagate to downstream systems and are either detected on their inputs, or pass through undetected and cause subsequent failures within the downstream system. By modeling the paths these faults can take and report the presence of interface faults to the CMC, it can filter out the cascaded faults and isolate the faulty LRU. In the event of an ambiguity that results in more than one possible LRU at fault, procedures are provided to the aircraft mechanic to assist in fault isolation.

Other model-based approaches include functional models and parametric models. Functional models attempt to model the actual operation of the vehicle and its subsystems, as opposed to modeling just their faults (as in the fault propagation approach). Parametric models use mathematical equations, created using live vehicle data collected during "nominal" operational scenarios, to predict the behavior of the vehicle in actual use. Both the functional and parametric model approaches consume input data in parallel with vehicle, process that data via the model and compares the model outputs against vehicle outputs. Resulting residual data between the model and vehicle outputs suggest the vehicle is performing off-nominal.

Similar to fault propagation models, the functional model approach provides the ability to determine the cause of the off-nominal performance, although the engineering resources required for creating and testing the model can be considerable approaching the amount of effort required creating and testing the actual vehicle. Further, providing enough computing resources to execute the model in real-time may not be feasible on-board the vehicle. Depending upon the complexity of the vehicle and its subsystems, the use of functional modeling may be cost-prohibitive.

Unlike fault propagation models and functional models, the parametric model approach does not specifically diagnose the cause of fault; rather it identifies how the vehicle performance has deviated from nominal. For example, engine thrust is 5% below nominal for current operating mode. Because the parametric model is developed based on nominal vehicle operation, it is typically developed after the vehicle has been in-service for some period of time and has stabilized in terms of operational reliability. Parametric modeling can also be viewed as a "black box" or "gray box" testing method in that knowledge of the relationship between inputs and outputs is needed, rather than intimate details of the inner-workings of the subsystems. For more in-depth information on parametric modeling, the reader is encouraged to consult the references and reading list at the end of this chapter.

22.4.4.3 Fielded Intelligent Reasoners

At the time of this writing, numerous intelligent reasoners have been fielded for use in the aviation and space industry. While not intended to be all-inclusive, listed in Table 22.5 is a sample of these products and their usage.

22.5 Examples of State-of-the-Art IVHM Systems

What follows is a brief overview of three IVHM systems, which reflect the state-of-the-art for aviation. Additional examples can also be found in the automotive and process control industries.

TABLE 22.5 Sample Listing of Fielded Intelligent Reasoners

Intelligent Reasoner	Type	Known Applications	Company Information
CMC	Fault propagation modeling	Boeing 777; Primus Epic (business jets, regional jets, helicopters)	Honeywell International (https://www.cas.honeywell.com/cmc)
TEAMS® Toolset	Multisignal dependency modeling (advanced form of fault propagation modeling)	Consult company Web site	Qualtech Systems Inc. (www.teamqsi.com)
eXpress™ Design Toolset	Dependency modeling (similar to fault propagation modeling)	Consult company Web site	DSI International (www.dsiintl.com)
Livingstone	Artificial intelligence based reasoner (mixture of functional and parametric modeling)	Deep Space One spacecraft; Earth Observing One (EO-1) satellite	NASA Ames Research Center (http://ic.arc.nasa.gov/projects/L2/doc)
BEAM And SHINE	Artificial intelligence based reasoner (mixture of functional and parametric modeling)	NASA Deep Space Missions (Voyager, Galileo, Magellan, Cassini and Extreme Ultraviolet Explorer)	NASA Jet Propulsion Laboratory (www.jpl.nasa.gov)

22.5.1   Honeywell Primus Epic® Aircraft Diagnostic Maintenance System

The Primus Epic avionics system represents the next generation in modular avionics systems. Receiving initial FAA certification and entering into service in 2003, Honeywell's Primus Epic system was designed to be scalable and extensible to aircraft and helicopters produced by many manufacturers for various markets, rather than targeting a specific aircraft. Today Primus Epic systems are flying on numerous business aircraft, regional aircraft and helicopters.

Integrated within the Primus Epic system, the Aircraft Diagnostic Maintenance System (ADMS) represents an evolution of several maintenance features used in previous systems. It is comprised of the CMC, Aircraft Condition Monitoring Function and the BIT functionality of the various member systems on the aircraft. Illustrated in Figure 22.14, the Primus Epic ADMS provides coverage of more than 200 aircraft subsystems, provided by both Honeywell and various third-party subsystem suppliers. ADMS serves as the maintenance access point to all subsystems via a point-and-click graphical user interface (GUI), allowing literally nose-to-tail coverage on most aircraft. ADMS performs root cause diagnostics to eliminate cascaded faults and provides correlation between flight deck effects and system faults. It is configurable via a separately loadable diagnostic database; provides fault information to the ground via aircraft data link; generates reports to the cockpit printer; and provides on-board data loading of aircraft subsystems.

22.5.2   Lockheed Martin Joint Strike Fighter

The avionics system of the new F-35 JSF is a mix of integrated and federated subsystems, featuring a triplex high-speed bus network that connects core processors and federated controllers, as shown in Figure 22.15. Core processors integrate applications that legacy aircraft performed separately such as vehicle and mission management. Core processors have no dedicated aircraft I/O; instead gathering such data off the bus network as supplied by remote I/O units. Federated controllers are used for specific applications that require dedicated, high-bandwidth control loops, such as flight controls.

In a system as complex as the JSF, diagnostic and prognostic requirements were contractually flowed down to all subsystem suppliers, with Lockheed Martin acting as the maintenance integrator. The JSF maintenance system is coupled with the United States Air Force integrated logistics system, enabling CBM activities as discussed earlier in this chapter.

22.5.3   Honeywell Health and Usage Monitoring for Helicopters

Health and Usage Monitoring Systems (HUMS) for helicopters enables the transition from traditional maintenance philosophy, which is largely based on fleet statistics and scheduled maintenance intervals, to a more efficient CBM philosophy that focuses on the usage of individual vehicles and their components. The Honeywell system monitors specific helicopter components and subsystems to form the basis for maintenance actions. Shown in Figure 22.16 is an example of the numerous sensors installed in the Bell 407 helicopter in support of HUMS. These sensors and the HUMS processing algorithms enable various types of monitoring including engine condition and performance; continuous vibration; engine exceedance; and rotor track and balance. In addition, HUMS provides ground-based tools that perform fleet data analysis. This combination of on-board and ground-based functionality allows more efficient, proactive use of maintenance assets and leads to higher fleet-readiness levels.

22.6   Future Trends in IVHM

The following is a brief overview of three IVHM systems currently under development, which reflect a glimpse into future IVHM technologies and applications. In the field of commercial aviation, new technologies continue to be investigated driven by the need to bring value to the customer, rather than just pure technology as in the past. These innovations find their roots in present day technologies, such

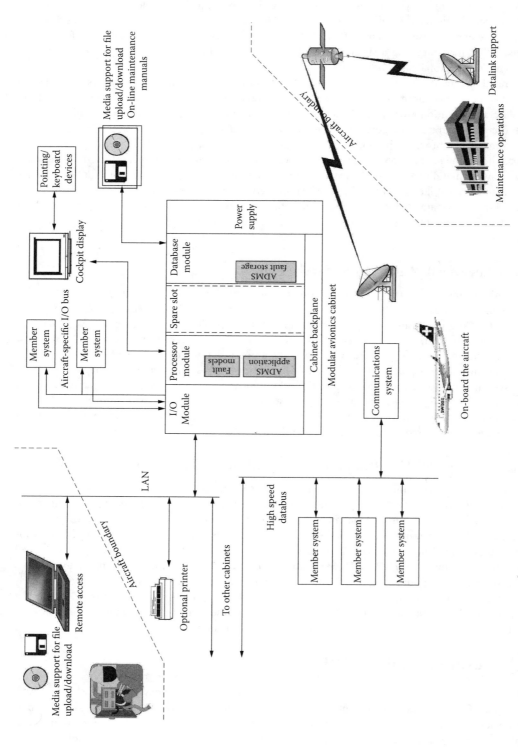

FIGURE 22.14 Honeywell Primus Epic Aircraft Diagnostic Maintenance System.

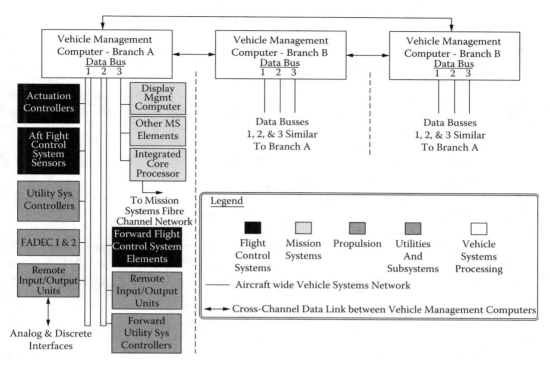

FIGURE 22.15 Lockheed Martin JSF.

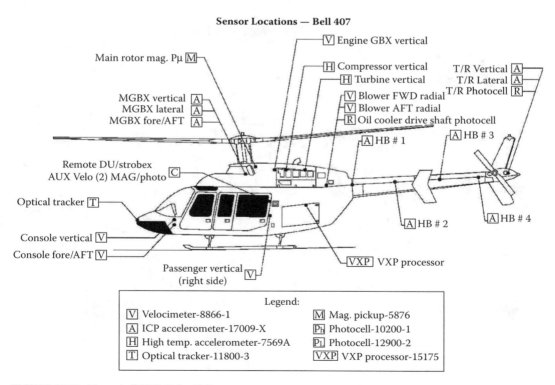

FIGURE 22.16 Honeywell HUMS for Helicopters.

as wireless communications and Internet accessibility, which are combined in new ways to provide value and service to customers [4].

If the field of space exploration, NASA's vision of the next-generation spacecraft depends upon using IVHM technologies in support of spacecraft automation, reconfigurability, mission planning and mission execution in addition to spacecraft maintenance. The innovations needed must focus upon the application of IVHM to safety-critical and mission-critical tasks, which pose substantially different challenges than a traditional maintenance-only role. It is in this area where more research and development will be needed to adapt commercial IVHM technologies to the applications of human spaceflight.

22.6.1   Boeing 787 Crew Information System/Maintenance System

The Boeing 787 Crew Information System (CIS) [4] will provide a networking infrastructure, as illustrated in Figure 22.17, enabling airborne applications to interact seamlessly with ground applications. Hosted within the CIS will be various applications, including the Maintenance System, Electronic Flight Bag, Software Distribution System, Crew Information Services, Flight Deck Printer and Wireless LAN support.

The lineage of the Boeing 787 Maintenance System portion can be traced to the approach used in the Boeing 777, which pioneered many of the maintenance concepts standardized by ARINC-624 (discussed earlier in this chapter). The maintenance system for 787 improves on many of these concepts by adopting the flexibility and power of today's wireless communication and Web-based technologies. Maintenance access will be provided via Internet browser technology hosted on laptop personal computers, which are members of the secure wireless network surrounding each aircraft. Using this network, maintenance access and the ability to upload software updates from secure Internet distribution servers will be provided as well as the ability to download maintenance data into the aircraft operator's maintenance and logistics network.

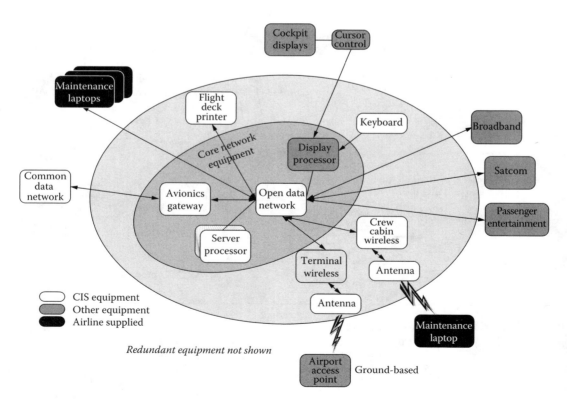

FIGURE 22.17   Boeing 787 CIS/Maintenance System [4].

22.6.2 Honeywell Sense and Respond Program

Current generation IVHM solutions are constrained by the fact that high fidelity access to many of the IVHM functions requires one to be in close proximity of the vehicle. Yet the majority of the expertise and resources best equipped to analyze, decide and act upon the outputs are widely distributed in time and space from the aircraft. Consider Formula 1 racing operation in which each race car is outfitted with hundreds of sensors wirelessly streaming data back to an operational and analytical hub located in the team van, parked in the infield. This data is analyzed by computers and vehicle experts who in turn forward instructions directly to the pit crew, the race strategists and the driver enabling real-time adjustments to improve performance and capability. Bulk data is also collected and forwarded immediately to automotive engineers back at the home office where the planning begins for vehicle design modifications and upgrades before the next race.

The innovation highlighted by this example is not a technological innovation; rather it is actually an operational and process innovation. It is through operational innovations such as this that commercial aviation will realize the next wave of improvements enabled by IVHM and information technology. Sensing and responding to anomalies in real time, coupled with the ability to engage a global community of resources and expertise to resolve issues will enable industry to move from reactive to proactive to predictive strategies.

Honeywell is currently developing their sense and respond solution, illustrated in Figure 22.18 designed to close the loop between vehicle, supplier and operator. In this system on-board diagnostic data will be collected from various aircraft subsystems (including a CMC, if so equipped) and automatically transmitted to a central data repository. From there the diagnostic data will be parsed, decoded, normalized and stored in a database that supports industry-standard queries using a Web-based service oriented architecture. Users will include both internal Honeywell organizations (engineering, reliability, product improvement and warranty services), as well as external customers (original equipment manufacturers (OEMs), aircraft owners and operators) in support of their manufacturing and fleet operations [4].

22.6.3 NASA Integrated Intelligent Vehicle Management

Integrated Intelligent Vehicle Management (IIVM) will provide the framework for manageable vehicle operations and quick response to system failures and space environmental events. IIVM research is currently being led by NASA Marshall Space Flight Center and coordinated with all other NASA centers as well as various industry and university participants. Target vehicles within the scope of the research

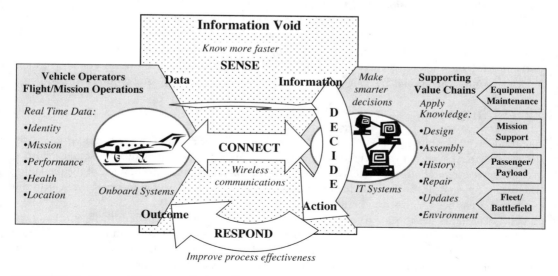

FIGURE 22.18 Honeywell "Sense and Respond" Program [4].

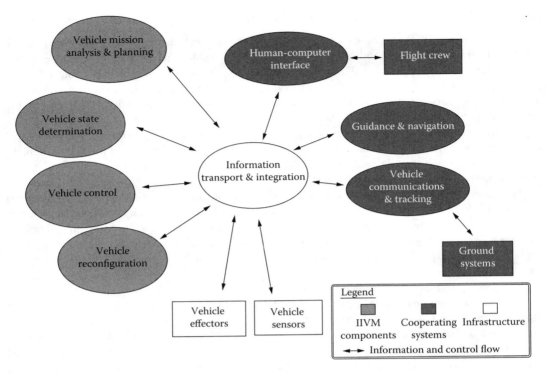

FIGURE 22.19 NASA IIVM.

include launch vehicles, interplanetary spacecraft and planetary landing craft. Both crewed and un-crewed vehicles are of interest although those carrying humans pose significant safety challenges as noted in the introduction to this section.

As a broad overview, IIVM will determine the current vehicle state and predicted future state(s), analyze vehicle mission objectives and plans, and then coordinate with the vehicle mission computer to execute vehicle commands to accomplish the mission. In the event of unexpected incidents or anomalies, IIVM will command reconfiguration of vehicle subsystems to ensure the ability to meet mission objectives. IIVM may also modify mission procedures, making optimum use of the available resources.

Illustrated in Figure 22.19 is a highly simplified view of IIVM and its relationship to other systems. Components of IIVM include Vehicle State Determination, Vehicle Mission Analysis and Planning, Vehicle Control and Systems Reconfiguration, all supported by an underlying infrastructure of Information Transport and Integration. Other systems that work in cooperation with IIVM include Guidance and Navigation, Vehicle Communications and Tracking, and interfaces to both the flight crew and ground systems.

22.7 Summary

The fundamental purpose of VHM is the determination of the health state of the target vehicle. Once determined, that knowledge can be used to ensure mission success by mitigating the cause of the failure(s) via system reconfiguration, or if necessary, modifying the mission to make the best use of the remaining healthy systems and still accomplish the mission objectives. VHM is later used to repair the vehicle and verify that it has been returned to nominal operational status.

VHM determines health state based on the use of diagnostics, prognostics and intelligent reasoners, which work in cooperation to detect faults, to determine the root cause and recommend the necessary mitigation actions. Following a layered approach, SHM performs health assessment at the subsystem level, providing that information to VHM who performs health assessment at the vehicle level. Depending upon the level of business enterprise integration, IVHM may be used to integrate vehicle health state into the design, production, planning, logistics and training processes of the enterprise.

IVHM is used in various applications, as shown by the examples in this chapter, spanning the range from noncritical maintenance-only usage to safety-critical/mission-critical decision-making. The IVHM design processes and architecture used must be appropriately matched to the level of criticality intended, to ensure a safe system.

Maximizing the effectiveness of IVHM requires that it be designed into the vehicle, rather than added on at some later date. Treating IVHM as a system engineering discipline is essential to designing such a system. Keeping the proper focus on IVHM is achieved by health management integration, whose ultimate goal is to ensure an optimal IVHM balance across the system, resulting in improved system safety, improved reliability and reduced life cycle costs.

Defining Terms

ACMF: Airplane Condition Monitoring Function
A/D: Analog to Digital (conversion)
ADMS: Aircraft Diagnostic and Maintenance System
AIAA: American Institute of Aeronautics and Astronautics
AIMS: Airplane Information Management System
ARINC: Aeronautical Radio Inc.
BEAM: Beacon-Based Exception Analysis for Multimissions
BIT[E]: Built-In Test [Equipment]
CAS: Crew Alerting System
CBM: Condition-Based Maintenance
CMC: Central Maintenance Computer
CPU: Central Processing Unit
CRC: Cyclic Redundancy Checking
D/A: Digital to Analog (conversion)
EDAC: Error Detect and Correct
FAA: Federal Aviation Administration
FDIR: Fault Detection, Isolation, and Recovery
GUI: Graphical User Interface
HUMS: Health and Usage Monitoring System
IEEE: Institute of Electrical and Electronics Engineers
IIVM: Integrated Intelligent Vehicle Management
I/O: Input/Output
IVHM: Integrated Vehicle Health Management
LAN: Local Area Network
LRM: Line Replaceable Module
LRU: Line Replaceable Unit
MUX: Multiplexer
NASA: National Aeronautics and Space Administration
OMS: Onboard Maintenance System
OSA-CBM: Open Systems Architecture for Condition-Based Maintenance
PCI: Peripheral Component Interconnect
RAM: Random Access Memory
ROM: Read Only Memory
Rx: Receiver
SHINE: Spacecraft Health Inference Engine
SHM: Subsystem Health Management
Tx: Transmitter

VHM: Vehicle Health Management

References

[1] Aaseng, G.B. (Honeywell International), Blueprint for an Integrated Vehicle Health Management System, IEEE 20th Digital Avionics Systems Conf., October 2001.

[2] Scandura, P.A. Jr. (Honeywell International), Integrated Vehicle Health Management as a System Engineering Discipline, IEEE 24th Digital Avionics Systems Conf., October 2005.

[3] Scandura, P.A. Jr. and Garcia-Galan, C. (Honeywell International), A Unified System to Provide Crew Alerting, Electronic Checklists and Maintenance Using IVHM, IEEE 23rd Digital Avionics Systems Conf., October 2004.

[4] Bird, G., Christensen, M., Lutz, D., and Scandura, P.A. Jr. (Honeywell International), Use of Integrated Vehicle Health Management in the Field of Commercial Aviation, NASA First International Forum on Integrated System Health Engineering and Management in Aerospace, November 2005.

[5] Open System Architecture for Condition Based Maintenance, www.osacbm.org.

[6] Engel, S.J., Gilmartin, B.J., Bongort, K., and Hess, A., Prognostics, The Real Issues Involved with Predicting Life Remaining," *Proc. IEEE Aerospace Conf.*, 6, March 2000, pp. 457–69.

[7] U.S. Department of Transportation, *Transportation Statistics Annual Report*, September 2004. [On-Line]. Available: www.bts.gov/publications/transportation_statistics_annual_report/2004

Further Information

In addition to the specific references listed in the previous section, the author recommends the Web sites and papers listed below for more information on various VHM-related topics.

Aerospace Standards

ARINC Avionics Maintenance Conference (www.arinc.com/amc)

Military OSA-CBM

Discenzo, F.; Nickerson, W., Mitchell, C.E., and Keller, K.J., Open Systems Architecture Enables Health Management for Next Generation System Monitoring and Maintenance, OSA-CBM Development Group.

Machinery Information Management Open Systems Alliance (MIMOSA) (www.mimosa.org)

OSA-CBM Online Training Manual (www.osacbm.org/Documents/Training/TrainingMaterial/TrainingWebsite)

Diagnostics and Prognostics

Aaseng, G.B.; Patterson-Hine, A., and Garcia-Galan, C., A Review of System Health State Determination Methods, presented at AIAA 1st Space Exploration Conference, January 31, 2005.

"Fault Diagnostics/Prognostics for Machine Health Maintenance," Four-day short course offered by Georgia Tech (www.pe.gatech.edu)

Park, H., Mackey, R., James, M., and Zak, M., BEAM: Technology for Autonomous Self-Analysis, presented at IEEE Aerospace Conference, Big Sky, MT, 2001.

"Prognostics and Health Management and Condition Based Maintenance," Two-day design course offered by Impact Technologies (www.impact-tek.com)

Parametric Modeling

Ganguli, S., Deo, S., and Gorinevsky, D., Parametric Fault Modeling and Diagnostics of a Turbofan Engine, IEEE CCA, September 2004, [On-Line]. Available: www.stanford.edu/~gorin/papers/CCA04gls.pdf.

NASA IIVM

Paris, D.E.; Watson, M.D., and Trevino, L.D., A Framework for Integration of IVHM Technologies for Intelligent Integration for Vehicle Management, presented at IEEE Aerospace Conference, Big Sky, MT, March 2005.

Paris, D.E.; Watson, M.D., and Trevino, L.D., An Intelligent Integration Framework for In-Space Propulsion Technologies for Integrated Vehicle Health Management, 41st AIAA/ASME/SAE/ASEE Joint Propulsion Conf. and Exhibit, Tucson, AZ, July 2005.

23

Boeing B-777: Fly-By-Wire Flight Controls

Gregg F. Bartley
Federal Aviation Administration

23.1 Introduction

Fly-By-Wire (FBW) primary flight controls have been been used in military applications such as fighter airplanes for a number of years. It has been a rather recent development to employ them in a commercial transport application. The 777 is the first commercial transport manufactured by Boeing which employs a FBW primary flight control system. This chapter will examine a FBW primary flight control system using the system on the 777 as an example. It must be kept in mind while reading this chapter that this is only a single example of what is currently in service in the airline industry. There are several other airplanes in commercial service made by other manufacturers that employ a different architecture for their FBW flight control system than described here.

A FBW flight control system has several advantages over a mechanical system. These include:

- Overall reduction in airframe weight.
- Integration of several federated systems into a single system.
- Superior airplane handling characteristics.
- Ease of maintenance.

- Ease of manufacture.
- Greater flexibility for including new functionality or changes after initial design and production.

23.2 System Overview

Conventional primary flight controls systems employ hydraulic actuators and control valves controlled by cables that are driven by the pilot controls. These cables run the length of the airframe from the cockpit area to the surfaces to be controlled. This type of system, while providing full airplane control over the entire flight regime, does have some distinct drawbacks. The cable-controlled system comes with a weight penalty due to the long cable runs, pulleys, brackets, and supports needed. The system requires periodic maintenance, such as lubrication and adjustments due to cable stretch over time. In addition, systems such as the yaw damper that provide enhanced control of the flight control surfaces require dedicated actuation, wiring, and electronic controllers. This adds to the overall system weight and increases the number of components in the system.

In a FBW flight control system, the cable control of the primary flight control surfaces has been removed. Rather, the actuators are controlled electrically. At the heart of the FBW system are electronic computers. These computers convert electrical signals sent from position transducers attached to the pilot controls into commands that are transmitted to the actuators. Because of these changes to the system, the following design features have been made possible:

- Full-time surface control utilizing advanced control laws. The aerodynamic surfaces of the 777 have been sized to afford the required airplane response during critical flight conditions. The reaction time of the control laws is much faster than that of an alert pilot. Therefore, the size of the flight control surfaces could be made smaller than those required for a conventionally controlled airplane. This results in an overall reduction in the weight of the system.
- Retention of the desirable flight control characteristics of a conventionally controlled system and the removal of the undesirable characteristics. This aspect is discussed further in the section on control laws and system functionality.
- Integration of functions such as the yaw damper into the basic surface control. This allows the separate components normally used for these functions to be removed.
- Improved system reliability and maintainability.

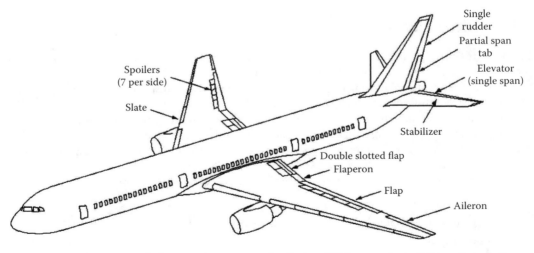

FIGURE 23.1   The primary flight control system on the Boeing 777 is comprised of the outboard ailerons, flaperons, elevator, rudder, horizontal stabilizer, and the spoiler/speedbrakes.

23.3 Design Philosophy

The philosophy employed during the design of the 777 primary flight control system maintains a system operation that is consistent with a pilot's past training and experience. What is meant by this is that however different the actual system architecture is from previous Boeing airplanes, the presentation to the pilot is that of a conventionally controlled mechanical system. The 777 retains the conventional control column, wheel, and rudder pedals, whose operations are identical to the controls employed on other Boeing transport aircrafts. The flight deck controls of the 777 are very similar to those of the Boeing 747-400, which employs a traditional mechanically controlled primary flight control system.

Because the system is controlled electronically, there is an opportunity to include system control augmentation and envelope protection features that would have been difficult to provide in a conventional mechanical system. The 777 primary flight control system has made full use of the capabilities of this architecture by including such features as:

- Bank angle protection
- Turn compensation
- Stall and overspeed protection
- Pitch control and stability augmentation
- Thrust asymmetry compensation

More will be said of these specific features later. What should be noted, however, is that none of these features limit the action of the pilot. The 777 design utilizes *envelope protection* in all of its functionality rather than *envelope limiting*. Envelope *protection* deters pilot inputs from exceeding certain predefined limits but does not prohibit it. Envelope *limiting* prevents the pilot from commanding the airplane beyond set limits. For example, the 777 bank angle protection feature will significantly increase the wheel force a pilot encounters when attempting to roll the airplane past a predefined bank angle. This acts as a prompt to the pilot that the airplane is approaching the bank angle limit. However, if deemed necessary, the pilot may override this protection by exerting a greater force on the wheel than is being exerted by the backdrive actuator. The intent is to inform the pilot that the command being given would put the airplane outside of its normal operating envelope, but the ability to do so is not precluded. This concept is central to the design philosophy of the 777 primary flight control system.

23.4 System Architecture

23.4.1 Flight Deck Controls

As noted previously, the 777 flight deck uses standard flight deck controls; a control column, wheel, and rudder pedals that are mechanically linked between the Captain's and First Officer's controls. This precludes any conflicting input between the Captain and First Officer into the primary flight control system. Instead of the pilot controls driving quadrants and cables, as in a conventional system, they are attached to electrical transducers that convert mechanical displacement into electrical signals.

A gradient control actuator is attached to the two control column feel units. These units provide the tactile feel of the control column by proportionally increasing the amount of force the pilot experiences during a maneuver with an increase in airspeed. This is consistent with a pilot's experience in conventional commercial jet transports.

Additionally, the flight deck controls are fitted with what are referred to as "backdrive actuators." As the name implies, these actuators backdrive the flight deck controls during autopilot operation. This feature is also consistent with what a pilot is used to in conventionally controlled aircraft and allows the pilot to monitor the operation of the autopilot via immediate visual feedback of the pilot controls that is easily recognizable.

23.4.2   System Electronics

There are two types of electronic computers used in the 777 primary flight control system: the Actuator Control Electronics (ACE), which is primarily an analog device, and the Primary Flight Computer (PFC), which utilizes digital technology. There are four ACEs and three PFCs employed in the system. The function of the ACE is to interface with the pilot control transducers and to control the primary flight control system actuation with analog servo loops. The role of the PFC is the calculation of control laws by converting the pilot control position into actuation commands, which are then transmitted to the ACE. The PFC also contains ancillary functions, such as system monitoring, crew annunciation, and all the primary flight control system onboard maintenance capabilities.

Four identical ACEs are used in the system, referred to as L1, L2, C, and R. These designations correspond roughly to the left, center, and right hydraulic systems on the airplane. The flight control functions are distributed among the four ACEs. The ACEs decode the signals received from the transducers used on the flight deck controls and the primary surface actuation. The ACEs convert the transducer position into a digital value and then transmit that value over the ARINC 629 data busses for use by the PFCs. There are three PFCs in the system, referred to as L, C, and R. The PFCs use these pilot control and surface positions to calculate the required surface commands. At this time, the command of the automatic functions, such as the yaw damper rudder commands, are summed with the flight deck control commands, and are then transmitted back to the ACEs via the same ARINC 629 data busses. The ACEs then convert these commands into analog commands for each individual actuator.

23.4.3   ARINC 629 Data Bus

The ACEs and PFCs communicate with each other, as well as with all other systems on the airplane, via triplex, bi-directional ARINC 629 flight controls data busses, referred to as L, C, and R. The connection from these electronic units to each of the data busses is via a stub cable and an ARINC 629 coupler. Each coupler may be removed and replaced without disturbing the integrity of the data bus itself.

23.4.4   Interface to Other Airplane Systems

The primary flight control system transmits and receives data from other airplane systems by two different pathways. The Air Data and Inertial Reference Unit (ADIRU), Standby Attitude and Air Data Reference Unit (SAARU), and the Autopilot Flight Director Computers (AFDC) transmit and receive data on the ARINC 629 flight controls data busses, which is a direct interface to the primary flight computers. Other systems, such as the Flap Slat Electronics Unit (FSEU), Proximity Switch Electronics Unit (PSEU), and Engine Data Interface Unit (EDIU) transmit and receive their data on the ARINC 629 systems data busses. The PFCs receive data from these systems through the Airplane Information Management System (AIMS) Data Conversion Gateway (DCG) function. The DCG supplies data from the systems data busses onto the flight controls data busses. This gateway between the two main sets of ARINC 629 busses maintains separation between the critical flight controls busses and the essential systems busses but still allows data to be passed back and forth.

23.4.5   Electrical Power

There are three individual power systems dedicated to the primary flight control system, which are collectively referred to as the Flight Controls Direct Current (FCDC) power system. An FCDC Power Supply Assembly (PSA) powers each of the three power systems. Two dedicated Permanent Magnet Generators (PMG) on each engine generate AC power for the FCDC power system. Each PSA converts the PMG alternating current into 28 V DC for use by the electronic modules in the Primary Flight Control System. Alternative power sources for the PSAs include the airplane Ram Air Turbine (RAT), the 28-V DC main airplane busses, the airplane hot battery buss, and dedicated 5 Ah FCDC batteries. During flight, the PSAs draw power from the PMGs. For on-ground engines-off operation or for in-flight failures of the PMGs, the PSAs draw power from any available source.

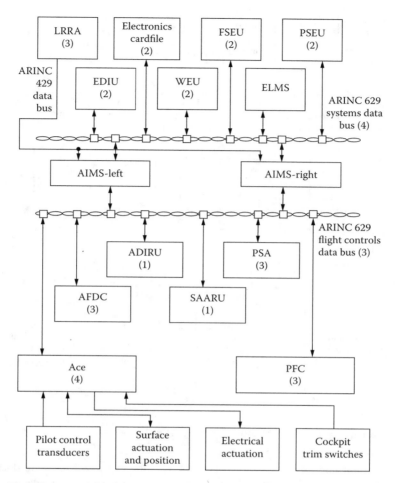

FIGURE 23.2 Block diagram of the electronic components of the 777 Primary Flight Control System, as well as the interfaces to other airplane systems.

23.5 Control Surface Actuation

23.5.1 Fly-by-Wire Actuation

The control surfaces on the wing and tail of the 777 are controlled by hydraulically powered, electrically signaled actuators. The elevators, ailerons, and flaperons are controlled by two actuators per surface, the rudder is controlled by three. Each spoiler panel is powered by a single actuator. The horizontal stabilizer is positioned by two parallel hydraulic motors driving the stabilizer jack-screw.

The actuation powering the elevators, ailerons, flaperons, and rudder have several operational modes. These modes, and the surfaces that each are applicable to, are defined below.

Active—Normally, all the actuators on the elevators, ailerons, flaperons, and rudder receive commands from their respective ACEs and position the surfaces accordingly. The actuators will remain in the active mode until commanded into another mode by the ACEs.

Bypassed—In this mode, the actuator does not respond to commands from its ACE. The actuator is allowed to move freely, so that the redundant actuator(s) on a given surface may position the

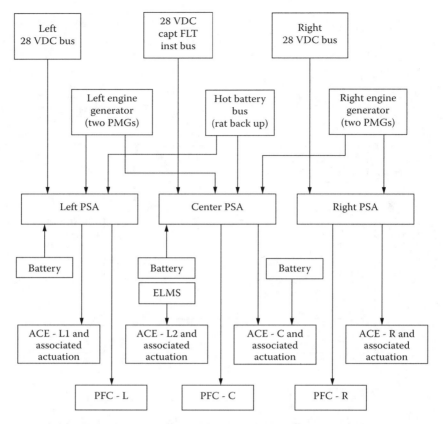

FIGURE 23.3   Block diagram of the 777 Fly-By-Wire Power Distribution System.

surface without any loss of authority, i.e., the actuator in the active mode does not have to overpower the bypassed actuator. This mode is present on the aileron, flaperon, and rudder actuators.

Damped—In this mode, the actuator does not respond to the commands from the ACE. The actuator is allowed to move, but at a restricted rate which provides flutter damping for that surface. This mode allows the other actuator(s) on the surface to continue to operate the surface at a rate sufficient for airplane control. This mode is present on elevator and rudder actuators.

Blocked—In this mode, the actuator does not respond to commands from the ACE, and it is not allowed to move. When both actuators on a surface (which is controlled by two actuators) have failed, they both enter the "Blocked" mode. This provides a hydraulic *lock* on the surface. This mode is present on the elevator and aileron actuators.

An example using the elevator surface illustrates how these modes are used. If the inboard actuator on an elevator surface fails, the ACE controlling that actuator will place the actuator in the "Damped" mode. This allows the surface to move at a limited rate under the control of the remaining operative outboard actuator. Concurrent with this action, the ACE also arms the "Blocking" mode on the outboard actuator on the same surface. If a subsequent failure occurs that will cause the outboard actuator to be placed in the "Damped" mode by its ACE, both actuators will then be in the "Damped" mode and have their "Blocking" modes armed. An elevator actuator in this configuration enters the "Blocking" mode, which hydraulically locks the surface in place for flutter protection.

23.5.2 Mechanical Control

Spoiler panel 4 and 11 and the alternate stabilizer pitch trim system are controlled mechanically rather than electrically. Spoilers 4 and 11 are driven directly from control wheel deflections via a control cable. The alternate horizontal stabilizer control is accomplished by using the pitch trim levers on the flight deck aisle stand. Electrical switches actuated by the alternate trim levers allow the PFCs to determine when alternate trim is being commanded so that appropriate commands can be given to the pitch control laws.

Spoiler panels 4 and 11 are also used as speedbrakes, both in the air and on the ground. The speedbrake function for this spoiler pair only has two positions: stowed and fully extended. The speedbrake commands for spoilers 4 and 11 are electrical in nature, with an ACE giving an *extend* or *retract* command to a solenoid-operated valve in each of the actuators. Once that spoiler pair has been deployed by a speedbrake command, there is no control wheel speedbrake command mixing, as there is on all the other fly-by-wire spoiler surfaces.

23.6 Fault Tolerance

"Fault tolerance" is a term that is used to define the ability of any system to withstand single or multiple failures which results in either no loss of functionality or a known loss of functionality or reduced level of redundancy while maintaining the required level of safety. It does not, however, define any particular method that is used for this purpose. There are two major classes of faults that any system design must deal with. These are:

- A failure which results in some particular component becoming totally inoperative. An example of this would be a loss of power to some electronic component, such that it no longer performs its intended function.

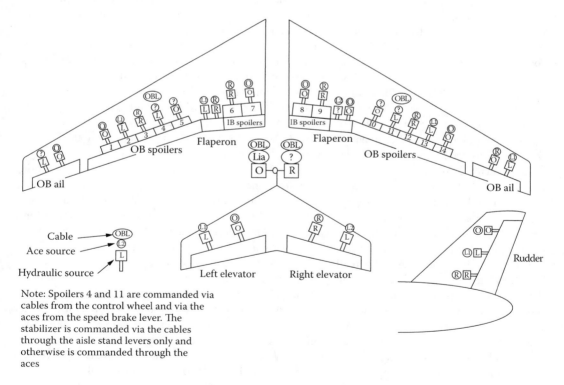

Note: Spoilers 4 and 11 are commanded via cables from the control wheel and via the aces from the speed brake lever. The stabilizer is commanded via the cables through the aisle stand levers only and otherwise is commanded through the aces

FIGURE 23.4 Schematic representation of the Boeing 777 Primary Flight Control System hydraulic power and electronic control functional distribution.

- A failure which results in some particular component remaining active, but the functionality it provides is in error. An example of this failure would be a Low Range Radio Altimeter whose output is indicating the airplane is at an altitude 500 ft above the ground when the airplane is actually 200 ft above the ground.

One method that is used to address the first class of faults is the use of redundant elements. For example, there are three PFCs in the 777 primary flight control system, each with three identical computing "lanes" within each PFC. This results in nine identical computing channels. Any of the three PFCs themselves can fail totally due to loss of power or some other failure which affects all three computing lanes, but the primary flight control system loses no functionality. All four ACEs will continue to receive all their surface position commands from the remaining PFCs. All that is affected is the level of available redundancy. Likewise, any single computing lane within a PFC can fail, and that PFC itself will continue to operate with no loss of functionality. The only thing that is affected is the amount of redundancy of the system. The 777 is certified to be dispatched on a revenue flight, per the Minimum Equipment List (MEL), with two computing lanes out of the nine total (as long as they are not within the same PFC channel) for 10 days and for a single day with one total PFC channel inoperative.

Likewise, there is fault tolerance in the ACE architecture. The flight control functions are distributed among the four ACEs such that a total failure of a single ACE will leave the major functionality of the system intact. A single actuator on several of the primary control surfaces may become inoperative due to this failure, and a certain number of spoiler symmetrical panel pairs will be lost. However, the pilot flying the airplane will notice little or no difference in handling characteristics with this failure. A total ACE failure of this nature will have much the same impact to the Primary Flight Control System as that of a hydraulic system failure.

The second class of faults is one that results in erroneous operation of a specific component of the system. The normal design practice to account for failures of this type is to have multiple elements doing the same task and their outputs voted or compared in some manner. This is sometimes referred to as a "voting plane." All critical interfaces into the 777 FBW Primary Flight Control System use multiple inputs which are compared by a voting plane. For interfaces that are required to remain operable after a first failure, at least three inputs must be used. For example, there are three individual Low Range Radio Altimeter (LRRA) inputs used by the PFCs. The PFCs compare all three inputs and calculates a mid-value select on the three values; i.e., the middle value LRRA input is used in all calculations which require radio altitude. In this manner, any single failure of an LRRA that results in an erroneous value will be discarded. If a subsequent failure occurs which causes the remaining two LRRA signals to disagree by a preset amount, the PFCs will throw out both values and take appropriate action in those functions which use these data.

Additionally, a voting plane scheme is used by the PFCs on themselves. Normally, a single computing lane within a PFC channel is declared as the "master" lane, and that lane is responsible for transmitting all data onto the data busses for use by the ACEs and other airplane systems. However, all three lanes are simultaneously computing the same control laws. The outputs of all three computing lanes within a single PFC channel are compared against each other. Any failure of a lane that will cause an erroneous output from that lane will cause that lane to be condemned as "failed" by the other two lanes.

Likewise, the outputs from all three PFC channels themselves are compared. Each PFC looks at its own calculated command output for any particular actuator, and compares it with the same command that was calculated by the other two PFC channels. Each PFC channel then does a mid-value select on the three commands, and that value (whether it was the one calculated by itself or by one of the other PFC channels) is then output to the ACEs for the individual actuator commands. In this manner, it is assured that each ACE receives identical commands from each of the PFC channels.

By employing methods such as those described above, it is assured that the 777 primary flight control system is able to withstand single or multiple failures and be able to contain those failures in such a manner that the system remains safe and does not take inappropriate action due to those failures.

23.7 System Operating Modes

The 777 FBW primary flight control system has three operating modes: normal, secondary, and direct. These modes are defined below:

Normal—In the "Normal" mode, the PFCs supply actuator position commands to the ACEs, which convert them into an analog servo command. Full functionality is provided, including all enhanced performance, envelope protection, and ride quality features.

Secondary—In the "Secondary" mode, the PFCs supply actuator position commands to the ACEs, just as in the "Normal" mode. However, functionality of the system is reduced. For example, the envelope protection functions are not active in the "Secondary" mode. The PFCs enter this mode automatically from the "Normal" mode when there are sufficient failures in the system or interfacing systems such that the "Normal" mode is no longer supported. An example of a set of failures that will automatically drop the system into the "Secondary" mode is total loss of airplane air data from the ADIRU and SAARU. The airplane is quite capable of being flown for a long period of time in the "Secondary" mode. It cannot, however, be dispatched in this condition.

Direct—In the "Direct" mode, the ACEs do not process commands from the PFCs. Instead, each ACE decodes pilot commands directly from the pilot controller transducers and uses them for the closed loop servo control of the actuators. This mode will automatically be entered due to total failure of all three PFCs, failures internal to the ACEs, loss of the flight controls ARINC 629 data busses, or some combination of these failures. It may also be selected manually via the PFC disconnect switch on the overhead panel in the flight deck. The airplane handling characteristics in the "Direct" mode closely match those of the "Secondary" mode.

23.8 Control Laws and System Functionality

The design philosophy employed in the development of the 777 primary flight control system control laws stresses aircraft operation consistent with a pilot's past training and experience. The combination of electronic control of the system and this philosophy provides for the feel of a conventional airplane, but with improved handling characteristics and reduced pilot workload.

23.8.1 Pitch Control

Pitch control is accomplished through what is known as a *maneuver demand* control law, which is also referred to as a C*U control law. C* (pronounced "C-Star") is a term that is used to describe the blending of the airplane pitch rate and the load factor (the amount of acceleration felt by an occupant of the airplane during a maneuver). At low airspeeds, the pitch rate is the controlling factor. That is, a specific push or pull of the column by the pilot will result in some given pitch rate of the airplane. The harder the pilot pushes or pulls on the column, the faster the airplane will pitch nose up or nose down. At high airspeeds, the load factor dominates. This means that, at high airspeeds, a specific push or pull of the column by the pilot will result in some given load factor.

The "U" term in C*U refers to the feature in the control law which will, for any change in the airspeed away from a referenced trim speed, cause a pitch change to return to that referenced airspeed. For an increase in airspeed, the control law will command the airplane nose up, which tends to slow the airplane down. For a decrease in airspeed, the control law causes a corresponding speed increase by commanding the airplane nose down. This introduces an element of speed stability into the airplane pitch control. However, airplane configuration changes, such as a change in the trailing edge flap setting or lowering the landing gear, will NOT result in airplane pitch changes, which would require the pilot to re-trim the airplane to the new configuration. Thus, the major advantage of this type of control law is that the nuisance-handling characteristics found in a conventional, mechanically controlled flight control system which increase the pilot workload are minimized or eliminated, while the desirable characteristics are maintained.

While in flight, the pitch trim switches on the Captain's and First Officer's control wheels do not directly control the horizontal stabilizer as they normally do on conventionally controlled airplanes. When the trim switches are used in flight, the pilot is actually requesting a new referenced trim speed. The airplane will pitch nose up or nose down, using the elevator surfaces, in response to that reference airspeed change to achieve that new airspeed. The stabilizer will automatically trim, when necessary, to offload the elevator surface and allow it to return to its neutral surface when the airplane is in a trimmed condition. When the airplane is on the ground, the pitch trim switches do trim the horizontal stabilizer directly. While the alternate trim levers (described previously) move the stabilizer directly, even in flight, the act of doing so will also change the C*U referenced trim speed such that the net effect is the same as would have been achieved if the pitch trim switches on the control wheels had been used. As on a conventional airplane, trimming is required to reduce any column forces that are being held by the pilot.

The pitch control law incorporates several additional features. One is called landing flare compensation. This function provides handling characteristics during the flare and landing maneuvers consistent with that of a conventional airplane, which would have otherwise been significantly altered by the C*U control law. The pitch control law also incorporates stall and overspeed protection. These functions will not allow the referenced trim speed to be set below a predefined minimum value or above the maximum operating speed of the airplane. They also significantly increase the column force that the pilot must hold to fly above or below those speeds. An additional feature incorporated into the pitch control law is turn compensation, which enables the pilot to maintain a constant altitude with minimal column input during a banked turn.

The unique 777 implementation of maneuver demand and speed stability in the pitch control laws means that:

- An established flight path remains unchanged unless the pilot changes it through a control column input, or if the airspeed changes and the speed stability function takes effect.
- Trimming is required only for airspeed changes and not for airplane configuration changes.

23.8.2 Yaw Control

The yaw control law contains the usual functionality employed on other Boeing jetliners, such as the yaw damper and rudder ratio changer (which compensates a rudder command as a function of airspeed). However, the 777 FBW rudder control system has no separate actuators, linkages, and wiring for these functions, as have been used in previous airplane models. Rather, the command for these functions are calculated in the PFCs and included as part of the normal rudder command to the main rudder actuators. This reduces weight, complexity, maintenance, and spares required to be stocked.

The yaw control law also incorporates several additional features. The gust suppression system reduces airplane tag wag by sensing wind gusts via pressure transducers mounted on the vertical tail fin and applying a rudder command to oppose the movement that would have otherwise been generated by the gust. Another feature is the wheel-rudder crosstie function, which reduces sideslip by using small amounts of rudder during banked turns.

One important feature in the yaw control is Thrust Asymmetry Compensation, or TAC. This function automatically applies a rudder input for any thrust asymmetry between the two engines which exceed approximately 10% of the rated thrust. This is intended to cancel the yawing moment associated with an engine failure. TAC operates at all airspeeds above 80 kn even on the ground during the take-off roll. It will not operate when the engine thrust reversers are deployed.

23.8.3 Roll Control

The roll control law utilized by the 777 primary flight control system is fairly conventional. The outboard ailerons and spoiler panels 5 and 10 are locked out in the faired position when the airspeed exceeds a value that is dependent upon airspeed and altitude. It roughly corresponds to the airplane "flaps up" maneuvering speed. As with the yaw damper function described previously, this function does not have a separate actuator, but is part of the normal aileron and spoiler commands. The bank angle protection feature in the roll control law has been discussed previously.

23.8.4 757 Test Bed

The control laws and features discussed here were incorporated into a modified 757 and flown in the summer of 1992, prior to full-scale design and development of the 777 primary flight control system. The Captain's controls remained connected to the normal mechanical system utilized on the 757. The 777 control laws were flown through the First Officer's controls. This flying testbed was used to validate the flight characteristics of the 777 fly-by-wire system, as was flown by Boeing, customer, and regulatory agency pilots. When the 777 entered into its flight test program, its handling characteristics were extremely close to those that had been demonstrated with the 757 flying testbed.

23.8.5 Actuator Force-Flight Elimination

One unique aspect of the FBW flight control system used on the 777 is that the actuators on any given surface are all fully powered at all times. There are two full-time actuators driving each of the elevator, aileron, and flaperon surfaces, just as there are three full-time actuators on the rudder. The benefit of this particular implementation is that each individual actuator was able to be sized smaller than it would have had to have been if each surface was going to be powered by a single actuator through the entire flight regime. In addition, there is not a need for any redundancy management of an active/standby actuation system. However, this does cause a concern in another area. This is a possible actuator force-fight condition between the multiple actuators on a single flight control surface.

Actuator force-fight is caused by the fact that no two actuators, position transducers, or set of controlling servo loop electronics are identical. In addition, there always will be some rigging differences of the multiple actuators as they are installed on the airplane. These differences will result in one actuator attempting to position a flight control surface in a slightly different position than its neighboring actuator. Unless addressed, this would result in a twisting moment upon the surface as the two actuators fight each other to place the surface in different positions. In order to remove this unnecessary stress on the flight control surfaces, the primary flight computer control laws include a feature which "nulls out" these forces on the surfaces.

Each actuator on the 777 primary flight control system includes what is referred to as a Delta Pressure, or Delta P, pressure transducer. These transducer readings are transmitted via the ACEs to the PFCs, which are used in the individual surface control laws to remove the force-fight condition on each surface. The PFCs add an additional positive or negative component to each of the individual elevator actuator commands, which results in the difference between the two Delta P transducers being zero. In this way, the possibility of any force-fight condition between multiple actuators on a single surface is removed. The surface itself, therefore, does not need to be designed to withstand these stresses, which would have added a significant amount of weight to the airplane.

23.9   Primary Flight Controls System Displays and Annunciations

The primary displays for the primary flight control system on the 777 are the Engine Indication and Crew Alerting System (EICAS) display and the Multi-Function Display (MFD) in the flight deck. Any failures that require flight crew knowledge or action are displayed on these displays in the form of an English language message. These messages have several different levels associated with them, depending upon the level of severity of the failure.

Warning (Red with accompanying aural alert): A nonnormal operational or airplane system condition that requires immediate crew awareness and immediate pilot corrective compensatory action.

Caution (Amber with accompanying aural alert): A nonnormal or airplane system condition that requires immediate crew awareness. Compensatory or corrective action may be required.

Advisory (Amber with no accompanying aural alert): A nonnormal operational or airplane system condition which requires crew awareness. Compensatory or corrective action may be required.

Status (White): No dispatch or Minimum Equipment List (MEL) related items requiring crew awareness prior to dispatch.

Also available on the MFD, but not normally displayed in flight, is the flight control synoptic page, which shows the position of all the flight control surfaces.

23.10   System Maintenance

The 777 primary flight control system has been designed to keep line maintenance to a minimum, but when tasks do need to be accomplished, they are straightforward and easy to understand.

23.10.1   Central Maintenance Computer

The main interface to the primary flight control system for the line mechanic is the Central Maintenance Computer (CMC) function of AIMS. The CMC uses the Maintenance Access Terminal (MAT) as its primary display and control. The role of the CMC in the maintenance of the primary flight control system is to identify failures present in the system and to assist in their repair. The two features utilized by the CMC that accomplish these tasks are maintenance messages and ground maintenance tests. Maintenance messages describe to the mechanic, in simplified English, what failures are present in the system and the components possibly at fault. The ground maintenance tests exercise the system, test for active and latent failures, and confirm any repair action taken. They are also used to unlatch any EICAS and maintenance messages that may have become latched due to failures.

The PFCs are able to be loaded with new software through the data loader function on the MAT. This allows the PFCs to be updated to a new software configuration without having to take them out of service.

23.10.2   Line Replaceable Units

All the major components of the system are Line Replaceable Units (LRU). This includes all electronics modules, ARINC 629 data bus couplers, hydraulic and electrical actuators, and all position, force, and pressure transducers. The installation of each LRU has been designed such that a mechanic has ample space for component removal and replacement, as well as space for the manipulation of any required tools.

Each LRU, when replaced, must be tested to assure that the installation was accomplished correctly. The major LRUs of the system (transducers, actuators, and electronics modules) have LRU replacement tests that are able to be selected via a MAT pull-down menu and are run by the PFCs. These tests are user-friendly and take a minimum amount of time to accomplish. Any failures found in an LRU replacement test will result in a maintenance message, which details the failures that are present.

23.10.3 Component Adjustment

The primary surface actuators on the 777 are replaced in the same manner as on conventional airplanes. The difference is how they are adjusted. Each elevator, aileron, flaperon, and rudder actuator has what is referred to as a null adjust transducer, which is rotated by the mechanic until the actuator is positioned correctly. For example, when a rudder actuator is replaced, all hydraulic systems are depressurized except for the one that supplies power to the actuator that has just been replaced. The null adjust transducer is then adjusted until the rudder surface aligns itself with a mark on the empennage, showing that the actuator has centered the rudder correctly.

The transducers used on the pilot controls are, for the most part, individual LRUs. However, there are some packages, such as the speedbrake lever position transducers and the column force transducers, which have multiple transducers in a single package. When a transducer is replaced, the primary flight controls EICAS maintenance pages are used to adjust the transducer to a certain value at the system rig point. There are CMC-initiated LRU replacement tests which check that the component has been installed correctly and that all electrical connections have been properly mated.

23.11 Summary

The Boeing 777 fly-by-wire primary flight control system uses new technology to provide significant benefits over that of a conventional system. These benefits include a reduction in the overall weight of the airplane, superior handling characteristics, and improved maintainability of the system. At the same time, the control of the airplane is accomplished using traditional flight deck controls, thereby allowing the pilot to fly the airplane without any specialized training when transferring from a more conventional commercial jet aircraft. The technology utilized by the 777 primary flight control system has earned its way onto the airplane, and is not just technology for technology's sake.

Defining Terms

ACE: Actuator Control Electronics
ADIRU: Air Data Inertial Reference Unit
ADM: Air Data Module (Static and Total Pressure)
AFDC: Autopilot Flight Director Computer
AIMS: Airplane Information Management System
ARINC: Aeronautical Radio Inc. (Industry Standard)
C: Center
C*U: Pitch Control Law utilized in the Primary Flight Computer
CMC: Central Maintenance Computer Function in AIMS
DCGF: Data Conversion Gateway Function of AIMS
EDIU: Engine Data Interface Unit
EICAS: Engine Indication and Crew Alerting System
ELMS: Electrical Load Management System
FBW: Fly-By-Wire
FCDC: Flight Controls Direct Current (power system)
FSEU: Flap Slat Electronic Unit
L: Left
L1:   Left 1
L2:   Left 2

LRRA: Low Range Radio Altimeter
LRU: Line Replaceable Unit
MAT: Maintenance Access Terminal
MEL: Minimum Equipment List
MFD: Multi-Function Display
MOV: Motor-Operated Valve
PCU: Power Control Unit (hydraulic actuator)
PFC: Primary Flight Computer
PMG: Permanent Magnet Generator
PSA: Power Supply Assembly
R: Right
RAT: Ram Air Turbine
SAARU: Standby Attitude and Air Data Reference Unit
TAC: Thrust Asymmetry Compensation
WEU: Warning Electronics Unit

24

Electrical Flight Controls, From Airbus A320/330/340 to Future Military Transport Aircraft: A Family of Fault-Tolerant Systems

Dominique Briere
Airbus—Retired

Christian Favre
Airbus

Pascal Traverse
Airbus

24.1 Introduction

The first electrical flight control system for a civil aircraft was designed by Aerospatiale and installed on the Concorde. This is an analog, full-authority system for all control surfaces. The commanded control surface positions are directly proportional to the stick inputs. A mechanical back-up system is provided on the three axes.

The first generation of electrical flight control systems with digital technology appeared on several civil aircraft at the start of the 1980s with the Airbus A310 program. These systems control the slats, flaps, and spoilers. These systems were designed with very stringent safety requirements (control surface runaway must be extremely improbable). As the loss of these functions results in a supportable increase in the crew's workload, it is possible to lose the system in some circumstances.

TABLE 24.1 Incremental Introduction of New Technologies

First Flight In:	1955	1969	1972	1978–1983	1983	1987
Servo-Controls, and Artificial Feel	x	x	x	x	x	--> x
Electro-Hydraulic Actuators		x	x	x	x	--> x
Command and Monitoring Computers		x	x	x	x	--> x
Digital Computers				x	x	--> x
Trim, Yaw Damper, Protection	x	x	x	x	x	--> x
Electrical Flight Controls		x		x	x	-->x
Side-Stick, Control Laws				x		--> x
Servoed Aircraft (Auto-pilot)	x	x	x	x	x	--> x
Formal System Safety Assessment		x	x	x	x	--> x
System Integration Testing	x	x	x	x	x	--> x
	Carevelle	Concorde	A300	Flight test Concorde A300	A310, A300–600	A320

The Airbus A320 (certified in early 1988) is the first example of a second generation of civil electrical flight control aircraft, rapidly followed by the A340 aircraft (certified at the end of 1992). These aircraft benefit from the significant experience gained by Aérospatiale in the technologies used for a fly-by-wire system (see Table 24.1). The distinctive feature of these aircrafts is that all control surfaces are electrically controlled and that the system is designed to be available under all circumstances.

This system was built to very stringent dependability requirements both in terms of safety (the system may generate no erroneous signals) and availability (the complete loss of the system is extremely improbable).

The overall dependability of the aircraft fly-by-wire system relies in particular on the computer arrangement (the so-called control/monitor architecture), the system tolerance to both hardware and software failures, the servo-control and power supply arrangement, the failure monitoring, and the system protection against external aggressions. It does this without forgetting the flight control laws which minimize the crew workload, the flight envelope protections which allow fast reactions while keeping the aircraft in the safe part of the flight envelope, and finally the system design and validation methods.

The aircraft safety is demonstrated by using both qualitative and quantitative assessments; this approach is consistent with the airworthiness regulation. Qualitative assessment is used to deal with design faults, interaction (maintenance, crew) faults, and external environmental hazard. For physical ("hardware") faults, both qualitative and quantitative assessments are used. The quantitative assessment covers the FAR/JAR 25.1309 requirement, and links the failure condition classification (minor to catastrophic) to its probability target.

This chapter describes the Airbus fly-by-wire systems from a fault-tolerant standpoint. The fly-by-wire basic principles are presented first, followed by the description of the main system features common to A320 and A340 aircraft, the failure detection and reconfiguration procedures, the A340 particularities, and the design, development, and validation procedures. Future trends in terms of fly-by-wire fault-tolerance conclude this overview.

24.2 Fly-by-Wire Principles

On aircraft of the A300 and A310 type, the pilot orders are transmitted to the servo-controls by an arrangement of mechanical components (rods, cables, pulleys, etc.). In addition, specific computers and actuators driving the mechanical linkages restore the pilot feels on the controls and transmit the autopilot commands (see Figure 24.1).

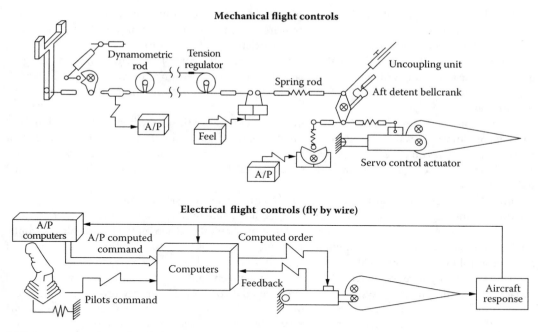

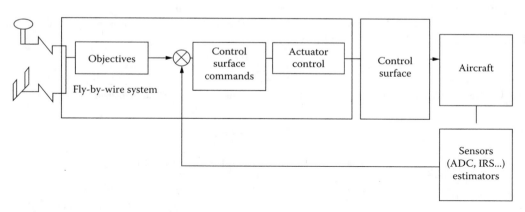

FIGURE 24.1 Mechanical and electrical flight control.

FIGURE 24.2 Flight control laws.

The term fly-by-wire has been adopted to describe the use of electrical rather than mechanical signalling of the pilot's commands to the flying control actuators. One can imagine a basic form of fly-by-wire in which an airplane retained conventional pilot's control columns and wheels, hydraulic actuators (but electrically controlled), and artificial feel as experienced in the 1970s with the Concorde program. The fly-by-wire system would simply provide electrical signals to the control actuators that were directly proportional to the angular displacement of the pilot's controls, without any form of enhancement.

In fact, the design of the A320, A321, A330, and A340 flight control systems takes advantage of the potential of fly-by-wire for the incorporation of control laws that provide extensive stability augmentation and flight envelope limiting [Favre, 1993]. The positioning of the control surfaces is no longer a simple reflection of the pilot's control inputs and conversely, the natural aerodynamic characteristics of the aircraft are not fed back directly to the pilot (see Figure 24.2).

The sidesticks, now part of a modern cockpit design with a large visual access to instrument panels, can be considered as the natural issue of fly-by-wire, since the mechanical transmissions with pulleys, cables, and linkages can be suppressed with their associated backlash and friction.

The induced roll characteristics of the rudder provide sufficient roll maneuverability of design a mechanical back-up on the rudder alone for lateral control. This permitted the retention of the advantages of the sidestick design, now rid of the higher efforts required to drive mechanical linkages to the roll surfaces.

Looking for minimum drag leads us to minimize the negative lift of the horizontal tail plane and consequently diminishes the aircraft longitudinal stability. It was estimated for the Airbus family that no significant gain could be expected with rear center-of-gravity positions beyond a certain limit. This allowed us to design a system with a mechanical back-up requiring no additional artificial stabilization.

These choices were obviously fundamental to establish the now-classical architecture of the Airbus fly-by-wire systems (Figures 24.3 and 24.4), namely a set of five full-authority digital computers controlling the three pitch, yaw, and roll axes and completed by a mechanical back-up on the trimmable horizontal stabilizer and on the rudder. (Two additional computers as part of the auto pilot system are in charge of rudder control in the case of A320 and A321 aircraft.)

Of course, a fly-by-wire system relies on the power systems energizing the actuators to move the control surfaces and on the computer system to transmit the pilot controls. The energy used to pressurize the servo-controls is provided by a set of three hydraulic circuits, one of which is sufficient to control the aircraft. One of the three circuits can be pressurized by a Ram air turbine, which automatically extends in case of an all-engine flame-out.

The electrical power is normally supplied by two segregated networks, each driven by one or two generators, depending on the number of engines. In case of loss of the normal electrical generation, an emergency generator supplies power to a limited number of fly-by-wire computers (among others). These computers can also be powered by the two batteries.

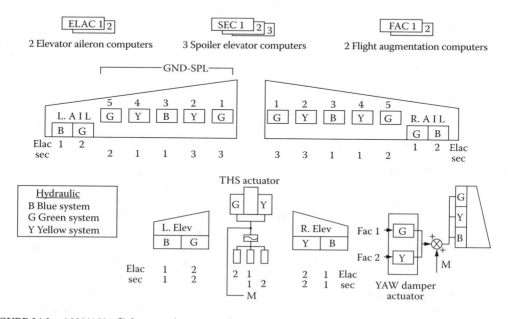

FIGURE 24.3   A320/A321 flight control system architecture.

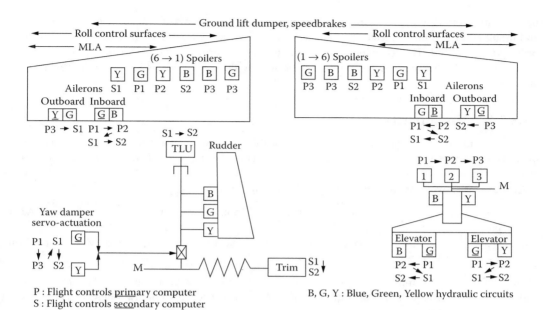

P : Flight controls <u>p</u>rimary computer
S : Flight controls <u>s</u>econdary computer

B, G, Y : Blue, Green, Yellow hydraulic circuits

FIGURE 24.4   A330/A340 flight control system architecture.

24.3   Main System Features

24.3.1 Computer Arrangement

24.3.1.1   Redundancy

The five fly-by-wire computers are simultaneously active. They are in charge of control law computation as a function of the pilot inputs as well as individual actuator control, thus avoiding specific actuator control electronics. The system incorporates sufficient redundancies to provide the nominal performance and safety levels with one failed computer, while it is still possible to fly the aircraft safely with one single computer active.

As a control surface runaway may affect the aircraft safety (elevators in particular), each computer is divided into two physically separated channels (Figure 24.5). The first one, the control channel, is

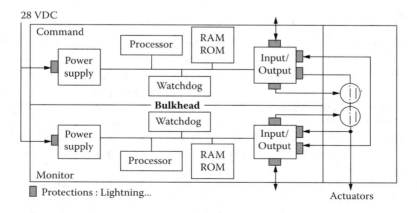

FIGURE 24.5   Command and monitoring computer architecture.

permanently monitored by the second one, the monitor channel. In case of disagreement between control and monitor, the computer affected by the failure is passivated, while the computer with the next highest priority takes control. The repartition of computers, servo-controls, hydraulic circuit, and electrical bus bars and priorities between the computers are dictated by the safety analysis including the engine burst analysis.

24.3.1.2 Dissimilarity

Despite the nonrecurring costs induced by dissimilarity, it is fundamental that the five computers all be of different natures to avoid common mode failures. These failures could lead to the total loss of the electrical flight control system.

Consequently, two types of computers may be distinguished:

1.2 ELAC (elevator and aileron computers) and 3 SEC (spoiler and elevator computers) on A320/A321 and,
2.3 FCPC (flight control primary computers) and 2 FCSC (flight control secondary computers) on A330/A340.

Taking the 320 as an example, the ELACs are produced by Thomson-CSF around 68010 microprocessors and the SECs are produced in cooperation by SFENA/Aerospatiale with a hardware based on the 80186 microprocessor. We therefore have two different design and manufacturing teams with different micro-processors (and associated circuits), different computer architectures, and different functional specifications. At the software level, the architecture of the system leads to the use of four software packages (ELAC control channel, ELAC monitor channel, SEC control channel, and SEC monitor channel) when, functionally, one would suffice.

24.3.1.3 Serve-Control Arrangement

Ailerons and elevators can be positioned by two servo-controls in parallel. As it is possible to lose control of one surface, a damping mode was integrated into each servo-control to prevent flutter in this failure case. Generally, one servo-control is active and the other one is damped. In case of loss of electrical control, the elevator actuators are centered by a mechanical feedback to increase the horizontal stabilizer efficiency.

Rudder and horizontal stabilizer controls are designed to receive both mechanical and electrical inputs. One servo-control per spoiler surface is sufficient. The spoiler servo-controls are pressurized in the retracted position in case of loss of electrical control.

24.3.1.4 Flight Control Laws

The general objective of the flight control laws integrated in a fly-by-wire system is to improve the natural flying qualities of the aircraft, in particular in the fields of stability, control, and flight domain protections. In a fly-by-wire system, the computers can easily process the anemometric and inertial information as well as any information describing the aircraft state. Consequently, control laws corresponding to simple control objectives could be designed. The stick inputs are transformed by the computers into pilot control objectives which are compared to the aircraft's actual state measured by the inertial and anemometric sensors. Thus, as far as longitudinal control is concerned, the sidestick position is translated into vertical load factor demands, while lateral control is achieved through roll rate, sideslip, and bank angle objectives.

The stability augmentation provided by the flight control laws improves the aircraft flying qualities and contributes to aircraft safety. As a matter of fact, the aircraft remains stable in case of perturbations such as gusts or engine failure due to a very strong spin stability, unlike conventional aircraft. Aircraft control through objectives significantly reduces the crew workload; the fly-by-wire system acts as the inner loop of an autopilot system, while the pilot represents the outer loop in charge of objective management.

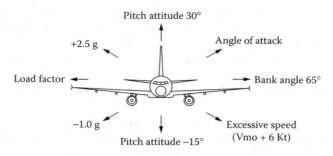

FIGURE 24.6   A320 flight envelope protections.

Finally, protections forbidding potentially dangerous excursions out of the normal flight domain can be integrated in the system (Figure 24.6). The main advantage of such protections is to allow the pilot to react rapidly without hesitation, since he knows that this action will not result in a critical situation.

24.3.1.5 Computer Architecture

Each computer can be considered as being two different and independent computers placed side by side (see Figure 24.5). These two (sub)computers have different functions and are placed adjacent to each other to make aircraft maintenance easier. Both command and monitoring channels of the computer are simultaneously active or simultaneously passive, ready to take control.

Each channel includes one or more processors, their associated memories, input/output circuits, a power supply unit, and specific software. When the results of these two channels diverge significantly, the links between the computer and the exterior world are cut by the channel or channels which detected the failure. The system is designed so that the computer outputs are then in a dependable state (signal interrupt via relays). Failure detection is mainly achieved by comparing the difference between the control and monitoring commands with a predetermined threshold. As a result, all consequences of a single computer fault are detected and passivated, which prevents the resulting error from propagating outside of the computer. This detection method is completed by permanently monitoring the program sequencing and the program correct execution.

Flight control computers must be robust. In particular, they must be especially protected against overvoltages and undervoltages, electromagnetic aggressions, and indirect effects of lightning. They are cooled by a ventilation system but must operate correctly even if ventilation is lost.

24.3.1.6 Installation

The electrical installation, in particular the many electrical connections, also comprises a common-point risk. This is avoided by extensive segregation. In normal operation, two electrical generation systems exist without a single common point. The links between computers are limited, the links used for monitoring are not routed with those used for control. The destruction of a part of the aircraft is also taken into account; the computers are placed at three different locations, certain links to the actuators run under the floor, others overhead, and others in the cargo compartment.

24.4   Failure Detection and Reconfiguration

24.4.1 Flight Control Laws

The control laws implemented in the flight control system computers have full authority and must be elaborated as a function of consolidated information provided by at least two independent sources in agreement.

Consequently, the availability of control laws using aircraft feedback (the so-called normal laws) is closely related to the availability of the sensors. The Airbus aircraft fly-by-wire systems use the information of three air data and inertial reference units (ADIRUs), as well as specific accelerometers and rate gyros. Moreover, in the case of the longitudinal normal law, analytical redundancy is used to validate the pitch rate information when provided by a single inertial reference unit. The load factor is estimated through the pitch rate information and compared to the available accelerometric measurements to validate the IRS data.

After double or triple failures, when it becomes impossible to compare the data of independent sources, the normal control laws are reconfigured into laws of the direct type where the control surface deflection is proportional to the stick input. To enhance the dissimilarity, the more sophisticated control laws with aircraft feedback (the normal laws) are integrated in one type of computer, while the other type of computer incorporates the direct laws only.

24.4.2   Actuator Control and Monitor

The general idea is to compare the actual surface position to the theoretical surface position computed by the monitoring channel. When needed, the control and monitor channels use dedicated sensors to perform these comparisons. Specific sensors are installed on the servovalve spools to provide an early detection capability for the elevators. Both channels can make the actuator passive. A detected runaway will result in the servo-control deactivation or computer passivation, depending on the failure source.

24.4.3   Comparison and Robustness

Specific variables are permanently compared in the two channels. The difference between the results of the control and monitoring channels are compared with a threshold. This must be confirmed before the computer is disconnected. The confirmation consists of checking that the detected failure lasts for a sufficiently long period of time. The detection parameters (threshold, temporization) must be sufficiently "wide" to avoid unwanted disconnections and sufficiently "tight" so that undetected failures are tolerated by the computer's environment (the aircraft). More precisely, all systems tolerance (most notably sensor inaccuracy, rigging tolerances, computer asynchronism) are taken into account to prevent undue failure detection, and errors which are not detectable (within the signal and timing thresholds) are assessed with respect to their handling quality and structural loads effect.

24.4.4   Latent Failures

Certain failures may remain masked a long time after their occurrence. A typical case is a monitoring channel affected by a failure resulting in a passive state and detected only when the monitored channel itself fails. Tests are conducted periodically so that the probability of the occurrence of an undesirable event remains sufficiently low (i.e., to fulfill [FAR/JAR 25] § 25.1309 quantitative requirement). Typically, a computer runs its self-test and tests its peripherals during the energization of the aircraft, and therefore at least once a day.

24.4.5   Reconfiguration

As soon as the active computer interrupts its operation relative to any function (control law or actuator control), one of the standby computers almost instantly changes to active mode with no or limited jerk on the control surfaces. Typically, duplex computers are designed so that they permanently transmit healthy signals which are interrupted as soon as the "functional" outputs (to an actuator, for example) are lost.

24.4.6 System Safety Assessment

The aircraft safety is demonstrated using qualitative and quantitative assessments. Qualitative assessment is used to deal with design faults, interaction (maintenance, crew) faults, and external environmental hazard. For physical ("hardware") faults, both a qualitative and a quantitative assessments are done. In particular, this quantitative assessment covers the link between failure condition classification (Minor to Catastrophic) and probability target.

24.4.7 Warning and Caution

It is deemed useful for a limited number of failure cases to advise the crew of the situation, and possibly that the crew act as a consequence of the failure. Nevertheless, attention has to be paid to keep the level of crew workload acceptable. The basic rule is to get the crews attention only when an action is necessary to cope with a failure or to cope with a possible future failure. On the other hand, maintenance personnel must get all the failure information.

The warnings and cautions for the pilots are in one of the following three categories:

- Red warning with continuous sound when an immediate action is necessary (for example, to reduce airplane speed).
- Amber caution with a simple sound, such that the pilot be informed although no immediate action is needed (for example, in case of loss of flight envelope protections an airplane speed should not be exceeded).
- Simple caution (no sound), such that no action is needed (for example, a loss of redundancy).

Priority rules among these warnings and cautions are defined to present the most important message first (see also [Traverse, 1994]).

24.5 A340 Particularities

The general design objective relative to the A340 fly-by-wire system was to reproduce the architecture and principles chosen for the A320 as much as possible for the sake of commonality and efficiency, taking account of the A340 particularities (long-range four-engine aircraft).

24.5.1 System

As is now common for each new program, the computer functional density was increased between the A320 and A330/A340 programs: The number of computers was reduced to perform more functions and control an increased number of control surfaces (Figure 24.3).

24.5.2 Control Laws

The general concept of the A320 flight control laws was maintained, adapted to the aircraft characteristics, and used to optimize the aircraft performance, as follows:

- The angle of attack protection was reinforced to better cope with the aerodynamic characteristics of the aircraft.
- The dutch roll damping system was designed to survive against rudder command blocking, thanks to an additional damping term through the ailerons, and to survive against an extremely improbable complete electrical failure thanks to an additional autonomous damper. The outcome of this was that the existing A300 fin could be used on the A330 and A340 aircraft with the associated industrial benefits.

- The take-off performance could be optimized by designing a specific law that controls the aircraft pitch attitude during the rotation.

- The flexibility of fly-by-wire was used to optimize the minimum control speed on the ground (VMCG). In fact, the rudder efficiency was increased on the ground by fully and asymmetrically deploying the inner and outer ailerons on the side of the pedal action as a function of the rudder travel: the inner aileron is commanded downwards, and the outer aileron (complemented by one spoiler) is commanded upwards.

- A first step in the direction of structural mode control through fly-by-wire was made on the A340 program through the so-called "turbulence damping function" destined to improve passenger comfort by damping the structural modes excited by turbulence.

24.6   Design, Development, and Validation Procedures

24.6.1   Fly-by-Wire System Certification Background

An airline can fly an airplane only if this airplane has a type certificate issued by the aviation authorities of the airline country. For a given country, this type certificate is granted when the demonstration has been made and accepted by the appropriate organization (Federal Aviation Administration in the U.S, Joint Aviation Authorities in several European countries, etc.) that the airplane meets the country's aviation rules and consequently a high level of safety. Each country has its own set of regulatory materials although the common core is very large. They are basically composed of two parts: the requirements on one part, and a set of interpretations and acceptable means of compliance in a second part. An example of requirement is "The aeroplane systems must be designed so that the occurrence of any failure condition which would prevent the continued safe flight and landing of the aeroplane is extremely improbable" (in Federal and Joint Aviation Requirements 25.1309, [FAR/JAR 25]). An associated part of the regulation (Advisory Circular from FAA, Advisory Material — Joint from JAA 25.1309) gives the meaning and discuss such terms as "failure condition," and "extremely improbable." In addition, guidance is given on how to demonstrate compliance.

The aviation regulatory materials are evolving to be able to cover new technologies (such as the use of fly-by-wire systems). This is done through special conditions targeting specific issues of a given airplane, and later on by modifying the general regulatory materials. With respect to A320/A330/A340 fly-by-wire airplane, the following innovative topics were addressed for certification (note: some of these topics were also addressing other airplane systems):

- Flight envelope protections
- Side-stick controller
- Static stability
- Interaction of systems and structure
- System safety assessment
- Lightning indirect effect and electromagnetic interference
- Integrity of control signal transmission
- Electrical power
- Software verification and documentation, automatic code generation
- System validation
- Application-specific integrated circuit

It is noteworthy that an integration of regulatory materials is underway which is resulting in a set of four documents:

- A document on system design, verification and validation, configuration management, quality assurance [ARP 4754, 1994]
- A document on software design, verification, configuration management, quality assurance [DO178B, 1992]
- A document on hardware design, verification, configuration management, quality assurance [DO254, 2000]
- A document on the system safety assessment process [ARP 4761, 1994]

24.6.2 The A320 Experience

24.6.2.1 Design

The basic element developed on the occasion of the A320 program is the so-called SAO specification (Spécification Assistée par Ordinateur), the Aerospatiale graphic language defined to clearly specify control laws and system logics. One of the benefits of this method is that each symbol used has a formal definition with strict rules governing its interconnections. The specification is under the control of a configuration management tool and its syntax is partially checked automatically.

24.6.2.2 Software

The software is produced with the essential constraint that it must be verified and validated. Also, it must meet the world's most severe civil aviation standards (level 1 software to [D0178A, 1985]–see also [Barbaste, 1988]). The functional specification acts as the interface between the aircraft manufacturer's world and the software designer's world. The major part of the A320 flight control software specification is a copy of the functional specification. This avoids creating errors when translating the functional specification into the software specification. For this "functional" part of the software, validation is not required as it is covered by the work carried out on the functional specification. The only part of the software specification to be validated concerns the interface between the hardware and the software (task sequencer, management of self-test software inputs/outputs). This part is only slightly modified during aircraft development.

To make software validation easier, the various tasks are sequenced in a predetermined order with periodic scanning of the inputs. Only the clock can generate interrupts used to control task sequencing. This sequencing is deterministic. A part of the task sequencer validation consists in methodically evaluating the margin between the maximum execution time for each task (worst case) and the time allocated to this task. An important task is to check the conformity of the software with its specification. This is performed by means of tests and inspections. The result of each step in the development process is checked against its specification. For example, a code module is tested according to its specification. This test is, first of all, functional (black box), then structural (white box).

Adequate coverage must be obtained for the internal structure and input range. The term "adequate" does not mean that the tests are assumed as being exhaustive. For example, for the structural test of a module, the equivalence classes are defined for each input. The tests must cover the module input range taking these equivalence classes and all module branches (among other things) as a basis. These equivalence classes and a possible additional test effort have the approval of the various parties involved (aircraft manufacturer, equipment manufacturer, airworthiness authorities, designer, and quality control).

The software of the control channel is different from that of the monitoring channel. Likewise, the software of the ELAC computer is different from that of the SEC computer (the same applies to the FCPC and FCSC on the A340). The aim of this is to minimize the risk of a common error which could cause control surface runaway (control/monitoring dissimilarity) or complete shutdown of all computers (ELAC/SEC dissimilarity).

The basic rule to be retained is that the software is made in the best possible way. This has been recognized by several experts in the software field both from industry and from the airworthiness authorities. Dissimilarity is an additional precaution which is not used to reduce the required software quality effort.

24.6.2.3 System Validation

Simulation codes, full-scale simulators and flight tests were extensively used in a complementary way to design, develop, and validate the A320 flight control system (see also [Chatrenet, 1989]), in addition to analysis and peer review.

A "batch" type simulation code called OSMA (Outil de Simulation des Mouvements Avion) was used to initially design the flight control laws and protections, including the nonlinear domains and for general handling quality studies.

A development simulator was then used to test the control laws with a pilot in the loop as soon as possible in the development process. This simulator is fitted with a fixed-base faithful replica of the A320 cockpit and controls and a visual system; it was in service in 1984, as soon as a set of provisional A320 aero data, based on wind tunnel tests, was made available. The development simulator was used to develop and initially tune all flight control laws in a closed-loop cooperation process with flight test pilots.

Three "integration" simulators were put into service in 1986. They include the fixed replica of the A320 cockpit, a visual system for two of them, and actual aircraft equipment including computers, displays, control panels, and warning and maintenance equipment. One simulator can be coupled to the "iron bird" which is a full-scale replica of the hydraulic and electrical supplies and generation, and is fitted with all the actual flight control system components including servojacks. The main purpose of these simulators is to test the operation, integration, and compatibility of all the elements of the system in an environment closely akin to that of an actual aircraft.

Finally, flight testing remains the ultimate and indispensable way of validating a flight control system. Even with the current state of the art in simulation, simulators cannot yet fully take the place of flight testing for handling quality assessment. On this occasion a specific system called SPATIALL (Système Pour Acquisition et Traitement d'Informations Analogiques ARINC et Logiques) was developed to facilitate the flight test. This system allows the flight engineer to:

- Record any computer internal parameter
- Select several preprogrammed configurations to be tested (gains, limits, thresholds, etc.)
- Inject calibrated solicitations to the controls, control surfaces, or any intermediate point.

The integration phase complemented by flight testing can be considered as the final step of the validation side of the now-classical V-shaped development/validation process of the system.

24.6.3 The A340 Experience

24.6.3.1 Design

The definition of the system requires that a certain number of actuators be allocated to each control surface and a power source and computers assigned to each actuator. Such an arrangement implies checking that the system safety objectives are met. A high number of failure combinations must therefore be envisaged. A study has been conducted with the aim of automating this process.

It was seen that a tool which could evaluate a high number of failure cases, allowing the use of capacity functions, would be useful and that the possibility of modeling the static dependencies was not absolutely necessary even though this may sometimes lead to a pessimistic result. This study gave rise to a data processing tool which accepts as input an arrangement of computers, actuators, hydraulic and electrical power sources, and also specific events such as simultaneous shutdown of all engines and, therefore, a high number of power sources. The availability of a control surface depends on the availability of a certain number of these resources. This description was made using a fault tree-type support as input to the tool.

The capacity function used allows the aircraft roll controllability to be defined with regard to the degraded state of the flight control system. This controllability can be approached by a function which measures the roll rate available by a linear function of the roll rate of the available control surfaces. It is then possible to divide the degraded states of the system into success or failure states and thus calculate the probability of failure of the system with regards to the target roll controllability.

The tool automatically creates failure combinations and evaluates the availability of the control surfaces and, therefore, a roll controllability function. It compares the results to the targets. These targets are, on the one hand, the controllability (availability of the pitch control surfaces, available roll rate, etc.) and, on the other hand, the reliability (a controllability target must be met for all failure combinations where probability is greater than a given reliability target). The tool gives the list of failure combinations which do not meet the targets (if any) and gives, for each target controllability, the probability of nonsatisfaction. The tool also takes into account a dispatch with one computer failed.

24.6.3.2 Automatic programming

The use of automatic programming tools is becoming widespread. This tendency appeared on the A320 and is being confirmed on the A340 (in particular, the FCPC is, in part, programmed automatically). Such a tool has SAO sheets as inputs, and uses a library of software packages, one package being allocated to each symbol. The automatic programming tool links together the symbol's packages.

The use of such tools has a positive impact on safety. An automatic tool ensures that a modification to the specification will be coded without stress even if this modification is to be embodied rapidly (situation encountered during the flight test phase for example). Also, automatic programming, through the use of a formal specification language, allows onboard code from one aircraft program to be used on another. Note that the functional specification validation tools (simulators) use an automatic programming tool. This tool has parts in common with the automatic programming tool used to generate codes for the flight control computers. This increases the validation power of the simulations. For dissimilarity reasons, only the FCPC computer is coded automatically (the FCSC being coded manually). The FCPC automatic coding tool has two different code translators, one for the control channel and one for the monitoring channel.

24.6.3.3 System validation

The A320 experience showed the necessity of being capable of detecting errors as early as possible in the design process, to minimize the debugging effort along the development phase. Consequently, it was decided to develop tools that would enable the engineers to actually fly the aircraft in its environment to check that the specification fulfils the performance and safety objectives before the computer code exists.

The basic element of this project is the so-called SAO specification, the Aerospatiale graphic language defined to clearly specify control laws and system logics and developed for A320 program needs. The specification is then automatically coded for engineering simulation purposes in both control law and system areas.

In the control law area, OCAS (Outil de Conception Assistée par Simulation) is a real-time simulation tool that links the SAO definition of the control laws to the already-mentioned aircraft movement simulation (OSMA). Pilot orders are entered through simplified controls including side-stick and engine thrust levels. A simplified PFD (primary flight display) visualizes the outputs of the control law. The engineer is then in a position to judge the quality of the control law that he has just produced, in particular with respect to law transition and nonlinear effects. In the early development phase, this same simulation was used in the full-scale A340 development simulator with a pilot in the loop.

In the system area, OSIME (Outil de SImulation Multi Equipement) is an expanded time simulation that links the SAO definition of the whole system (control law and system logic) to the complete servo-control modes and to the simulation of aircraft movement (OSMA). The objective was to simulate the whole fly-by-wire system including the three primary computers (FCPC), the two secondary computers (FCSC), and the servo-controls in an aircraft environment.

This tool contributed to the functional definition of the fly-by-wire system, to the system validation, and to the failure analysis. In addition, the behavior of the system at the limit of validity of each

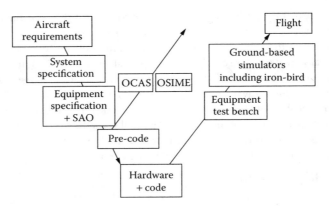

FIGURE 24.7   Validation methodology.

parameter, including time delays, could be checked to define robust monitoring algorithms. Non-regression tests have been integrated very early into the design process to check the validity of each new specification standard.

Once validated, both in the control law and system areas using the OCAS and OSIME tools, a new specification standard is considered to be ready to be implemented in the real computers (automatic coding) to be further validated on a test bench, simulator, and on the aircraft (Figure 24.7).

24.7   Future Trends

The fly-by-wire systems developed on the occasion of the A320, A321, A340, and A330 programs now constitute an industrial standard for commercial applications and are well adapted to future military transport aircraft, thanks to the robustness of the system and its reconfiguration capabilities. What are the possible system evolutions? Among others, are the following:

1. New actuator concepts are arising. In particular, systems using both electrical and hydraulic energy within a single actuator were developed and successfully tested on A320 aircraft. This is the so-called electrical back-up hydraulic actuator or EBHA. This actuator can be used to design flight control systems that survive the total loss of hydraulic power, which is a significant advantage for a military transport aircraft particularly in the case of battle damage.
2. The hardware dissimilarity of the fly-by-wire computer system and the experience with A320 and A340 airline operation will probably ease the suppression of the rudder and trimmable horizontal stabilizer mechanical controls of future aircraft.
3. The integration of new functions, such as structural mode control, may lead to increased dependability requirements, in particular if the loss of these functions is not allowed.
4. Finally, future flight control systems will be influenced by the standardization effort made through the IMA concept (integrated modular avionics) and by the "smart" concept where the electronics destined to control and monitor each actuator are located close to the actuator.

References

ARP 4754, 1994. *System Integration Requirements.* Society of Automotive Engineers (SAE) and European Organization for Civil Aviation electronics (EUROCAE).

ARP 4761, 1994. *Guidelines and tools for Conducting the Safety Assessment Process on Civil Airborne Systems and Equipment.* Society of Automotive Engineers (SAE) and European Organization for Civil Aviation Electronics (EUROCAE).

Barbaste, L. and Desmons, J. P., 1988. Assurance qualité du logiciel et la certification des aéronefs/ Expérience A320. ler séminaire EOQC sur la qualité des logiciels, April 1988, Brussels, pp. 135–146.

Chatrenet, D., 1989. Simulateurs A320 d'Aérospatiale: leur contribution à la conception, au développement et à la certification. *INFAUTOM 89*, Toulouse.

DO178A. 1985. *Software Considerations in Airborne Systems and Equipment Certification.* RTCA and European Organization for Civil Aviation Electronics (EUROCAE).

DO178B, 1992. *Software Considerations in Airborne Systems and Equipment Certification.* RTCA and European Organization for Civil Aviation Electronics (EUROCAE).

DO254, 1995. *Design Assurance Guidance for Complex Electronic Hardware Used in Airborne Systems.* RTCA and by European Organization for Civil Aviation Electronics (EUROCAE).

FAR/JAR 25. *Airworthiness Standards: Transport Category Airplanes.* Part 25 of "Code of Federal Regulations, Title 14, Aeronautics and Space," for the Federal Aviation Administration, and "Airworthiness Joint Aviation Requirements — large aeroplane" for the Joint Aviation Authorities.

Favre, C., 1993. Fly-by-wire for commercial aircraft — the Airbus experience. *Int. J. Control,* special issue on "Aircraft Flight Control".

Traverse, P., Brière, D., and Frayssignes, J. J., 1994. Architecture des commande de vol électriques Airbus, reconfiguration automatique et information équipage. *INFAUTOM 94*, Toulouse.

Index

A

B

V

W